T0315093

ADVANCES IN ELECTRIC POWER AND ENERGY SYSTEMS

ADVANCES IN ELECTRIC POWER AND ENERGY SYSTEMS
Load and Price Forecasting

Edited by

Mohamed E. El-Hawary

IEEE PRESS SERIES ON POWER ENGINEERING

IEEE PRESS

WILEY

Published by John Wiley & Sons, Inc., Hoboken, New Jersey.
Published simultaneously in Canada.

For general information on our other products and services or for technical support, please contact our Customer Care Department within the United States at (800) 762-2974, outside the United States at (317) 572-3993 or fax (317) 572-4002.

Wiley also publishes its books in a variety of electronic formats. Some content that appears in print may not be available in electronic formats. For more information about Wiley products, visit our web site at www.wiley.com.

Library of Congress Cataloging-in-Publication Data is available.

ISBN: 978-1-118-17134-9

Printed in the United States of America.

10 9 8 7 6 5 4 3 2 1

Contents

Preface and Acknowledgments

My goal in writing this book is to present a sampling of leading-edge works treating forecasting problems in the operation of electric power systems throughout the world.

The book's audience consists mainly of practicing professionals, regulators, planners, and consultants engaged in the electric power business. In addition, senior undergraduate and graduate students and researchers will find this book to be useful in their work. Background requirements include some mathematical notions from algebra and calculus.

The core of this book (Chapters 2–6) consists of two chapters dealing with power system load forecasting and three chapters on electricity price forecasting. Chapters 7 and 8 are unique treatments of estimation of post-storm restoration times in electric power distribution systems and river flow forecasting based on autonomous neural network models using a nonparametric approach. While Chapter 1 is usually devoted to charting the course of the book, I prepared Chapter 1 to offer background material for the two main forecasting issues considered. Each chapter is a self-contained treatment of its subject matter.

I am indebted to our Editor, Ms. Mary Hatcher. Her patience and constant encouragement contributed to the evolution of this book. My family is the source of the continuing motivation to complete this manuscript.

MOHAMED E. EL-HAWARY

Halifax, Nova Scotia, Canada

Contributors

Alexandre P. Alves da Silva is with GE Global Research, Brazil Technology Center, Rio de Janeiro, Brazil. He received his PhD degree in Electrical Engineering from the University of Waterloo, Canada. During 1999, he was a Visiting Professor in the Department of Electrical Engineering, University of Washington, Seattle, WA. He has authored and co-authored more than 200 papers on intelligent systems application to power systems. Professor Alves da Silva was the TPC Chairman of the First Brazilian Conference on Neural Networks in 1994 and of the International Conference on Intelligent System Applications to Power Systems (ISAP) in 1999, the first time it was held in Brazil. He is an IEEE fellow.

Tatiyana V. Apanasovich is an Associate Professor of Statistics at George Washington University, Washington, DC. Her research concerns measurement error models, non/semiparametric regression, and spatial statistics.

Rachel A. Davidson is a Professor in the Department of Civil and Environmental Engineering and a core faculty member in the Disaster Research Center at the University of Delaware. She received a PhD from Stanford University. Davidson conducts research on natural disaster risk modeling and civil infrastructure systems. Her work involves developing new engineering models to better characterize the impact of future natural disasters, and using that understanding to support decisions to help reduce future losses. She is a fellow and past president of the Society for Risk Analysis.

Vitor Hugo Ferreira is currently with Universidade Federal Fluminense (UFF), Niterói, Brazil. He received his BSc degree in Electrical Engineering from the Federal University of Itajubá in 2002, and MSc and DSc degrees in Electrical Engineering from the Federal University of Rio de Janeiro (COPPE/UFRJ) in 2005 and 2008, respectively, all in Brazil. His research interests include time series forecasting and neural networks. He is the Chairman of Electrical Engineering Department at UFF.

Carolina García-Martos is an Associate Professor at the Escuela Tecnica Superior de Ingenieros Industriales (ETSII), Universidad Politécnica de Madrid (UPM), Spain. She received her degree in Industrial Engineering and her PhD in Applied Statistics from the Universidad Politécnica de Madrid in July 2005 and June 2010, respectively.

She has received several awards for her PhD thesis: Extraordinary Prize from the UPM, the Loyola de Palacio Best PhD Prize on EU Energy Policy given by the European University Institute (EUI), Florence School of Regulation and Loyola de Palacio Chair (Third Prize Winner), the "ELECNOR" PhD Thesis Award and the PhD Thesis Special Mention by the Professional Association of Industrial Engineers of Madrid (Spain). She has published her works in *Technometrics*, *IEEE Transactions on Power Systems*, *Applied Energy*, *Energy Economics*, and *Wiley Encyclopedia of Electrical and Electronics Engineering*, among others.

Antonio Gómez-Expósito received an Industrial Engineering degree, major in electrical engineering, and a Doctor of Engineering degree in Power Engineering from the University of Seville, Spain, in 1982 and 1985, respectively. He is currently the Endesa Red Industrial Chair Professor with the University of Seville. His research interests include optimal power system operation, state estimation, digital signal processing, and control of flexible ac transmission system devices.

Catalina Gómez-Quiles received an engineering degree from the University of Seville, Spain, and an MSc degree from McGill University, Montreal, Canada, both in Electrical Engineering. In 2012, she got a PhD degree from the University of Seville. Her research interests relate to mathematical and computer models for power system analysis.

Yunhe Hou received his BE and PhD degrees in Electrical Engineering from the Huazhong University of Science and Technology, Wuhan, China, in 1999 and 2005, respectively. He was a post-doctoral research fellow at Tsinghua University, Beijing, China, and a post-doctoral researcher at Iowa State University, Ames, IA, and the University College Dublin, Dublin, Ireland. He was also a Visiting Scientist at the Massachusetts Institute of Technology, Cambridge, MA. He is currently an Associate Professor with the Department of Electrical and Electronic Engineering at the University of Hong Kong, Hong Kong.

Chen-Ching Liu received his PhD from the University of California, Berkeley, CA. He is Boeing Distinguished Professor at Washington State University, Pullman, WA, and Visiting Professor at University College Dublin, Ireland. He was a Palmer Chair Professor at Iowa State University and a Professor at the University of Washington, Seattle, WA. Dr. Liu received the IEEE PES Outstanding Power Engineering Educator Award in 2004. He was recognized with a Doctor Honoris Causa from University Politehnica of Bucharest, Romania. Professor Liu is a fellow of the IEEE and member of the Washington State Academy of Sciences.

Haibin Liu is with State Key Laboratory of Power Equipment and System Security and New Technology, College of Electrical Engineering, Chongqing University, China.

Paras Mandal is an Assistant Professor of Electrical and Computer Engineering and Director of Power and Renewable Energy Systems (PRES) Lab at the University of

Texas at El Paso (UTEP). He received his ME and BE degrees from Thailand and India, respectively, and a PhD degree from the University of the Ryukyus, Japan. His research interests include AI application to forecasting problems, renewable energy systems, power systems operations and markets, power system optimization, and smart grid.

José L. Martínez-Ramos received a PhD degree in Electrical Engineering from the University of Sevilla, Spain. Since 1990, he has been with the Department of Electrical Engineering, University of Sevilla, where he is currently a Professor. His primary areas of interest are active and reactive power optimization and control and power system analysis.

Patrick E. McSharry is a senior research fellow at the Smith School of Enterprise and the Environment, faculty member of the Oxford Man Institute of Quantitative Finance at Oxford University, Visiting Professor at the Department of Electrical and Computer Engineering, Carnegie Mellon University, and will lead the new World Bank-funded African Centre of Excellence in Data Science (ACE-DS) based in Rwanda. He is a fellow of the Royal Statistical Society, a senior member of the IEEE, and a senior academic member of the Willis Research Network. He takes a multidisciplinary approach to developing quantitative techniques for data science, decision-making, and risk management. His research focuses on Big Data, forecasting, predictive analytics, machine learning, and the analysis of human behavior. He has published over 100 peer-reviewed papers, participated in knowledge exchange programs, and consults for national and international government agencies and the insurance, finance, energy, telecoms, environment, and healthcare sectors. Patrick received a first-class honours BA in Theoretical Physics and an MSc in Engineering from Trinity College Dublin and a DPhil in Mathematics from Oxford University.

Michael Negnevitsky is a Professor/Chair in Power Engineering and Computational Intelligence and the Director of the Centre for Renewable Energy and Power Systems (CREPS) at the University of Tasmania, Hobart, Australia. He received a PhD degree from Byelorussian University of Technology, Minsk, Belarus. He was a senior research fellow and a Senior Lecturer in the Department of Electrical Engineering, Byelorussian University of Technology, Minsk.

Jesús M. Riquelme-Santos received his PhD degree in Electrical Engineering from the University of Sevilla, Spain. Since 1994, he has been with the Department of Electrical Engineering, University of Seville, where he is currently an Associate Professor. His areas of interest are active and reactive power optimization and control, power system analysis, and power quality and forecasting techniques.

José C. Riquelme received a PhD degree in Computer Science from the University of Sevilla, Spain. Since 1987, he has been with the Department of Computer Science, University of Sevilla, where he is currently an Associate Professor. His primary areas of interest are data mining, knowledge discovery in databases, and machine learning techniques.

Julio Rodríguez has a BA in Mathematics from the Universidad Autónoma de Madrid and a PhD in Mathematics from the Universidad Carlos III de Madrid. He was a Visiting Associate Professor at the Graduate School of Business (GSB) at the University of Chicago in 2002, and an Associate Professor of Statistics at the Universidad Politecnica de Madrid, from 2003 to 2005. Since 2006, he is an Associate Professor of Econometrics and Statistical Methods at the Departamento de Análisis Económico: Economía Cuantitativa, Universidad Autónoma de Madrid. His areas of interest are applications in electricity markets, time series, multivariate analysis, graphical methods, and functional data. He has published his work in *Journal of the American Statistical Association*, *Technometrics*, *International Journal of Forecasting*, and *Journal of Multivariate Analysis*, among others.

Harold Salazar received a PhD degree in Electrical Engineering and an MS degree in Economics from Iowa State University, Ames, IA. He is currently a Professor at the Technological University of Pereira (Universidad Tecnológica de Pereira), Colombia. He is also a consultant for the National Energy and Gas Regulatory Commission of Colombia (CREG in Spanish) and for the Power Market Operator of Colombia.

María Jesús Sánchez is an Associate Professor of Statistics at the Escuela Técnica Superior de Ingenieros Industriales Universidad Politécnica de Madrid. She obtained her degree in Electrical Engineering and her PhD in Applied Statistics both from the Universidad Politécnica de Madrid. Her research areas of interest are outliers in time series, Kriging models, reliability of electric power generating systems, load and prices forecasting models, and dimensionality reduction techniques with application to liberalized electricity markets forecasting. She has published her works in *Technometrics*, *Computational Statistics and Data Analysis*, *Reliability Engineering and System Safety*, and *IEEE Transactions on Power Systems*, among others.

Tomonobu Senjyu is a Professor with the Department of Electrical and Electronics Engineering, University of the Ryukyus. He received her BS and MS degrees in Electrical Engineering from the University of the Ryukyus, Okinawa, and a PhD degree in Electrical Engineering from Nagoya University, Nagoya, Japan. His research interests are in the areas of power system optimization and operation, electricity market, intelligent systems, power electronics, renewable energy, and smart grid.

Anurag K. Srivastava is an Associate Professor in the School of Electrical Engineering and Computer Science at Washington State University, Pullman, WA. In 2005, he received his PhD degree from Illinois Institute of Technology (IIT), Chicago, IL.

James W. Taylor is a Professor of Decision Science at the Saïd Business School of the University of Oxford. His research is in the area of time series forecasting, and he teaches analytics courses for the Oxford MBA and Executive MBA Programmes. He has a PhD in Time Series Forecasting from the London Business School. He is a former Associate Editor of the *International Journal of Forecasting* and *Management Science*.

Alicia Troncoso is a Professor in the School of Engineering at Pablo de Olavide University in Seville, Spain. She received a PhD degree in Computer Science from the University of Seville, Spain, in 2005. Presently, she is an Associate Professor at the University Pablo de Olavide, Seville. Her primary areas of interest are time series analysis, control and forecasting, and optimization techniques.

Chapter 1

Introduction

Mohamed E. El-Hawary

PRELUDE

Since time immemorial, human communities have been preoccupied with foreseeing the future and events that lie ahead. In the past, shamans, soothsayers, oracles, and comets foretold the future on the basis of prevailing ideologies, customs, and past observations of past events. Present-day science of forecasting is based on skillfully blending statistical principles, ingenious deductions and observations and measurements of the causal interactions between social, economic, and physical quantities and of the underlying processes. For an insightful and highly readable historical treatment of the evolution of forecasting, Reference [1] is recommended.

The intent of this introductory chapter is to offer the reader a brief discussion of selected technologies and issues of forecasting in electric power systems, and specifically load and electricity forecasting contributions in Chapters 2–8.

FORECASTING: GENERAL CONSIDERATIONS

The term forecasting refers to the process of making statements about events whose actual outcomes (typically) have not yet been observed. Moreover, forecasting is a decision-making tool that deals with predicting future events, and the proper presentation and use of forecasts to help in budgeting, planning, and estimating future growth of a quantity (or quantities.) Prediction is a similar, but more general term. Both might refer to formal statistical methods. In the simplest terms, forecasting aids in predicting future outcomes based on past events and expert insights. It is generally accepted that while forecasts are rarely perfect, they are more accurate for grouped data than for individual items and for shorter than longer time periods. In

Advances in Electric Power and Energy Systems: Load and Price Forecasting, First Edition.
Edited by Mohamed E. El-Hawary.
© 2017 by The Institute of Electrical and Electronics Engineers, Inc. Published 2017 by John Wiley & Sons, Inc.

hydrology, the terms "forecast" and "forecasting" are sometimes reserved for estimates of values at certain specific future times, while the term "prediction" is used for more general estimates, such as the number of times floods will occur over a long period.

There are diverse applications of forecasting, the most familiar applications are in the area of weather conditions, where temperature, precipitation, wind, barometric pressure are to be forecast. Typical applications of forecasting in the business domain include *Supply Chain Management* which makes sure that the necessary productive resources (capital, labor, component parts, and the like) are always available to manufacture the required output to meet consumer demand and from that estimate determine the necessary resources to produce the forecasted amount of output. In other words, forecasts make sure that the right product is at the right place at the right time. Another business application of forecasting is *Inventory Control* which aims to maximize profits, efficient inventory management is required so as not to tie up idle inventory unnecessarily. Alternatively, accurate forecasting will help retailers reduce excess inventory and therefore increase profit margins. Accurate forecasting will also help retailers to meet consumer demand.

It is important to note that forecasts may be conditional in the sense that, if policy A is adopted then X will take place. The field of forecasting relies on judgment, uses intuition and experience in addition to quantitative or statistical methods. Quantitative forecasting relies on identifying repeated patterns in data, so it may take some time to see the same pattern repeat more than once. Combining judgment and quantitative forecasting gets the best results. The most trustworthy forecasts combine both methods to support their strengths and counteract their weaknesses.

FORECASTING IN ELECTRIC POWER SYSTEMS

Forecasting of electric power system variables is vital for many operational and planning functions. Historically, forecasting power system load has been a dominant application in the electric utility business. In this regard, forecasting the demand for water and gas in a corresponding utility occupy the same prominent position in the utilities business.

The advent of power system competition and deregulation has introduced new requirements for forecasting additional quantities, with varying degrees of importance. Forecasting the electrical energy price in power markets is most relevant along with other applications such as:

1. Energy price forecasting and bidding strategy in power system markets
2. Day-ahead prediction of residual capacity of energy storage unit of microgrid in islanded state
3. Reservoir inflow forecasting
4. Flood forecasting

The introduction of renewable energy sources, has introduced new forecasting challenges such as:

1. Wind power forecasting
2. Photovoltaic and solar power forecasting
3. Marine currents for tidal, wave, and river turbines

In scanning the literature on forecasting in electric power systems, we encounter new challenges such as:

1. Electric power consumption forecast of life of energy sources
2. Power quality prediction

Our conversation will focus on advanced load and price forecasting approaches.

LOAD FORECASTING IN ELECTRIC POWER SYSTEMS

Some may argue that electricity load forecasting has reached a state of maturity, but the advent of electricity markets and the progress in renewable energy sources have changed the nature of electricity production and consumption. In their now classic review paper, Gross and Galiana [2] defined electric power systems' short-term load forecasting (STLF) as dealing with prediction times of the order of minutes, hours, or possibly half hour up to 168 hours or a few weeks for the short-term problem. It is natural to recognize that from a mathematical forecasting (prediction) technique point of view, this qualification is not essential because the techniques apply equally to longer-term forecasts of months and even longer. Since the load forecasts play a crucial role in the composition of these prices, they have become vital for the electricity industry. In the pre-competitive era, the basic variable of interest in STLF has been, typically, the hourly integrated total system load. In a competitive market, the participants such as power producers, independent system operators, and power aggregators determine the "sampling" frequency and hence the prediction duration. The term load may mean peak daily system load, or system load values at pre-defined times of the day, or the hourly (or weekly) system energy, or individual bus loads or energy levels. Short-term forecasting is closer to operations, while the long-term function is closer to system planning applications.

The forecasting models vary because the factors affecting the prediction vary.

The active power generation of the system follows the active power load at all times. Whole units must be brought on line or taken out on an hourly basis, and prediction of load over such intervals is essential. Unit commitment and spinning reserve allocation need STLF based on 24 hour predictions. Security assessment relies on a priori knowledge of the expected values of bus loads from 15 minutes to a few hours to allow detecting vulnerable situations and taking corrective measures.

Regular maintenance scheduling requires load forecasts of 1 or 2 weeks to maintain a predetermined reliability level.

The behavior of an electric power system load depends on factors such as time, weather, and small random disturbances reflecting the inherent statistical nature of the load because not every user is affected in the same way by the time and weather effects. Time factors include weekly periodicity and seasonal variations. Temperature, humidity, intensity of light, wind speed, precipitation, and cloud cover are reasonably important weather-related variables known to modify power consumption. Annual load growth or decline is a factor that used to be relatively easy to identify, reflecting primarily the growth of sales of new electric equipment relative to preceding years.

Other attributes can be suggested, such as the geographically distributed nature of the load and the possible decomposition into residential, commercial, municipal, and industrial type loads.

Predicting load involves developing a model describing its behavior based on possibly abstract rules discerned by experienced operators, or it may be a concrete mathematical model. The rules used by the operator in load forecasting may extrapolate past load behavior correlated with expected future weather, and are conceptually not different from a mathematical model.

Establishing mathematical load forecasting models involves modeling, identification, and performance analysis. The mathematical models hypothesize the structure of a model relating load to the effects influencing its behavior based on physical observations. The identification step determines the values of those free parameters of the model which result in the closest "fit" of the load behavior generated by the model to the actual observed load behavior. The last step tests the validity of the model to forecast load. If the last step indicates that the hypothesis of the modelling step was inadequate, one returns to the model step for a modification of the model structure and a repetition of the next two steps is necessary.

ELECTRICITY PRICE FORECASTING IN ELECTRIC POWER SYSTEMS

In competitive electricity markets, participants, such as generators, power suppliers, investors, and traders, require accurate electricity price forecasts to maximize their profits. Forecasting loads and prices in electricity markets are mutually intertwined activities, and errors in load forecasting will propagate to price forecasting. Unlike load forecasting, electricity price forecasting is much more complex because of the unique characteristics, uncertainties in operation, as well as the bidding strategies of market participants. The main features that make it so specific include the nonstorability of power, which implies that prices depend strongly on the power demand. Electricity prices depend on fuel prices, generation unit operation costs, weather conditions, and probably the most theoretically significant factor, the balance between overall system supply and demand. Another characteristic is the seasonal behavior of the electricity price at different levels (daily, weekly, and annual seasonality.) Because

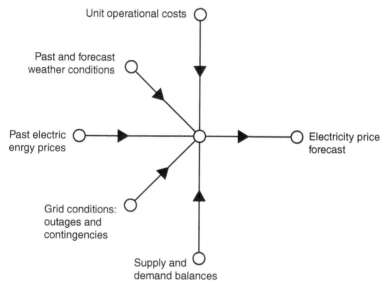

Figure 1-1   Factors influencing electricity prices.

electricity cannot be stored and needs constant balance between demand and supply, the price of electricity is volatile, which causes high risks to market participants.

Electricity price forecasts may be classified into three time horizons: short-term, medium-term, and long-term forecasts. In spot markets, short-term (mainly one day-ahead) price forecasts allow market participants to maximize their profits. The hourly price series are nonstationary and volatile, exhibiting multiple seasonality, spikes, and high frequency due to unexpected incidents such as transmission congestion, transmission and generation contingencies. Mid-term price forecasts are required for negotiating bilateral contracts between suppliers and consumers.

Long-term price forecasts guide decision-making on transmission expansion and enhancement, generation augmentation, distribution planning, and regional energy exchange. The main factors influencing electricity prices are shown in Fig. 1-1. Some additional factors are may be unavailable and may need forecasting as well.

TIME SERIES ANALYSIS

A time series is a sequence of data points $x(k)$, $k = 0, 1, 2, \ldots$, measured typically over regular successive regular time intervals such as hourly, daily, monthly. The measurements taken during an event in a time series are arranged in a proper chronological order and the function $x(k)$ is treated as a random variable. A time series made of the records of a single variable is called univariate, but if records of more than one variable are considered, it is referred to as multivariate. A time series can be continuous or discrete. In a continuous time, series observations are measured at every instant of time, whereas a discrete time series contains observations measured at discrete points of time.

A time series in general is affected by the following four main components, which are separable from the observed data.

Trend: The general tendency of a time series to increase, decrease, or remain invariant over a long period of time is termed trend. A trend is a long-term movement in a time series.

Cyclical: The cyclical variation in a time series describes the medium-term changes in the series, caused by circumstances, which repeat in cycles. The duration of a cycle extends over longer period of time, usually 2 or more years.

Seasonal: Seasonal variations in a time series are fluctuations within an interval during the season. The important factors causing seasonal variations are climate and weather conditions, customs, traditional habits, etc.

Irregular components: Irregular or random variations in a time series are caused by unpredictable influences, which are not regular and also do not repeat in a particular pattern.

Multiplicative and Additive Models

Considering the effects of the four components, two different types of representations (models) are generally used for a time series.

$$\text{Multiplicative model: } Y(k) = T(k) \times S(k) \times C(k) \times I(k).$$
$$\text{Additive model: } Y(k) = T(k) + S(k) + C(k) + I(k).$$

Here $Y(k)$ is the observation and $T(k)$, $S(k)$, $C(k)$, and $I(k)$ are respectively the trend, seasonal, cyclical, and irregular variation at time k.

The multiplicative model assumes that the four components of the time series are not necessarily independent and they can affect each other. On the other hand, the additive model assumes that the four components are independent of each other.

A component of a time series of random variables is independent and identically distributed (i.i.d.) if each random variable has the same probability distribution as the others and all are mutually independent.

Occam's Razor, the Principle of Parsimony

According to the principle of parsimony (or Occam's razor), an adequate representation of the underlying time series data is achieved by using the model with the smallest possible number of parameters. In other words, out of a number of suitable models, one should consider the simplest one, while still maintaining an accurate description of inherent properties of the time series.

In time series forecasting, past observations are used to establish an appropriate mathematical model (or representation), which captures the underlying data generating process for the series. This model is then used to predict future events. This approach is particularly useful when there is not much knowledge about the statistical pattern followed by the successive observations or when there is a lack of a satisfactory explanatory model.

The Stationarity Concept

The concept of stationarity of a stochastic process can be visualized as a form of statistical equilibrium. The statistical properties (such as mean and variance) of a stationary process do not depend upon time. It is a necessary condition for building a time series model that is useful for future forecasting purposes. Further, with this assumption, the mathematical complexity of the fitted model is decreased. In general, there are two fundamental types of stationary processes.

The Autoregressive (AR) Process

The autoregressive (AR) model is the simplest stationary process model which represents the dependency of the values of a time series on its past. The AR model generalizes the idea of regression to represent the linear dependence between a dependent variable $y(z_t)$ and an explanatory variable $x(z_{t-1})$. In general, an AR(p) has direct effects on observations separated by p lags and the direct effects of the observations separated by more than p lags are zero.

The AR processes have a relatively "long" memory, since the current value of a series is correlated with all previous ones, although with decreasing coefficients. This property means that we can write an AR process as a linear function of all its innovations, with weights that tend to zero with the lag. The AR processes cannot represent short memory series, where the current value of the series is only correlated with a small number of previous values.

Moving Average Processes

A family of processes that have the "very short memory" property is the moving average (MA) processes. The MA processes are a function of a finite, and generally small, number of its past innovations. A first-order moving average, MA(1), is defined by a linear combination of the last two innovations. Generalizing the idea of an MA(1), we can write processes whose current value depends not only on the last innovation but on the last q innovations. Thus the MA(q) process is obtained.

Based on the two fundamental models AR and MA, useful combinations are obtained.

Autoregressive Moving Average Processes

The ARMA (p,q) processes combine the properties of AR (p) and MA (q) processes and allow us to represent a time series using few parameters, those processes whose first coefficients can be any, whereas subsequent ones decay according to simple rules. autoregressive moving average (ARMA) processes give us a very broad and flexible family of stationary stochastic processes useful in representing many practical time series. It is always assumed that there are no common roots between the autoregressive and moving average components of an ARMA process.

Integrated Processes and ARIMA Models

A process can be nonstationary in the mean, in the variance, or in other characteristics of the distribution of variables. When the level of the series is not stable in time, in particular showing increasing or decreasing trends, we say that the series is not stable in the mean. When the variability or autocorrelations change with time, we say that the series is not stationary in the variance or autocovariance.

The most important nonstationary processes are the integrated ones, which have the basic property that stationary processes are obtained when they are differentiated. Nonstationary processes are useful when describing the behavior of many climatological or financial time series. When a nonstationary process in the mean is differentiated and a stationary process is obtained, the original process is known as an integrated one.

Seasonality and Seasonal ARIMA Models

Seasonality is a particular case of nonstationarity. A time series presents a seasonal pattern when the mean is not constant and evolves according to a cyclical pattern. For example, data coming from electricity markets present a variable mean, and it varies depending on the hour of the day and also depending on the day of the week. It is an example of data with two seasonal patterns: daily and weekly.

ARTIFICIAL NEURAL NETWORKS

Artificial neural network (ANN) methods have been promoted as tools to solve a large class of forecasting problems. An ANN contains simple processing units which are called neurons designed to replicate the action of the brain conducting a specific task [3]. The neurons are arranged in a distinct layered topology which consists of an input layer, one or more hidden layers, and an output layer. Each neuron follows a learning procedure to produce a biased weighted sum of its inputs and pass this activation level through a transfer function which maybe a sigmoid or tan-sigmoid function to generate its output.

The multiple layers of neurons with nonlinear activation functions allow the network to learn relationships between the input and the output of the network. ANN models may differ with regard to combinations of different numbers of hidden layers, different numbers of units in each layer and different types of transfer functions. Most applications, in compliance with the principle of parsimony, use three layers for the forecasting task.

The training process is crucial to the success of the ANN in the forecasting task. It employs a set of examples of proper network behavior. In training, the weights define a vector in a multidimensional space. The weights and biases are updated iteratively with the objective of minimizing the mean squared error between the actual output and the desired output for each pattern on the training set. The iterative process is repeated until an acceptable error is achieved. The training process is called backpropagation (BP). Once the model has acquired the knowledge, new data are then tested for forecasting.

$$E_{\mathrm{p}} = \frac{1}{N} \sum_{i=1}^{N} (t_{\mathrm{p}i} - o_{\mathrm{p}i})^2 \tag{1-1}$$

where $t_{\mathrm{p}i}$ is the target at ith pattern, $o_{\mathrm{p}i}$ is the predicted value of the network's output at ith pattern, and N is the number of training set examples.

Thus the parameters of data flow from the input neurons, forward through any hidden layer neurons, eventually reaching the output layer neurons. All of the layers are fully interconnected with each other by weights as shown in the typical structure of an ANN model shown in Fig. 1-2.

Early BP algorithms used gradient or conjugate gradient descent algorithms requiring the transfer function of each neuron to be differentiable making them slow to train and sensitive to the initial guess which could possibly be trapped at a local minimum.

Particle swarm optimization (PSO) was successfully employed to train neural networks resulting in better training performance, faster convergence rate, and enhanced

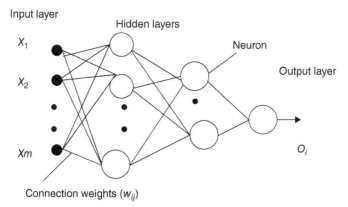

Figure 1-2   Typical backpropagation network.

predictions than BP-based ANN. However, the PSO algorithm has attractive features over other conventional algorithms in that it can be computationally inexpensive, easily implemented, and does not require gradient information of the objective function. PSO also has disadvantage; its parameters affect overall algorithm performance and stability, and must be selected depending on the experience of designer. A modified adaptive PSO (MAPSO) method was subsequently used to resolve all of these issues [4–6].

Radial Basis Function Networks

Evans et al. [7] credit Moody and Darken [8] for proposing neural networks referred to as radial basis function (RBF) networks that employ locally tuned neurons to perform function mapping. Keller et. al [9] suggested that the RBF network is a strong and viable alternative to multilayer perceptrons. While in the latter, the function approximation is defined by a nested set of weighted sums, in an RBF network, the approximation is defined by a single weighted sum. Building RBF networks choosing a Gaussian function as the RBF has many desirable features.

OVERVIEW OF CHAPTERS

This book deals with advances in forecasting technologies in electric power system applications. In addition to power system load forecasting, topics this book treats electricity price forecasting and a chapter dealing with storm-caused outage duration forecasting. The treatments are authoritative by forecasting experts from around the globe. We next offer an overview of the chapters that follow.

Chapters 2 and 3 offer sophisticated treatments of load forecasting, while Chapters 4–6 deal with innovative approaches to price forecasting.

Chapter 2 by Taylor and McSharry deals with univariate methods used to forecast intraday series of power system load for lead times up to a day ahead. While load is often modeled in terms of weather variables, univariate methods can be sufficient for short lead times because meteorological variables tend to change smoothly in the short term, and this will be reflected in the load series itself. A notable feature of an intraday load series is the presence of both an intraweek and an intraday seasonal cycle.

In this chapter, the authors present five methods designed to accommodate this characteristic. These methods are the ARMA modeling; periodic autoregressive (AR) modeling; exponential smoothing for double seasonality; a recently proposed alternative exponential smoothing formulation; and a method based on principal component analysis (PCA) of the daily load profiles. The chapter includes a comparison of the post-sample forecast accuracy of the methods using load data from 10 European countries. The results show a similar ranking of methods across the series, with exponential smoothing for double seasonality consistently outperforming the rest. The ARMA and PCA methods also performed well, but the results were disappointing for the other two methods.

Chapter 3 entitled "Application of the Weighted Nearest Neighbor Method to Power System Forecasting Problems" is by Gómez-Expósito, Troncoso-Lora, Riquelme-Santos, Gómez-Quiles, Martínez-Ramos, and Riquelme-Santos. The weighted nearest neighbors (WNN) approach is a data mining (DM) pattern classification technique based on the similarity of individual members of a population. The learning rule of a WNN classifier is that members of a population are surrounded by individuals who have similar properties. As a result, the nearest neighbors decision rule assigns to an unclassified sample point the classification of the nearest of a set of previously classified points. Unlike most statistical methods, the WNN method considers the training set as the model itself.

In this chapter a forecasting methodology, based on the WNN technique is implemented to forecast power system variables characterized by daily and weekly repetitive patterns, such as energy demand and prices. The chapter discusses ways of tuning model parameters such as time series window length, the number of neighbors chosen for the prediction, and the way the weighting coefficients are computed.

Three case studies are used to illustrate the potential of the WNN method. The first case involves energy demand auctioned every hour in the day-ahead Spanish electricity market. When applied to the weekdays of the year 2005, this technique provides a monthly relative prediction error ranging from 2.4% to 4.37%, which compares well with that obtained by an AR model (4.65%).

The second case involves the more irregular time series of hourly marginal prices of the day-ahead Spanish electricity market. In order to allow comparisons with some previously published methods, forecasting results corresponding to the market of mainland Spain for the entire year of 2002 are reported, yielding an average monthly error which is close to 8%. The WNN prediction accuracy is generally better, on average, than that of other more sophisticated techniques such as ANN, GARCH, and ARIMA (with and without applying the Wavelet transform).

The third case involves the hourly demand of a faculty building at the University of Seville during 2005. The time series is more volatile, as there are many ill-defined factors affecting the electricity consumption, which means that larger prediction errors are to be expected. Average weekly errors ranging from 4.4% to over 20% are provided by the WNN approach, depending on whether the weekly demand follows a regular pattern or not. This compares well with an AR model, yielding weekly errors from 3.4% to 75% for the same test periods. Because exogenous variables, such as weather conditions or equipment availability, are not explicitly included as an influential variable, unlike in other competing methods, good results provided by the WNN method are somewhat unexpected.

We shift our emphasis to price forecasting. Chapter 4, which deals with "Electricity Prices as a Stochastic Process," is by Hou, Liu, and Salazar. The authors' point of departure is that the new competitive environment of the power industry led to a radical change in how prices of electricity are established. Generators' bids, load demand, and power grid conditions are the primary factors that set the prices. While the mechanisms that determine electricity prices are known, random events such as contingencies and congestive conditions of transmission lines introduce uncertainties. Therefore, it is not easy to develop analytical models for electricity prices that can be used by market or system planners and investors in their decision-making process.

The chapter summarizes a set of stochastic process models that can be used for electricity prices according to the purpose of modeling. Significant features of electricity prices from various markets, such as mean reversion, seasonality, volatility, and spikes, are discussed. The nonstorable property of electricity, the need to constantly balance supply and demand, and available capabilities of the transmission network, constitute the physical constraints of electricity markets.

Continuous time stochastic models are widely used for modeling financial assets and derivatives. In this chapter, commonly used continuous time stochastic models such as Brownian motion, mean reversion process, geometric Brownian motion, geometric mean reversion process, are described in detail. In these models, deterministic components account for regularities in the behavior of electricity prices. The stochastic components describe the uncertainties involved in electricity prices. Among these models, the geometric mean reversion process is well accepted for modeling of electricity prices, especially with the seasonal pattern of prices. With price-dependent volatility, geometric mean reversion process is superior to other models. Moreover, the behavior of underlying electricity prices is nonlinear, implying the structural nonstationarity of fundamentals. This suggests that spot prices may be better modeled by a set of adaptive regime-switching models than from a single specification.

Time series models are used to understand the underlying context of data points, or to make predictions. Among these time series models, the stationary time series model, ARMA, is the foundation. For the system with covariance stationary property, ARMA-class models can be used to model the data series well. These models can be viewed as a special class of linear stochastic difference equations. In electricity price modeling, familiar approaches include ARMA with time-varying parameters and ARMA with exogenous variables. Usually, time-varying parameters are used to describe the seasonal pattern of prices; historical and forecasted loads, fuel price, and time factors served as exogenous variables. For the short-term price modeling, fluctuation of covariance is not prominent. Therefore, these models are good for price fitting and forecasting. However, the nonstationary property is an essential characteristic of electricity prices. ARCH and GARCH models are employed to deal with the issue. A distinctive feature of the models is that they recognize that volatilities and correlations are not constant. Similarly, these models can be extended with time-varying parameters and exogenous variables.

Establishing an adequate methodology to compute accurate short-run forecasts is the basis of every profit maximization strategy, and makes possible scheduling power generation units. This is the topic of Chapter 5 "Short-Term Forecasting of Electricity Prices Using Mixed Models" by García-Martos, Rodríguez, and Sánchez. The chapter establishes a methodology to compute accurate one-step-ahead forecasts. The main idea is to develop a powerful forecasting tool based on combining several prediction models whose forecasting performance has been verified and compared over a long period of time. A tutorial review of the theoretical basis of the topics used to develop the models presented in the chapter (time series analysis, locally weighted regression, and design of experiments) is included.

The novelty of the authors' approach is that they consider that for each day the price data available consists of a vector of 24 hourly prices, instead of a unique time series with several seasonalities (daily, weekly, etc.), as had been assumed in most earlier work. Therefore, considering what is known as the *parallel approach* allows one to take advantage of the homogeneity of these 24 time series. The objective is to select the model that leads to smaller prediction errors (among two proposed models) as well as the appropriate length of the time series used to build the forecasting model.

The results have been obtained by means of a computational experiment, estimating and calculating forecasts for every combination of models and lengths under study. Then, the results obtained are carefully and descriptively analyzed. Moreover, nonparametric estimates of the conditional mean of the errors taking into account the time evolution is given using locally weighted regression. The authors used a design of experiments with several factors to arrive at a proposal of a combination of models and lengths which establishes a mixed model which combines the advantages of the new models discussed. The most appropriate length in terms of smaller prediction error is selected. Finally, some numerical results for the Spanish market are shown.

Chapter 6 deals with "Electricity Price Forecasting Using Neural Networks and Similar Days" by Mandal, Srivastava, Senjyu, and Negnevitsky. This chapter presents an application of artificial neural network (ANN) assisted by similar days (SD) method to predict day-ahead electricity price. Publicly available data pertaining to the PJM electricity market are used to demonstrate the efficiency and accuracy of the proposed ANN and similar days approach. In the similar days method, the price curves are forecasted by using the information of the days being similar to that of the forecast day. Two procedures are analyzed, that is, prediction based on averaging the prices of similar days and prediction based on averaging the prices of similar days plus neural network refinement. The factors impacting the electricity price forecasting, including time factors, load factors, and historical price factors, are discussed. Comparison of forecasting performance of the proposed ANN model with that of forecasts obtained from the similar days method is presented. To give a better insight about the overall accuracy of the results obtained in this study, this chapter further presents a comparative analysis with existing literatures. This study also explores a new technique of recursive neural network (RNN) integrated with SD. RNN is a multistep approach based on one output node, which uses the previous prediction as input for the subsequent forecasts. Mean absolute percentage error (MAPE), mean absolute error (MAE), and forecast mean square error (FMSE) values obtained from the forecasting results demonstrate that the proposed models function reasonably well to predict short-term electricity price at different levels (daily, weekly, and annual seasonality).

Chapter 7, "Estimation of Post-Storm Restoration Times for Electric Power Distribution Systems," is by Davidson, Liu, and Apanasovich. Better estimation ahead of time of how long storm-caused power outages are expected to last would allow a utility company to better inform its customers, the public, and regulators; and would allow those people to better plan for the time without power. This chapter introduces a new method for estimating the time at which electric power will be restored after a major storm. The method was applied for hurricanes and ice storms for three major electric power companies on the East Coast of the United States. Using an unusually large dataset that includes the companies' experiences with six hurricanes and eight ice storms, accelerated failure time models were fitted and used to predict the duration of each probable outage in a storm. By aggregating those individual estimated outage durations and accounting for variable outage start times, restoration curves were then estimated for each county in the companies' service areas. The method can be applied as storm approaches, before damage assessments are available from the field. Results of model applications using test data suggest they have promising predictive ability. The real-life post-storm restoration process is described and alternative approaches to restoration modeling are reviewed and compared.

There are several opportunities for future research, both in improving the method introduced herein, and in building on the method and its application. A key limitation of the current method is that, because the necessary data could not be found, the outage duration models do not include potentially important tree-related or tree trimming covariates or covariates describing restoration resources used (e.g., number of line crews). Including such covariates would improve the models' predictive power and allow evaluation of the effect of changing tree trimming practices or the amount of resources.

Chapter 8, "A Nonparametric Approach for River Flow Forecasting Based on Autonomous Neural Network Models" is by Ferreira and Alves da Silva. Their work presents a forecasting method based on appropriate techniques for controlling ANN complexity with simultaneous selection of explanatory input variables via a combination of filter and wrapper techniques. In order to automatically minimize the out-of-sample prediction error, the three levels of Bayesian inference applied to multilayered perceptrons (MLPs) have been exploited. This training method includes complexity control terms in their objective function, which allow autonomous modeling and adaptation.

REFERENCES

1. D.N. McCloskey, "The art of forecasting, ancient to modern times," *Cato Journal*, vol. 12, no. 1, pp. 23–43, 1992.

2. G. Gross and F.D. Galiana, "Short-term load forecasting," *Proceedings of IEEE*, vol. 75, no. 12, 1558–1573, 1987.

3. S. Haykin, *Neural Networks: A Comprehensive Foundation*, Prentice-Hall, Upper Saddle River, NJ, 1999.

4. Z. Zhixia and L. Caiwu, "Application of the Improved Particle Swarm Optimizer to Vehicle Routing and Scheduling Problems," Proceedings of 2007 IEEE International Conference on Grey Systems and Intelligent Services, Nanjing, pp. 1150–1152, 18–20 Nov 2007.

5. J.-C. Cheng, T.-J. Su, M.-Y. Huang, C.-Y. Juang, "Artificial Neural Networks Design Based on Modified Adaptive Particle Swarm Optimization," Second International Conference on Next Generation Information Technology (ICNIT), pp. 201–206, July 2011.

6. Z.A. Bashir and M.E. El-Hawary, "Load and Locational Marginal Pricing Prediction in competitive Electrical Power Environment Using Computational Intelligence," *Canadian Conference on Electrical and Computer Engineering (CCECE'09)*, 2009.

7. P. Evans, K.C. Persaud, A.S. McNeish, R.W. Sneath, N. Hobson, and N. Magan, "Evaluation of a radial basis function neural network for the determination of wheat quality from electronic nose data," *Sensors and Actuators B*, vol. 69, pp. 348–358, 2000.

8. J. Moody and C. Darken, "Fast learning in networks of locally-tuned processing units," *Neural Computing*, vol. 1, no. 2, pp. 281–294, 1989.

9. J.M. Keller, D. Liu, and D.B. Fogel, "Radial-Basis Function Networks," Chapter in *Fundamentals of Computational Intelligence: Neural Networks, Fuzzy Systems, and Evolutionary Computation*, Wiley-IEEE Press, 2016.

Chapter 2

Univariate Methods for Short-Term Load Forecasting

James W. Taylor and Patrick E. McSharry

INTRODUCTION

Electricity load forecasting is of great importance for the management of power systems. Transmission systems operators require accurate load forecasts for managing the generation and distribution of electrical power. While load forecasts are required to balance demand and supply, they also play a crucial role in determining electricity prices. The appropriate modeling technique depends on the specific application and the forecast horizon. Long-term forecasts of peak electricity load are needed for capacity planning and maintenance scheduling [1]. Medium-term load forecasts are required for power system operation and planning [2]. Short-term load forecasts are required for the control and scheduling of power systems. Short-term forecasts are also required by transmission companies when a self-dispatching market is in operation. There are several such markets in Europe and the United States. For example, in Great Britain, one-hour-ahead forecasts are a key input to the balancing market, which operates on a rolling one-hour-ahead basis to balance supply and demand after the closure of bilateral trading between generators and suppliers [3, 4]. More generally, error in predicting electricity load has significant cost implications for companies operating in competitive power markets [5].

It is well recognized that meteorological variables, such as temperature, wind speed, and cloud cover, have a very significant influence on electricity load (see References [6–8]). However, in online short-term forecasting systems, multivariate modeling is usually considered impractical [9]. In such systems, the lead times considered are less than a day ahead, and univariate methods can be sufficient because the meteorological variables tend to change smoothly, which will be captured in

Advances in Electric Power and Energy Systems: Load and Price Forecasting, First Edition.
Edited by Mohamed E. El-Hawary.

17

the load series itself. Univariate models are often used for predictions up to about 3–6 hours ahead, and, due to the expense or unavailability of weather forecasts, univariate methods are sometimes used for longer lead times. By investigating the performance of a range of univariate methods over different forecast horizons, it is possible to ascertain the value of employing multivariate information such as weather variables.

In a recent study [10], methods for short-term load forecasting are reviewed and two intraday load time series are used to compare a variety of univariate methods. One of the aims of this chapter is to validate the results of that study. It concluded that a double seasonal version of Holt–Winters exponential smoothing was the most accurate method, with a new approach based on principal component analysis (PCA) also performing well. Using time series of intraday electricity load from 10 European countries, we empirically compare the better methods identified in Reference [10] and also the following two new candidate methods: an intraday cycle exponential smoothing method (see Reference [11]) and a new periodic autoregressive (AR) approach, which we believe has not previously been considered for electricity load forecasting. All of the methods are specifically formulated to deal with the double seasonality that typically arises in load data. This seasonality involves intraday and intraweek seasonal cycles. This chapter is based on the study described in Reference [12].

Artificial neural networks (ANNs) have featured prominently in the load forecasting literature (see References [13] and [14]). Their nonlinear and nonparametric features have been useful for multivariate modeling in terms of weather variables. However, their usefulness for univariate short-term load prediction is less obvious. Indeed, the results in Reference [10] for the ANN were not competitive. Although we accept that a differently specified neural network may be useful for univariate load modeling, in this chapter, for simplicity, we do not include an ANN. We would hope that the better performing methods in our study can serve as benchmarks in future studies with ANNs.

The chapter is structured as follows: Intraday Load Data describes the electricity load data; the section entitled Univariate Methods for Load Forecasting describes the methods included in the study; Empirical Forecasting Study presents the post-sample results of a comparison of the short-term forecasts with lead times up to one day ahead; Extensions of the Methods describes the extensions of the methods; and Summary and Concluding Comments summarizes the results and concludes the chapter.

INTRADAY LOAD DATA

Our dataset consisted of intraday electricity loads from 10 European countries for the 30 week period from Sunday April 3, 2005 to Saturday October 29, 2005. We obtained half-hourly data for six of the countries, and hourly data for the remaining four. The first 20 weeks of each series were used to estimate parameters and the remaining 10 weeks to evaluate post-sample accuracy of forecasts up to 24 hours ahead. For the half-hourly series, this implies 6720 observations for estimation and

3360 for evaluation. For the hourly series, 3360 observations were used for estimation and 1680 for evaluation. We represent the lengths of the intraday and intraweek cycles as s_1 and s_2, respectively. For the half-hourly load series, $s_1 = 48$ and $s_2 = 336$, and for the hourly series, $s_1 = 24$ and $s_2 = 168$.

Electricity load on "special days," such as bank holidays, is very different from normal days and can give rise to problems with online forecasting systems. For certain bank holidays in the year, the adjacent days also exhibit unusual load behavior and can be classed as special days. In practice, interactive facilities tend to be used for special days, allowing operator experience to supplement or override the standard forecasting system. As the emphasis of our study is online prediction, prior to fitting and evaluating the forecasting methods, we chose to smooth out the special days leaving the natural periodicities of the data intact. In our study, we smoothed out these unusual observations by taking averages of the observations from the corresponding period in the two adjacent weeks. An alternative approach of explicitly modeling special days is described in Reference [15].

Table 2-1 lists the 10 countries along with the mean electricity load for the 30 week period and in order to give a feel for the size of the countries, we also list their respective populations.

The 10 load series are plotted in Figs. 2-1 to 2-3. We use three graphs in order to avoid overlapping between any two series. (The three graphs do not represent a classification into three distinct groups.) It is interesting to note common features in certain of the series. The series for Norway, Great Britain, and Sweden show a smooth swing throughout the year, which would seem to be the annual cycle with the colder winter months demanding more electricity. This pattern also exists in the series for France, although it is not so smooth. The pattern appears to be slightly reversed for the warmer countries Spain, Portugal, and Italy, where demand is at its annual peak in the summer. A feature present in the series for France, Spain, and Italy is the sizeable drop in demand in August, which is due to the summer vacation. An

Table 2-1   Mean Load and Population for the 10 Countries

	Mean Load (GW)	Population (million)	Mean Load (W) Per Capita
Half-hourly			
Belgium	9.4	10.4	906
Finland	8.3	5.2	1588
France	47.8	60.9	786
Great Britain	36.0	58.9	611
Ireland	2.9	4.1	711
Portugal	5.2	10.6	492
Hourly			
Italy	32.7	58.1	563
Norway	12.0	4.6	2599
Spain	25.5	40.4	630
Sweden	14.5	9.0	1608

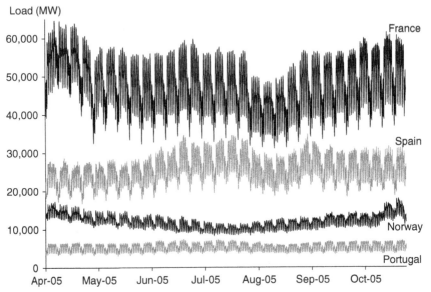

Figure 2-1   Electricity load (MW) in France, Norway, Portugal, and Spain from Sunday April 3, 2005 to Saturday October 29, 2005.

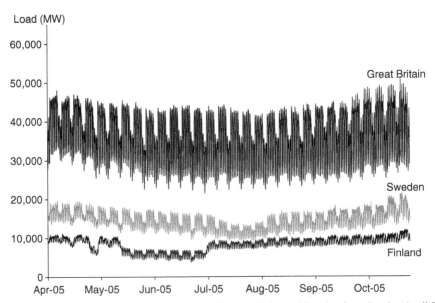

Figure 2-2   Electricity load (MW) in Finland, Great Britain, and Sweden from Sunday April 3, 2005 to Saturday October 29, 2005.

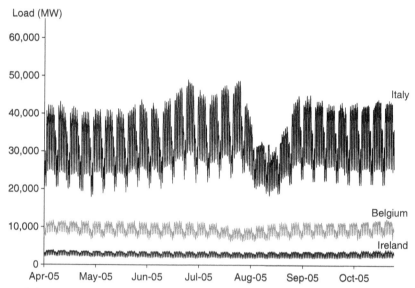

Figure 2-3   Electricity load (MW) in Belgium, Ireland, and Italy from Sunday April 3, 2005 to Saturday October 29, 2005.

interesting feature in the data for Finland is the temporary level shift in the first half of the series. This was due to the inactivity in the paper industry, which was caused by a large conflict in contract negotiations between the workers and employers.

Figure 2-4 shows the series for the two countries with highest electricity demand, France and Great Britain, for the fortnight in the middle of the 30 week period. This

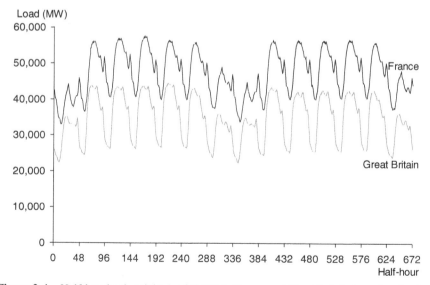

Figure 2-4   Half-hourly electricity load (MW) in France and Great Britain from Sunday June 10, 2005 to Saturday June 23, 2005.

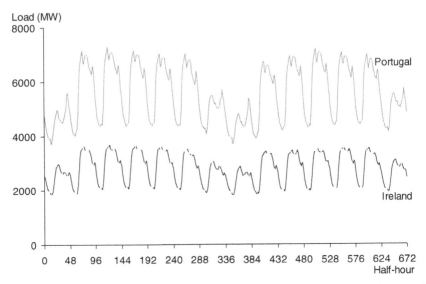

Figure 2-5   Half-hourly electricity load (MW) in Ireland and Portugal from Sunday June 10, 2005 to Saturday June 23, 2005.

graph shows a within-day seasonal cycle of duration $s_1 = 48$ periods and a within-week seasonal cycle of duration $s_2 = 336$ periods. The weekdays show similar patterns of load, whereas Saturday and Sunday have different levels and profiles. The intraweek and intraday features in Fig. 2-4 are typical of those in all 10 series. This is supported by Fig. 2-5, which shows the series for the two countries with lowest electricity demand, Ireland and Portugal. In addition to the differences in the levels of the four series plotted in Figs. 2-4 and 2-5, there are clear differences between their seasonal patterns. However, the features of intraday and intraweek seasonality are common to all the series, and so should be accommodated in a sophisticated univariate forecasting method.

UNIVARIATE METHODS FOR LOAD FORECASTING

Simplistic Benchmark Methods

In our empirical comparison of forecasting methods, we implemented two naïve benchmark methods. The first was a seasonal version of the random walk, which takes as a forecast the observed value for the corresponding period in the most recent occurrence of the seasonal cycle. With two seasonal cycles, it seems sensible to focus on the longer cycle, so that the prediction is constructed simply as the observed value for the corresponding period in the previous week. The forecast function is written as

$$\hat{y}_t(k) = y_{t+k-s_2}$$

where y_t is the load in period t, and k is the forecast lead time ($k \leq s_2$).

The second simplistic benchmark that we used was the simple average of the corresponding observations in each of the previous four weeks. For this method, the forecast function is given by the following expression:

$$\hat{y}_t(k) = (y_{t+k-s_2} + y_{t+k-2s_2} + y_{t+k-3s_2} + y_{t+k-4s_2})/4$$

The number of past periods to use in this simple average was made arbitrarily. A larger number would offer greater stability, but would not enable swift adaptation to changes in the series. The use of a simple average could also be criticized, as there is appeal in putting decreasing weights on older observations. The idea of using all observations and a decreasing weighting scheme is the common motivation for using exponential smoothing methods [16]. We consider a sophisticated version of exponential smoothing later in this section.

Seasonal ARMA

A forecasting method that has remained popular over the years, and appears in many load forecasting studies as a sophisticated benchmark, is ARMA modeling. The multiplicative seasonal autoregressive moving average (ARMA) model for a series with just one seasonal pattern can be written as

$$\varphi_p(L)\Phi_P(L^s)(y_t - c) = \theta_q(L)\Theta_Q(L^s)\varepsilon_t$$

where c is a constant term; s is the length of the seasonal cycle; L is the lag operator; ε_t is a white noise error term; and ϕ_p, Φ_P, θ_q, and Θ_Q are polynomial functions of orders p, P, q, and Q respectively. The model is often expressed as ARMA(p,q) $\times$ $(P,Q)_s$. Box et al. [17, p. 333] comment that the model can be extended to the case of multiple seasonalities. Double seasonal ARMA models are often used as benchmarks in load forecasting studies [18–20]. The multiplicative double seasonal ARMA model can be written as

$$\varphi_p(L)\Phi_{P_1}(L^{s_1})\Omega_{P_2}(L^{s_2})(y_t - c) = \theta_q(L)\Theta_{Q_1}(L^{s_1})\Psi_{Q_2}(L^{s_2})\varepsilon_t$$

In comparison with the ARMA model for single seasonality, there are several new terms in the double seasonal formulation; Φ_{P_1}, Ω_{P_2}, Θ_{Q_1}, and Ψ_{Q_2} are polynomial functions of orders P_1, P_2, Q_1, and Q_2 respectively. The model can be expressed as ARMA(p, q) $\times (P_1, Q_1)_{s_1} \times (P_2, Q_2)_{s_2}$.

For each of the 10 load series, we followed the Box–Jenkins methodology to identify the most suitable model based on the estimation sample of 20 weeks. We considered differencing, but the resultant models had weaker diagnostics than models fitted with no differencing. We estimated the models using maximum likelihood with the likelihood function based on the standard Gaussian assumption. We considered lag polynomials up to order 3. In their study of half-hourly data, Laing and Smith [18] write that forecasting performance is relatively insensitive to the order of the

Table 2-2   Orders of the Double Seasonal ARMA Model for Each of the 10 Load Series

	p	q	P_1	Q_1	s_1	P_2	Q_2	s_2
Half-hourly								
Belgium	[3]	3	3	[2, 3]	48	3	3	336
Finland	1	[1, 3]	3	[1, 3]	48	3	3	336
France	3	[2]	3	[3]	48	3	[1, 3]	336
Great Britain	3	1	[2, 3]	3	48	[1, 3]	[1, 3]	336
Ireland	2	0	3	[2, 3]	48	3	3	336
Portugal	3	0	3	[2, 3]	48	3	3	336
Hourly								
Italy	[1, 3]	2	[1, 3]	[1, 2]	24	3	[1, 3]	168
Norway	3	2	[1, 3]	[1, 3]	24	3	[2, 3]	168
Spain	3	2	3	[2, 3]	24	3	3	168
Sweden	3	2	3	3	24	3	[1, 3]	168

lag polynomials, and polynomials of order greater than two are rarely necessary. Experimentation with several of our series indicated that polynomials of order 3 were sufficient. We based model selection on the Schwarz Bayesian information criterion (BIC), with the requirement that all parameters were significant (at the 5% level). BIC measures the fit of a model as the sum of a term involving the log of the maximized likelihood and a penalty term for the number of parameters in the model [17, pp. 200–202].

The orders of the resulting double seasonal ARMA models are presented in Table 2-2. Our notation in the table is such that, if the order of the lag polynomial is given as 3, the polynomial includes linear, quadratic, and cubic terms. However, if the order is given as [3], this indicates that the lag polynomial includes only a cubic term and no linear or quadratic terms. As an example, we present here the resulting model for the British series

$$
\begin{aligned}
(1 &- 2.03L + 1.22L^2 - 0.18L^3)(1 - 0.42L^{2\times48} - 0.50L^{3\times48}) \\
&\times (1 - 0.74L^{336} - 0.22L^{3\times336})(y_t - 35,757) \\
= (1 &- 0.84L)(1 + 0.29L^{48} - 0.13L^{2\times48} - 0.33L^{3\times48}) \\
&\times (1 - 0.55L^{336} - 0.15L^{3\times336})\varepsilon_t
\end{aligned}
$$

For the French series, the optimal model was the following:

$$
\begin{aligned}
(1 &- 1.00L - 0.65L^2 + 0.66L^3)(1 - 0.29L^{48} - 0.15L^{2\times48} - 0.49L^{3\times48}) \\
&\times (1 - 0.58L^{336} - 0.09L^{2\times336} - 0.27L^{3\times336})(y_t - 47,760) \\
= (1 &- 0.58L^2)(1 - 0.35L^{3\times48}) \\
&\times (1 - 0.30L^{336} - 0.22L^{3\times336})\varepsilon_t
\end{aligned}
$$

In a recent load-forecasting study [21], the approach taken is to fit a separate model for each hour of the day. Each hourly model consists of a deterministic component and a seasonal autoregressive integrated moving average (ARIMA) component. Fitting a separate model for each hour of the day reduces the forecasting problem by eliminating the need to model the intraday seasonal cycle. The focus in Reference [21] is to forecast lead times of one day to one week ahead. For shorter lead times, which is the focus in our study, the use of separate models for each hour of the day is less appealing because the approach does not capture the level of the load at, or just prior to, the forecast origin.

Periodic AR

In intraday electricity load time series, the intraday seasonal cycle is usually reasonably similar for the five weekdays, but quite different for the weekends. This implies that the autocorrelation at a lag of one day is time-varying across the days of the week. Such time-variation cannot be captured in the seasonal ARMA model described in the previous section. A class of models that can capture this feature is periodic ARMA models. In these models, the parameters are allowed to change with the seasons [22]. Such models have been shown to be useful for modeling economic data [23]. In the electricity context, periodic models have been considered with some success in studies investigating methods for forecasting intraday net imbalance volume [4] and daily electricity spot prices [24].

To assess the potential for periodic ARMA models, we examined whether the autocorrelation at a specified lag exhibited variation across the periods of the day or the week. For example, for the half-hourly French series, Fig. 2-6 shows how the autocorrelation at lag $s_1 = 48$ varies across $s_2 = 336$ half-hours of the week. The first period of the x-axis corresponds to the first period on a Sunday. The autocorrelation

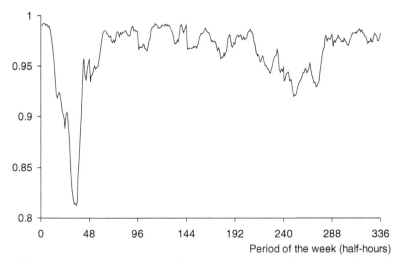

Figure 2-6   For the in-sample period of the French series, lag 48 autocorrelation estimated separately for each period of the week.

values were calculated from just the 20-week in-sample period. Although the sample size is not sufficiently large to conclude with confidence, the variation in the auto-correlation in this plot, and in similar graphs for the other series, suggested to us that there was some appeal in estimating periodic ARMA models for our data.

Studies have shown that periodic MA terms are unnecessary (see Reference [22], p. 28), and so for simplicity, in our work, we considered only periodic AR models. More specifically, we estimated models with periodicity in the coefficient of AR terms of lag s_1. The formulation for this method is presented in the following expressions:

$$(1 - \varphi_1 L)(1 - \varphi_{s_1}(t)L^{s_1})(1 - \varphi_{s_2}L^{s_2})(y_t - c) = \varepsilon_t$$

where

$$\varphi_{s_1}(t) = \omega + \sum_{i=1}^{4} \left(\begin{array}{c} \lambda_i \sin\left(2i\pi \frac{d(t)}{s_1}\right) + v_i \cos\left(2i\pi \frac{d(t)}{s_1}\right) \\ + \kappa_i \sin\left(2i\pi \frac{w(t)}{s_2}\right) + \upsilon_i \cos\left(2i\pi \frac{w(t)}{s_2}\right) \end{array} \right)$$

Here, $d(t)$ and $w(t)$ are repeating step functions that number the periods within each day and week, respectively. For example, for the half-hourly series, $d(t)$ counts from 1 to 48 within each day, and $w(t)$ counts from 1 to 336 within each week. ω, λ_i, v_i, κ_i, υ_i, φ_1, and φ_{s_2} are constant parameters. The periodic parameter, $\varphi_{s_1}(t)$, uses a flexible fast Fourier form similar to that employed in an analysis of the volatility in intraday financial returns [25]. For simplicity, we arbitrarily chose to sum from $i = 1$ to 4 for all 10 series. The parameters were estimated using maximum likelihood.

Exponential Smoothing for Double Seasonality

Exponential smoothing has found widespread use in automated applications, such as inventory control. For series with seasonality, the Holt–Winters method is often used. The multiplicative formulation for this method is presented in the following expressions:

$$\hat{y}_t(k) = l_t \, d_{t-s+k}$$
$$l_t = \alpha \, (y_t / d_{t-s}) + (1 - \alpha)l_{t-1}$$
$$d_t = \omega(y_t / l_{t-1}) + (1 - \omega) \, d_{t-s}$$

where l_t is the smoothed level; d_t is a seasonal index; α and ω are smoothing parameters; s is the length of the seasonal cycle; and $\hat{y}_t(k)$ is the k step-ahead forecast made from forecast origin t (where $k \leq s$). The Holt–Winters method has been adapted in order to accommodate the two seasonal cycles in electricity load series [26]. This involves the introduction of an additional seasonal index and an extra smoothing equation for the new seasonal index. The multiplicative formulation for the double seasonal Holt–Winters method is given in the following expressions:

$$\hat{y}_t(k) = l_t\, d_{t-s_1+k}\, w_{t-s_2+k} + \varphi^k(y_t - (l_{t-1}\, d_{t-s_1}\, w_{t-s_2})) \tag{2-1}$$

$$l_t = \alpha\,(y_t/(d_{t-s_1}\, w_{t-s_2})) + (1-\alpha)\, l_{t-1} \tag{2-2}$$

$$d_t = \delta\,(y_t/(l_t\, w_{t-s_2})) + (1-\delta)\, d_{t-s_1} \tag{2-3}$$

$$w_t = \omega(y_t/(l_t\, d_{t-s_1})) + (1-\omega)\, w_{t-s_2} \tag{2-4}$$

In this formulation, d_t and w_t are the seasonal indices for the intraday and intraweek seasonal cycles, respectively; and ω is a new smoothing parameter. The forecast function in expression (2-1) is valid for $k \le s_1$. The term involving the parameter φ in the forecast function is a simple adjustment for first-order autocorrelation that substantially improves the accuracy of the method [26]. A trend term was included in the original formulation, but we found it not to be of use for our 10 series.

An alternative formulation is the additive version of the method, which is presented in the following expressions:

$$\hat{y}_t(k) = l_t + d_{t-s_1+k} + w_{t-s_2+k} + \varphi^k(y_t - (l_{t-1} + d_{t-s_1} + w_{t-s_2}))$$

$$l_t = \alpha(y_t - d_{t-s_1} - w_{t-s_2}) + (1-\alpha)l_{t-1}$$

$$d_t = \delta(y_t - l_t - w_{t-s_2}) + (1-\delta)\, d_{t-s_1}$$

$$w_t = \omega(y_t - l_t - d_{t-s_1}) + (1-\omega)\, w_{t-s_2}$$

In our empirical work, this formulation led to results very similar to the multiplicative formulation of expressions (2-1) to (2-4), and so, for simplicity, we do not consider the additive formulation further in this chapter.

An important point to note regarding the double seasonal Holt–Winters exponential smoothing approach is that, in contrast to ARMA modeling and the majority of other approaches to short-term load forecasting, there is no model specification required. This gives the method a strong appeal in terms of simplicity and robustness.

The initial smoothed values for the level and seasonal components are estimated by averaging the early observations. The parameters are estimated in a single procedure by minimizing the sum of squared one step-ahead in-sample errors. We constrained the parameters to lie between 0 and 1, and estimated them by minimizing the sum of squared in-sample forecast errors (SSE). We applied various optimization algorithms and found that their success depended on the choice of initial values for the parameters. To address this, we followed an optimization procedure similar to that described for a somewhat different type of model in Reference [27]. For each model, we first generated 10^4 vectors of parameters from a uniform random number generator between 0 and 1. For each of the vectors, we then evaluated the SSE. The 10 vectors that produced the lowest SSE values were used, in turn, as the initial vector in a quasi-Newton algorithm. Of the 10 resulting vectors, the one producing the lowest SSE value was chosen as the final parameter vector.

Table 2-3 Parameters of the Holt–Winters Method for Double Seasonality for Each of the 10 Load Series

	α	δ	ω	φ
Half-hourly				
Belgium	0.043	0.146	0.175	0.820
Finland	0.000	0.083	0.153	0.996
France	0.004	0.249	0.231	0.987
Great Britain	0.002	0.316	0.168	0.970
Ireland	0.009	0.227	0.153	0.910
Portugal	0.094	0.201	0.210	0.771
Hourly				
Italy	0.039	0.271	0.281	0.944
Norway	0.039	0.126	0.151	0.863
Spain	0.036	0.193	0.217	0.871
Sweden	0.022	0.223	0.134	0.928

The resultant parameters for the 10 load series are presented in Table 2-3. For many of the series, the value of φ is very high and the value of α is very low, indicating that the adjustment for first-order autocorrelation has, to a large degree, made the smoothing equation for the level redundant. It is also interesting to note that, for a given series, the values are similar for δ and ω, the two smoothing parameters for the seasonal indices. We also implemented a version of the method with the optimized values of δ and ω constrained to be identical, and with $\alpha = 0$ so that the level was set as a constant value equal to the mean of the in-sample observations. This formulation delivered predictions only marginally poorer than the full method given in expressions (2-1) to (2-4). This is somewhat surprising, given that this reformulation of the method involves just two parameters.

From a theoretical perspective, exponential smoothing methods can be considered to have a sound basis as they have been shown to be equivalent to a class of state space models [28]. The double seasonal Holt–Winters formulation of expressions (2-1) to (2-4) can be expressed as a single source of error state-space model. This model can be used as the basis for producing prediction intervals. The motivation that led us to consider periodic AR models prompted us to also consider periodicity in the parameters of the double seasonal Holt–Winters method. Disappointingly, this did not lead to improved accuracy, and so, for simplicity, we do not report these further results in Empirical Forecasting Study. The issue of periodicity is addressed in the next section in an alternative exponential smoothing formulation that has recently been proposed.

Intraday Cycle Exponential Smoothing

A feature of the double seasonal Holt–Winters method is that it assumes the same intraday cycle for all days of the week, and that updates to the smoothed intraday cycle are made at the same rate for each day of the week. An alternative form of exponential smoothing for double seasonality is presented by Gould et al. in Reference [11]. It allows the intraday cycle for different days to be represented by different seasonal components. In addition, it allows the different seasonal components to be updated at different rates by using different smoothing parameters.

Our implementation of the method of Gould et al. involves the seasonality being viewed as consisting of the same intraday cycle for the five weekdays and a distinct intraday cycle for Saturday and another for Sunday. The days of the week are thus divided into three types: weekdays, Saturdays, and Sundays. By contrast with double seasonal Holt–Winters, there is no representation in the formulation for the intraweek seasonal cycle. Due to its focus on intraday cycles, we term this method "intraday cycle exponential smoothing." For any period t, the latest estimated values of the three distinct intraday cycles are given as c_{1t}, c_{2t}, and c_{3t}, respectively. The formulation requires three corresponding dummy variables, x_{1t}, x_{2t}, and x_{3t}, defined as follows:

$$x_{jt} = \begin{cases} 1 & \text{if time period } t \text{ occurs in a day of type } j \\ 0 & \text{otherwise} \end{cases}$$

Gould et al. present their approach in the form of a state-space model, and we follow this convention in our presentation of the model in expressions (2-5) to (2-7):

$$y_t = l_{t-1} + \sum_{i=1}^{3} x_{it}\, c_{i,t-s_1} + \varepsilon_t \tag{2-5}$$

$$l_t = l_{t-1} + \alpha \varepsilon_t \tag{2-6}$$

$$c_{it} = c_{i,t-s_1} + \left(\sum_{j=1}^{3} \gamma_{ij} x_{jt} \right) \varepsilon_t \qquad (i = 1,\, 2,\, 3) \tag{2-7}$$

where l_t is the smoothed level; ε_t is an error term; and α and the γ_{ij} are the smoothing parameters. (Expressions (2-6) and (2-7) can be rewritten as recursive expressions, which is the more widely used form for exponential smoothing methods.) We found that the results substantially improved with the inclusion of the adjustment for first-order autocorrelation that was used in expression (2-4) of the double seasonal Holt–Winters method. In Empirical Forecasting Study, we report only the results for this improved form of the intraday cycle method.

As with the double seasonal Holt–Winters method, we estimated the initial smoothed values for the level and seasonal components by averaging the early observations. The parameters were estimated by minimizing the sum-of-squared one-step-ahead in-sample errors. All parameters were constrained to lie between 0 and 1, and the optimization was performed using the same procedure described for the exponential method for double seasonality in Exponential Smoothing for Double Seasonality.

Table 2-4   Parameters of the Restricted Intraday Cycle Exponential Smoothing Method for Each of the 10 Load Series

	α	δ	ω	φ
Half-hourly				
Belgium	0.187	0.088	0.210	0.671
Finland	0.187	0.083	0.208	0.668
France	0.143	0.210	0.340	0.915
Great Britain	0.026	0.228	0.363	0.955
Ireland	0.000	0.168	0.241	0.948
Portugal	0.246	0.136	0.253	0.634
Hourly				
Italy	0.127	0.135	0.282	0.852
Norway	0.037	0.069	0.232	0.865
Spain	0.076	0.096	0.280	0.855
Sweden	0.077	0.133	0.344	0.890

The γ_{ij} can be viewed as a 3×3 matrix of parameters that enables the three types of intraday cycle to be updated at different rates. It also enables intraday cycle of type i to be updated even when the current period is not in a day of type i.

Several restrictions have been proposed for the matrix of γ_{ij} parameters (see Reference [11]). We included in our empirical study, two forms of the method; one involved estimation of the matrix of γ_{ij} parameters with the only restriction being that the parameters lie between 0 and 1, and the other involved the additional restrictions of common diagonal elements and common off-diagonal elements. Gould et al. note that these additional restrictions lead to the method being identical to the double seasonal Holt–Winters method of expressions (2-1) to (2-4), provided that seven distinct intraday cycles are used, instead of three, as in our study. In our discussion of the post-sample forecasting results in Empirical Forecasting Study, we refer to this second form of the model as the restricted form.

In Table 2-4, we present the parameters for the restricted form of the method. There are broad similarities between these values and those in Table 2-3 for the Holt–Winters method for double seasonality. For each series, the error autocorrelation parameter, φ, is substantially larger than the other parameters, and for most series, the smoothing parameter for the level, α, is the smallest parameter. One difference between the two methods is that, for each series, the value of α tends to be larger for the intraday cycle exponential smoothing method.

A Method Based on Principal Components Analysis

PCA provides a means of reducing the dimension of a multivariate data set to a smaller set of orthogonal variables, which are linear combinations of the original variables. PCA employs an orthogonal linear transformation to transform the data to a new coordinate system such that the first coordinate known as the first principal component captures the greatest fraction of the total variance. Similarly, the nth principal component captures the nth largest fraction of the variance. These components are called principal components because they are linearly uncorrelated and provide an efficient means of explaining the variation in the data. Dimension reduction is enabled by retaining only the first few components which are often capable of explaining a large fraction of the variability. PCA is often used to deal with collinear data where multiple variables are co-correlated, whereby the approach combines all collinear data into a small number of orthogonal variables, which can then be employed for regression analyses.

A forecasting method based on PCA has recently been proposed for short-term intraday load forecasting [10]. The method can be viewed as a development of the approach that uses a separate model for each of the s_1 periods of the day (see References [21] and [29]). In applying PCA, each period of the intraday cycle is viewed as one of a set of correlated variables. Instead of having to model each intraday period, the forecasting task is simplified to one that requires the modeling of just the reduced set of uncorrelated variables. Note that it would be relatively straightforward to extend the method to the multivariate case, if weather related variables were available.

In this section, we present only an overview of the method, as details are provided in Reference [10]. The method proceeds by arranging the intraday data as an ($n_d \times s_1$) matrix, Y, where n_d is the number of days in the estimation sample. Each column of this matrix constitutes a time series of daily observations for a particular period of the day. PCA is used to extract the main underlying components in these columns, and thus reduce the problem from having to forecast s_1 daily series to forecasting daily series for just the main components. To be more specific, PCA is applied to the columns of Y to deliver components that are columns of a new ($n_d \times s_1$) matrix. For each component, a regression model is built using day of the week dummies and quadratic trend terms. The models are then used to deliver a day-ahead forecast for each component. Load predictions are created by projecting forecasts of the components back onto the Y space. The method is refined, and speeded up, by focusing attention on just the *principal* components. Cross-validation is used to optimize two parameters: the number of principal components and the length of the training period used in the PCA. In our study, the cross-validation employed the first half of the 20-week in-sample period for estimation and the second half for evaluation. We set, as optimal parameters, those delivering the minimum sum of squared one-step-ahead forecast errors for the training data.

The errors resulting from this method exhibit autocorrelation. As for the Holt–Winters exponential smoothing method, the method benefits by the addition of an AR model of the error process. It is worth noting that in Reference [29], an error model was also employed in order to correct for serial correlation resulting from the use of separate models for each hour of the day. Let $E_t(k)$ be the prediction error associated with a k step-ahead forecast made from origin t. The error-correction model is of the following form:

$$E_t(k) = \alpha_0(k) + \alpha_1(k)E_{t-s_1}(k) + \alpha_2(k)E_{t-1}(1)$$

Table 2-5   The Optimal Number of Training Weeks and the Number of
Principal Components for the 10 Load Series for the PCA Method

	Number of Training Weeks	Number of Principal Components
Half-hourly		
Belgium	7	12
Finland	9	9
France	4	9
Great Britain	8	12
Ireland	8	10
Portugal	8	9
Hourly		
Italy	3	5
Norway	8	6
Spain	10	7
Sweden	8	6

where $\alpha_l(k)$ are parameters estimated separately for each lead time, k, using LS regression applied to the estimation sample. Finally, with this model specification that now includes the error correction term, cross-validation is used to optimize the number of principal components and the length of the training period used in the PCA. For each of our 10 load series, the optimal values are presented in Table 2-5.

EMPIRICAL FORECASTING STUDY

We evaluated post-sample forecasting performance from the various methods using the mean absolute percentage error (MAPE) and the mean absolute error (MAE). Having calculated the MAPE for each method at each forecast horizon, we then summarized each method's performance by averaging the MAPE across the 10 load series. For the half-hourly series, MAPE values were available for 48 half-hour lead times, while, for the hourly data, forecasts were obviously only available for 24 hourly lead times. In order to average across all the series, we focused only on the 24 hourly lead times. The resulting mean MAPE values are presented in Fig. 2-7.

The first point to note is that Fig. 2-7 does not show the results for the second simplistic benchmark method that involved the simple average of the corresponding observations in each of the previous four weeks. The results for this method were poorer than those for the seasonal random walk, and so, for simplicity, we opted to omit the results from the figure. Turning to the more sophisticated methods, the figure shows the double seasonal Holt–Winters method performing the best, followed by the PCA method and then the seasonal ARMA. Of the two versions of the intraday cycle exponential smoothing method, the restricted form appears to be better, which is consistent with the results in Reference [11]. However, the results for both forms of

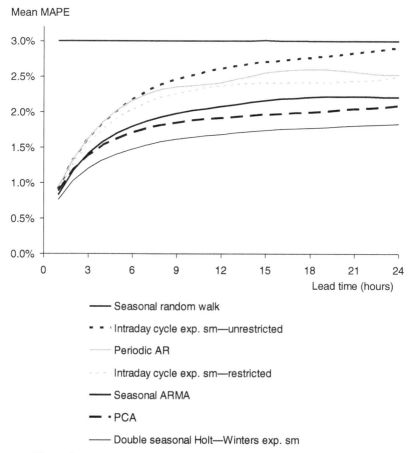

Mean MAPE

Figure 2-7   Mean MAPE plotted against lead time for the 10 load series.

this method are disappointing. This is also the case for the periodic AR method. Our view is that there is strong potential for the use of some form of periodic model but a longer time series may be needed to estimate the periodicity in the parameters. With regard to Fig. 2-6, the use of only 20 weeks of data implies that only 19 observations were available to estimate the intraweek cyclical pattern in the autocorrelation for a given lag. In a similar way, 20 weeks of data are perhaps too little to provide adequate estimates of periodic model parameters.

In addition to the MAPE, we also calculated MAE for each of the methods and for each series. Averaging MAE values across the 10 series did not seem sensible because the values tended to be substantially higher for the series corresponding to higher levels of electricity load. In view of this, for each method, we summarized the MAE performance across the 10 series by averaging, for each lead time, the ratio of the method's MAE to the MAE of the seasonal random-walk benchmark method. We present the results for this "relative MAE" measure in Fig. 2-8. The figure shows that,

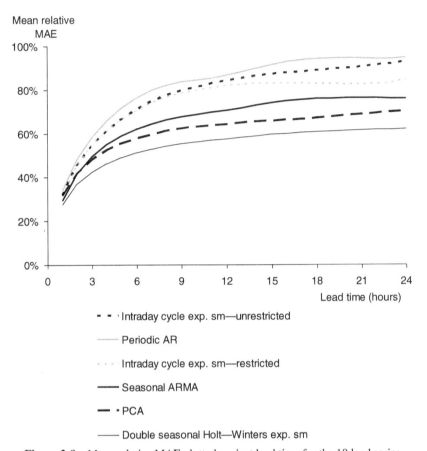

Figure 2-8 Mean relative MAE plotted against lead time for the 10 load series.

on the whole, the relative performances of the methods according to this measure were similar to those for the mean MAPE. The only noticeable difference between the results for the two measures is that the performance of the intraday cycle exponential smoothing methods improved relative to the periodic AR method when evaluated using the relative MAE measure.

We also evaluated the methods using the root mean squared percentage error and root mean squared error, but we do not report these results because the relative performances of the methods for these measures were similar to those for the MAPE and the relative MAE.

We should also comment that the rankings of the methods were really quite stable across the 10 series, and that the double seasonal Holt–Winters method was consistently the most accurate regardless of the error measure used for evaluation.

As we explained in Introduction, univariate methods tend to be used most often for predicting load up to lead times of just a few hours ahead. In Fig. 2-9, we focus more closely on the post-sample three-hour-ahead results for the three methods that

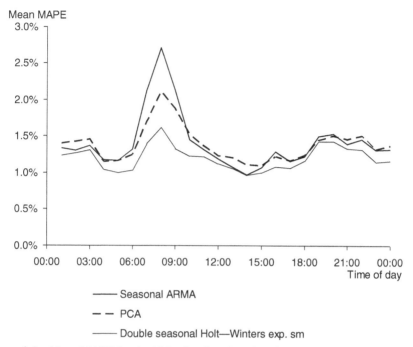

Figure 2-9   Mean MAPE for the 10 load series plotted against time of day for three-hour-ahead prediction.

performed the best in Figs. 2-7 and 2-8. Figure 2-9 shows the three-hour-ahead mean MAPE results plotted against time of day. The largest mean MAPE values occur for all three methods at around 8 a.m., and this is because at around this time of the day, the load tends to be changing more rapidly than at other periods of the day. The plot shows the double seasonal Holt–Winters method dominating at almost all periods of the day. The results for the other two methods are much closer, with the seasonal ARMA method matching the PCA method except for the periods around 8 a.m.

EXTENSIONS OF THE METHODS

Very Short-Term Forecasting with Minute-by-Minute Data

To enable the real-time scheduling of electricity generation as well as load–frequency control, online load forecasts are required for lead times as short as 10–30 minutes [30–32]. Such predictions have been termed *very short-term* forecasts. In a recent study [33], minute-by-minute British electricity load was used to compare the accuracies for very short-term forecasting of a variety of methods including nonseasonal methods, double seasonal ARMA, the Holt–Winters method for double seasonality, intraday cycle exponential smoothing, and a method based on regression models with weather forecasts as input.

For very short lead times, the most accurate methods were found to be the double seasonal Holt–Winters and the restricted intraday cycle exponential smoothing methods with parameters estimated using 30-minute-ahead in-sample forecast errors. Interestingly, the weather-based method was not competitive for very short-term prediction. The results for prediction beyond 30 minutes ahead showed that the weather-based approach was superior for lead times longer than about four hours. However, combining this method with the double seasonal Holt–Winters method led to the best results beyond about 1 hour ahead. This suggests that the double seasonal Holt–Winters method should be used for very short-term prediction, and that it can be used within a combined forecast to improve the accuracy of the weather-based method for longer lead times.

Recently Developed Exponentially Weighted Methods

French and British intraday load data was used in [34] for a comparison of five forecasting methods recently developed in [35]. The first of these methods involves exponential smoothing of both the total load for a week and the partition of this weekly total across the periods of the week. The second method is discount weighted regression (DWR) with trigonometric terms. This is a development of the use of exponentially weighted regression, which was applied to load data in [36]. The development provided by DWR is the use of more than one discount factor. The third method uses DWR to fit a time-varying regression spline to the seasonal cycles. The fourth method is an alternative form of time-varying spline, which uses exponential smoothing to model the spline at the knots. The fifth method aims to reduce the dimensionality of the modeling of intraday data by transforming the data using singular value decomposition (SVD), which is a procedure related to PCA, and then applying a form of exponential smoothing. All five methods were outperformed by a simpler SVD-based exponential smoothing formulation developed in [34]. In [35], SVD is essentially performed on the intraday cycle, while in the new method, SVD is applied to the intraweek cycle, which leads to a simpler and potentially more efficient model formulation. The empirical work in [34] showed that the methods that had not previously been applied to load data, were not able to outperform double seasonal Holt-Winters and intraday cycle exponential smoothing.

Triple Seasonal Methods for Multiple Years of Intraday Data

In this chapter, we have presented ARMA and exponential smoothing methods designed in order to capture the intraday and intraweek seasonal cycles evident in our 30-week samples of intraday load data. A time series of multiple years of intraday load data shows strong annual fluctuations in load, which motivates consideration of methods that are able to model the intraday, intraweek, and intrayear seasonal cycles. This is the focus in Reference [37], where double seasonal ARMA and exponential smoothing methods are extended for triple seasonality. An empirical analysis was performed using French and British data consisting of six years of half-hourly observations. The triple seasonal versions of the two methods were found to be more accurate than the double seasonal methods. For both series, there was little difference between the results of the ARMA and exponential smoothing methods for triple seasonality, and a simple average combination of the two methods led to the greatest accuracy.

SUMMARY AND CONCLUDING COMMENTS

In this chapter, we have used 10 time series of intraday electricity load to compare empirically, a number of univariate short-term forecasting methods. One of the aims of the chapter has been to validate the findings in Reference [10] using a substantially larger dataset. In addition to the methods that performed well in that study, we have also considered the intraday cycle exponential smoothing method (see Reference [11]) and a new periodic AR approach. Our results confirm the findings in Reference [10]. All the sophisticated methods outperformed the two naïve benchmark methods, and the best performing method was double seasonal Holt–Winters exponential smoothing, followed by the method based on PCA, and then seasonal ARMA. The results for the new intraday exponential smoothing method and for the periodic AR model were a little disappointing. We suspect that the performance of the periodic AR model may improve with the use of a longer time series. Our reasoning is that just 20 weeks of data may well be insufficient to capture the intraweek periodicity in a parameter. The same comment can also be applied to the intraday cycle exponential smoothing method because the approach involves the estimation of a relatively complex parameterization, which enables a form of periodic smoothing parameter.

The success of the double seasonal Holt–Winters exponential smoothing method is impressive, particularly in view of the method's simplicity. Ongoing work is aiming to gain insight into the method, with particular focus on the implication of including the autoregressive error correction term within the formulation. In terms of advising practitioners, the double seasonal Holt–Winters method would seem to be very attractive as it is simple to understand and implement and it has been shown to be accurate for short-term load prediction. Furthermore, the method is also appealing because of the existence of an underlying statistical model, which enables the calculation of prediction intervals. Finally, we should acknowledge that, if weather predictions are available, weather-based load forecasting methods may well be more accurate beyond about 3 to 6 hours ahead. However, for shorter lead times, the better of the univariate methods considered in this chapter should be competitive. In addition, the univariate methods have strong appeal, in terms of robustness, for online load prediction.

ACKNOWLEDGMENTS

We are grateful to the editor for his helpful comments. The authors are grateful to a number of people and organizations for supplying the data. These include Mark O'Mallley (University College Dublin, Ireland), Shanti Majithia (National Grid, UK),

Juan Toro (Transmarket, Spain), Rui Pestana (REN, Portugal), Maarit Uusitalo (Fingrid, Finland), Mikkel Sveen (Markedskraft), Samuele Grillo (University of Genova, Italy), and national transmission system operators, including Eirgrid in Ireland, RTE in France, Elia in Belgium, and Terna in Italy.

REFERENCES

1. P.E. McSharry, S. Bouwman, and G. Bloemhof, "Probabilistic forecasts of the magnitude and timing of peak electricity demand," *IEEE Transactions Power Systems*, vol. 20, pp. 1166–1172, 2005.

2. E. Gonzalez-Romera, M.A. Jaramillo-Moran, and D. Carmona-Fernandez, "Monthly electric energy demand forecasting based on trend extraction," *IEEE Transactions on Power Systems*, vol. 21, pp. 1946–1953, 2006.

3. M.P. Garcia and D.S. Kirschen, "Forecasting system imbalance volumes in competitive electricity markets," *IEEE Transactions on Power Systems*, vol. 21, pp. 240–248, 2006.

4. J.W. Taylor, "Density forecasting for the efficient balancing of the generation and consumption of electricity," *International Journal of Forecasting*, vol. 22, pp. 707–724, 2006.

5. D.W. Bunn, "Forecasting loads and prices in competitive power markets," *Proceedings of the IEEE*, vol. 88, pp. 163–169, 2000.

6. J.W. Taylor and R. Buizza, "Neural network load forecasting with weather ensemble predictions," *IEEE Transactions on Power Systems*, vol. 17, pp. 626–632, 2002.

7. J.W. Taylor and R. Buizza, "Using weather ensemble predictions in electricity demand forecasting," *International Journal of Forecasting*, vol. 19, pp. 57–70, 2003.

8. H. Ching-Lai, S.J. Watson, and S. Majithia, "Analyzing the impact of weather variables on monthly electricity demand," *IEEE Transactions on Power Systems*, vol. 20, pp. 2078–2085, 2005.

9. D.W. Bunn, "Short-term forecasting: A review of procedures in the electricity supply industry," *Journal of the Operational Research Society*, vol. 33, pp. 533–545, 1982.

10. J.W. Taylor, L.M. de Menezes, and P.E. McSharry, "A comparison of univariate methods for forecasting electricity demand up to a day ahead," *International Journal of Forecasting*, vol. 22, pp. 1–16, 2006.

11. P.G. Gould, A.B. Koehler, F. Vahid-Araghi, R.D. Snyder, J.K. Ord, and R.J. Hyndman, "Forecasting time-series with multiple seasonal patterns," *European Journal of Operational Research*, vol. 191, pp. 207–222, 2008.

12. J.W. Taylor and P.E. McSharry, "Short-term load forecasting methods: An evaluation based on European data," *IEEE Transactions on Power Systems*, vol. 22, pp. 2213–2219, 2007.

13. H.S. Hippert, C.E. Pedreira, and R.C. Souza, "Neural networks for short-term load forecasting: A review and evaluation," *IEEE Transactions on Power Systems*, vol. 16, pp. 44–55, 2001.

14. K. Methaprayoon, W.-J. Lee, S. Rasmiddatta, J.R. Liao, and R.J. Ross, "Multistage artificial neural network short-term load forecasting engine with front-end weather forecast," *IEEE Transactions on Power Systems*, vol. 43, pp. 1410–1416, 2007.

15. J.R. Cancelo, A. Espasa, and R. Grafe, "Forecasting from one day to one week ahead for the Spanish system operator," *International Journal of Forecasting*, vol. 24, pp. 588–602, 2008.

16. S. Makridakis, S.C. Wheelwright, and R.J. Hyndman, *Forecasting Methods and Applications*, 3rd edition, Wiley, NY, 1998.

17. G.E.P. Box, G.M. Jenkins, and G.C. Reinsel, *Time Series Analysis: Forecasting and Control*, 3rd edition, Prentice Hall, Englewood Cliffs, NJ, 1994.

18. W.D. Laing and D.G.C. Smith, "A Comparison of Time Series Forecasting Methods for Predicting the CEGB Demand," Proceedings of the Ninth Power Systems Computation Conference, 1987.

19. G.A. Darbellay and M. Slama, "Forecasting the short-term demand for electricity – Do neural networks stand a better chance?," *International Journal of Forecasting*, vol. 16, pp. 71–83, 2000.

20. R. Weron, *Modelling and Forecasting Electric Loads and Prices*, Wiley, Chichester, UK, 2006.

21. L.J. Soares and M.C. Medeiros, "Modeling and forecasting short-term electricity load: A comparison of methods with an application to Brazilian data," *International Journal of Forecasting*, vol. 24, pp. 630–644, 2008.

22. P.H. Franses and R. Paap, *Periodic Time Series Models*, Oxford University Press, Oxford, UK, 2004.

23. D.R. Osborn, S. Heravi, and C.R. Birchenhall, "Seasonality and the order of integration for consumption," *Oxford Bulletin of Economics and Statistics*, vol. 50, pp. 361–377, 1988.

24. S.J. Koopman, M. Ooms, and M.A. Carnero, "Periodic seasonal Reg-ARFIMA-GARCH models for daily electricity spot prices," *Journal of the American Statistical Association*, vol. 102, pp. 16–27, 2007.

25. T.G. Andersen and T. Bollerslev, "DM-dollar volatility: Intraday activity patterns, macroeconomic announcements and longer run dependencies," *Journal of Finance*, vol. 53, pp. 219–265, 1998.

26. J.W. Taylor, "Short-term electricity demand forecasting using double seasonal exponential smoothing," *Journal of Operational Research Society*, vol. 54, pp. 799–805, 2003.

27. R.F. Engle and S. Manganelli, "CAViaR: Conditional autoregressive value at risk by regression quantiles," *Journal of Business and Economic Statistics*, vol. 22, pp. 367–381, 2004.

28. R.J. Hyndman, A.B. Koehler, R.D. Snyder, and S. Grose, "A state space framework for automatic forecasting using exponential smoothing methods," *International Journal of Forecasting*, vol. 18, pp. 439–454, 2002.

29. R. Ramanathan, R. Engle, C.W.J. Granger, F. Vahid-Araghi, and C. Brace, "Short-run forecasts of electricity loads and peaks," *International Journal of Forecasting*, vol. 13, pp. 161–174, 1997.

30. W. Charytoniuk and M.-S. Chen, "Very short-term load forecasting using artificial neural networks," *IEEE Transactions on Power Systems*, vol. 15, pp. 263–268, 2000.

31. K. Liu, S. Subbarayan, R.R. Shoults, et al., "Comparison of very short-term load forecasting techniques," *IEEE Transactions on Power Systems*, vol. 11, pp. 877–882, 1996.

32. D.J. Trudnowski, W.L. McReynolds, and J.M. Johnson, "Real-time very short-term load prediction for power-system automatic generation control," *IEEE Transactions on Control Systems Technology*, vol. 9, pp. 254–260, 2001.

33. J.W. Taylor, "An evaluation of methods for very short-term load forecasting using minute-by-minute British data," *International Journal of Forecasting*, vol. 24, pp. 645–658, 2008.

34. J.W. Taylor, "Short-term load forecasting with exponentially weighted methods," *IEEE Transactions on Power Systems*, vol. 27, pp. 458–464, 2012.

35. J.W. Taylor, "Exponentially weighted methods for forecasting intraday time series with multiple seasonal cycles," *Int. J. Forecast.*, vol. 26, pp. 627–646, 2010.

36. W.R. Christiaanse, "Short-term load forecasting using general exponential smoothing," *IEEE Trans. Power App. Syst.*, vol. PAS-90, pp. 900–910, 1971.

37. J.W. Taylor, "Triple seasonal methods for short-term load forecasting," *European Journal of Operational Research*, vol. 204, pp. 139–152, 2010.

Chapter 3

Application of the Weighted Nearest Neighbor Method to Power System Forecasting Problems

Antonio Gómez-Expósito, Alicia Troncoso, Jesús M. Riquelme-Santos, Catalina Gómez-Quiles, José L. Martínez-Ramos, and José C. Riquelme

INTRODUCTION

A time-series database consists of sequences of values or events obtained over repeated measurements of time. Normally, the values are measured at equal time intervals. Time-series databases are popular in many applications, such as stock market analysis, economic and sales forecasting, process and quality control, and observation of natural phenomena (temperature, earthquake, pollution) or medical treatments.

The main goal of time series analysis is not only to forecast (i.e., to predict the future values), but also to find correlation relationships within time series; find similar or regular patterns; trends, and outliers. These last goals can be considered as time series modeling (i.e., to gain insight into the mechanisms or underlying causes that generate time series).

The time-series dynamics can be classified into four categories:

- Long-term or trend dynamics: general direction in which a time series is moving over a long interval of time.
- Cyclic dynamics or cycle variations: long-term oscillations about a trend line or curve, for example, business cycles.

Advances in Electric Power and Energy Systems: Load and Price Forecasting, First Edition. Edited by Mohamed E. El-Hawary.

- Seasonal dynamics or seasonal variations, that is, almost identical patterns that a time series appears to follow during corresponding months of successive years.
- Irregular or random variations.

Time-series modeling is referred to as the decomposition of a time series into these four basic dynamics, either as additive or as multiplicative model.

Classical methods for time-series analyses are often divided into two classes: frequency-domain methods and time-domain methods. The frequency methods are based on identifying periodic oscillations to describe the temporal evolution in terms of harmonics. The techniques used are Fourier transforms, spectral analysis and, more recently, wavelet analysis. Time-domain methods have a model-free subset consisting of the examination of auto-correlation and cross-correlation analysis. The well-known Box–Jenkins methods: parametric autoregressive (AR), moving average (MA), and autoregressive and moving average (ARMA) are time-domain methods.

Other division of time-series techniques is based on the concept of stationary or nonstationary sequences. Before analyzing the structure of a time-series model one must make sure if the series is stationary with respect to the variance and mean, that is, statistical stationarity of a time series implies that the marginal probability distribution is time independent. Often the raw data are transformed to become stationary, for example, if they are a linear combination of a stationary process and one or more processes exhibiting a trend. ARMA methods are based on the stationarity of the time series. However, many time series (especially from economic data) do not satisfy the stationarity conditions. Then these time series are called nonstationary and should be re-expressed such that they become stationary with respect to the variance and the mean. The Box–Jenkins ARIMA and the techniques based on the generalized autoregressive conditional heteroscedasticity (GARCH) are the most popular nonstationary methods.

In the last few years, machine learning techniques, such as artificial neural networks (ANNs), have been applied to time-series analyses owing to their relatively good performance in forecasting and pattern recognition. ANNs are trained to learn the nonlinear relationships between the input variables (mainly past values of the time series and other key variables) and historical patterns of the time series. More complex arrangements, combining ANNs with fuzzy logic, have been recently proposed.

Weighted nearest neighbors (WNNs) algorithms are data mining (DM) techniques for pattern classification that are based on the similarity of the individuals of a population. The members of a population are surrounded by individuals who have similar properties. This simple idea is the learning rule of the WNN classifier. The nearest neighbors decision rule assigns the classification of the nearest set of previously classified points to an unclassified sample point. Unlike most statistical methods, which elaborate a model from the information available in the data base, the WNN method considers the training set as the model itself.

In this chapter, a forecasting methodology, based on the WNN technique, is described in detail. This technique provides a very simple approach to forecast power system variables characterized by daily and weekly repetitive patterns, such as energy demand and prices. Three case studies are used in this chapter to illustrate the potential of the WNN method: (a) the hourly energy demand in the Spanish power system; (b) the hourly marginal prices of the day-ahead Spanish electricity market; and (c) the hourly demand of a particular customer (a faculty building owned by the University of Seville).

BACKGROUND

Data Mining Techniques and Time Series Analysis

Data Mining   With the significant progress in computing and related technologies and the expansion of its use in different aspects of life, large amounts of information are being collected and stored in digital databases. Extracting knowledge from this huge volume of data is a challenge in itself. Data mining (DM) is an attempt to find meaning in the amount of digital information that can be stored in a database. Similarly, the term KDD (Knowledge Discovery in Databases), coined in 1989, refers to the entire process of extracting knowledge from a database and put the emphasis on the useful knowledge that can be discovered from the data.

In fact, the terms DM and KDD are often confused as synonymous. It is generally accepted that the DM is a particular step in the KDD process based on algorithms to extract specific patterns of data. Other steps in the KDD process are the preparation, selection and cleaning of the data, the incorporation of prior knowledge, and the interpretation of the results. The concept of DM also overlaps with the concepts of automatic learning and statistics. In general, statistics is the first science that historically extracted information from the data applying mathematics. When computers were used, the concept of machine learning arose. Subsequently, with the increase in the size and structure of data, the DM concept was introduced.

Data mining can be defined as the process of building a model adjusted to the data to obtain knowledge. Therefore, we can distinguish two steps in DM: on the one hand, the choice of the model, and, on the other hand, the adjustment to the data.

The choice of the model is determined primarily by two conditions: the type of data and the objective of the DM process. Regarding the relationship between the model and the objective, different models are available for different objectives, that is, neural networks may be used for classification problems, regression trees for regression problems, and so on. Besides, there are models that are easy to explain to the user such as the association rules, and models that are difficult to explain such as the neural networks.

The second step is to conduct a "learning phase" with data available to adjust the model to a particular problem, for example, a neural network will require a particular configuration for a particular application and to adjust the weights of the connections, a regression function requires the coefficients to be computed, and a k-nearest neighbors (kNN) method will require to set a metric and a value of k, and so on.

Inherent to the learning phase is the concept of a "fitness function" as a way of evaluating the goodness of the model. In addition, related to the fitness function, a phenomenon known as over-fitting appears, that is, the model fits well to the data used for training but is not able to recognize similar patterns in new data.

Regarding the search for the values that maximize the fitness function, there are many possible approaches: from the classical mathematical analysis to heuristics provided by operational research, or evolutionary algorithms, taboo search, etc. Because this optimization process is present in all DM approaches, both are often confused, presenting, for example, the evolutionary algorithm used to adjust the model as a DM approach.

Finally, another factor to take into account is the treatment of the uncertainty generated by the model, introducing the use of alternative approaches such as fuzzy logic or rough sets. Linked to this is a final concept, *soft computing*, to refer to the set of computational techniques (fuzzy logic, probabilistic reasoning, etc.) that make it possible to develop learning tools.

The Nearest Neighbor Rule As pointed out earlier, DM techniques build a model by adjusting its parameters in the training phase, and later use this model to classify new samples of data. Thus, methods such as decision trees, Naïve Bayes, and neural networks are referred to as "eager learners."

In contrast, "lazy learner" approaches simply store the training samples (some or all) and only when it is necessary to classify a test sample a generalization is performed based on its similarity to the stored training set. These methods are also referred to as "instance-based learners." The main drawbacks of lazy learners are that they can be computationally expensive and require efficient storage techniques. However, they admit easily incremental learning and are well-suited to parallel implementations. Examples of lazy learners are the k-nearest neighbor rule and case-based reasoning.

The basic concepts of the technique known as Nearest Neighbor (NN) were established by Fix and Hodges in 1951 and 1953. The characteristics of nonparametric discrimination problems against linear discrimination analysis or other parametric procedures with known statistical distributions were presented, insisting on the fact that nonparametric discrimination has asymptotically optimum properties for large sample sets.

In 1967, Cover and Hart [1, 2] formally established the Nearest Neighbor Rule. Also they studied the bounds on the error rates of this rule: thus the lower bound is the Bayes error and the upper bound is at most twice the Bayes error. This result can be interpreted as follows: half of all the available information in an infinite training sample set is contained in the nearest neighbor.

Other variants of the kNN rule were proposed subsequently by others. As an example, instead of using the traditional majority-vote criterion, a threshold k_1 ($0 < k_1 < k$) was introduced in 1976 so that sample y was assigned to class 1 if the number of its kNN belonging to class 1 is greater than k_1, and assigned to class 2 otherwise. Also in 1976, the weighted kNN rule with the idea that the nearby samples would be most important in the decision than those farther away was established [3].

Finally, alternative kNN techniques have been proposed to both improve computational efficiency and avoid an exhaustive search in the neighborhood calculation in practical implementations using alternative metrics [4–6].

Demand and Price Forecasting in Power Systems

Demand Forecasting Methods Demand forecasting methods can be grouped into long-term (year), medium-term (month), and short-term models (hour, day, or week). The long-term and medium-term load forecasts are required mainly for generation planning, transmission expansion planning, and hydrothermal coordination. The short-term load forecast is required for the operation of any power system, including generation scheduling and security assessment.

In addition, forecasting errors in load predictions result in increased operating costs, requiring the start-up of too many units and unnecessarily high levels of reserves. Several studies have been developed to assess the economic impact of demand-forecasting errors [7–9].

The system demand time series is influenced by a large number of variables: economic, time, weather, social, and random effects. In this sense, some studies were focused on identifying nonlinear relationships between possible input variables [10–12]. As a result, it was demonstrated that a little set of input variables can identify the forecast model without sacrificing accuracy, the most import factors being seasonal effects, the day of the week, and temperature. However, factors like humidity, wind speed, and precipitation can be as important as the temperature in certain areas.

In practice, short-term load forecasting is affected by a variety of nonlinear factors and, consequently, a lot of techniques have been applied with many variants tuned to particular systems [13]. Proposed techniques include regression models [14, 15], Kalman filtering [16], Box–Jenkins models [17], expert systems [18], fuzzy inference [19], ANNs [20], and chaos time series analysis [21].

Several methods for load forecasting have been successfully applied in the last decades. Regression techniques have become the most popular demand forecasting functions, after the time-series approach was used with satisfactory results. Besides, Box–Jenkins models have also been extensively applied, for example, the approach used by National Grid in the United Kingdom, based on a combination of two Box–Jenkins models and a spectral method [22]. However, some new forecasting models have been recently introduced as expert systems, ANN and fuzzy systems. This is mainly due to the ability of these techniques to learn complex and nonlinear relationships that are difficult to model with conventional models. As an example, the Electric Power Research Institute's (EPRI) Artificial Neural Network Short-Term Load Forecaster (ANNSTLF) [23] was adopted by 40 North American utilities in 1999. Examples of practical approaches based on the combination of several techniques can be found in References [24–27].

Recently, data mining techniques based on the kNN have been successfully applied to the next-day load forecasting problem [28, 29].

Energy Price Forecasting Methods   In competitive electricity markets, prediction tools have become important for participating agents to be able to develop their bidding strategies in order to maximize the profit obtained by trading energy. Such techniques, traditionally applied to load forecasting, have recently focused on predicting the hourly energy prices of pool-based electricity markets. Electricity price models can be broadly classified into two sets [30], namely production cost models and statistical models. Production cost models try to simulate the operation of the system, taking into account the strategic behavior of the involved agents. The main drawback of simulation methods is the large amount of information required, which is difficult to obtain in deregulated markets.

On the other hand, statistical models predict price evolution based on historically observed relationships, without explicitly modeling the underlying physical processes. This category contains a diversity of methods ranging from the simplest "black-box" time-series methods, using only previous price as input data, to more complex structural forecasting models that include explanatory (causal) variables such as load demand, fuel prices, and generation availability.

In turn, time-series methods can be grouped as follows: classical time-series methods [31–34], and automated learning techniques [30, 35–38] (the reader is referred to Reference [30] for an excellent taxonomy of electricity price models). The main advantage of classical statistical methods, including regression models, ARIMA, transfer function models, etc., is their relative simplicity. However, owing to the non-linear nature of the price prediction problem, it is difficult to obtain accurate and realistic models for such methods.

In the last few years, machine learning techniques, such as ANNs, have been applied to energy price prediction owing to their relatively good performance in load forecasting and load pattern recognition [35, 39]. ANNs are trained to learn the nonlinear relationships between the input variables (mainly past values of prices and other key variables affecting the prices) and historical patterns of energy prices. More complex arrangements, combining ANNs with fuzzy logic, have been recently proposed [40].

Alternative data mining techniques such as the WNNs approach have also been successfully applied to next-day energy price forecasting [39, 41].

WEIGHTED NEAREST NEIGHBORS METHODOLOGY

For simplicity of presentation, as the three applications considered in this chapter have to do with the day-ahead prediction of hourly magnitudes, the theoretical description of the WNN methodology will be tailored to this particular context.

Given the time series of the variable of interest (demand, price, wind speed, etc.), recorded in the past up to day d, the problem consists of predicting the 24 hourly values of such magnitude corresponding to day $d + 1$.

Let $P_i \in R^{24}$ be a vector composed of the 24 hourly values corresponding to an arbitrary day i

$$P_i = [p_1, p_2, \dots, p_{24}] \qquad (3\text{-}1)$$

Then, the associated vector $WP_i \in R^{24m}$ is defined by gathering the values contained in a window composed of m consecutive days, from day i backward, as follows:

$$WP_i = [p_{i-(m-1)}, \cdots, p_{i-1}, p_i] \qquad (3\text{-}2)$$

where m is a parameter to be determined. Note that, when $m = 1$ the vector WP_i reduces to P_i. For any couple of days, i and j, the distance between the associated windows of data can be defined

$$\text{dist}(i,j) = \| WP_i - WP_j \| \qquad (3\text{-}3)$$

where $\| \cdot \|$ represents a suitable vector norm (the Euclidean and the so-called Manhattan norms have been used in this work).

The WNN method first identifies the kNN of day d, where k is a number to be determined and "neighborhood" in this context is measured according to the distance (3-3) above. This leads to the neighbor set

$$NS = \{\text{set of } k \text{ days}, q_1, \ldots, q_k, \text{ closest to day } d\} \qquad (3\text{-}4)$$

in which q_1 and q_k refer to the first and kth neighbors respectively, in order of distance.

According to the WNN methodology, the 24 hourly values of day $d + 1$ are predicted by linearly combining the values of the k days succeeding those in NS, that is,

$$\hat{P}_{d+1} = \frac{1}{\sum\limits_{i \in NS} \alpha_i} \cdot \sum\limits_{i \in NS} \alpha_i P_{i+1} \qquad (3\text{-}5)$$

where the weighting factors α_i can be obtained by any of the schemes described in Determination of the Weighting Coefficients subsection to follow.

Figure 3-1 illustrates the basic idea behind the WNN methodology. It is assumed that, if WP_i is close to WP_d, then P_{i+1}, already known, should be also similar to the unknown vector P_{d+1}.

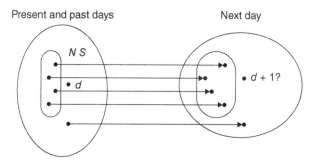

Figure 3-1   Illustration of the WNN approach.

In order to find candidate neighbors, a window of m days is simply slid along the entire list of values contained in the data base [42].

Determination of the Weighting Coefficients

The standard way of computing the weighting factors α_i, as described in the original formulation of the WNN method, is by means of

$$\alpha_i = \frac{\text{dist}(q_k, d) - \text{dist}(i, d)}{\text{dist}(q_k, d) - \text{dist}(q_1, d)} \qquad (3\text{-}6)$$

Obviously, α_i is null when $i = q_k$ (furthest neighbor) and is equal to one when $i = q_1$ (nearest neighbor), which means that only $k - 1$ neighbors are actually used to determine the next-day values. Note also that, although the $24m$ values contained in WP_i are used to determine if i is a candidate neighbor of d, and consequently the values of α_i, only the 24 components of P_{i+1} are relevant in determining P_{d+1}.

Two other schemes to obtain α_i have been developed for the applications described below. The goal is to let all selected neighbors appear in the linear combination (3-5).

Under the first scheme, the weighting coefficients are chosen to be inversely proportional to the distance (IPD) between WP_d and WP_i for $i \in NS$. For each selected neighbor i, the respective coefficient is then given by

$$\alpha_i = \frac{\dfrac{1}{\text{dist}(i, d)}}{\displaystyle\sum_{j=1}^{k} \dfrac{1}{\text{dist}(j, d)}} \qquad (3\text{-}7)$$

The second scheme is based on the assumption that the data window WP_d can be expressed as a linear combination of the selected neighbors' windows WP_i times the weighting coefficients α_i, plus a vector of residuals

$$[WP_d] = [WP_1 \quad WP_2 \quad \cdots \quad WP_k] \begin{bmatrix} \alpha_1 \\ \alpha_2 \\ \vdots \\ \alpha_k \end{bmatrix} + \varepsilon \qquad (3\text{-}8)$$

This leads to a least-squares (LS) problem in which the coefficients α_i are those that minimize the Euclidean norm of ε, that is,

$$\begin{bmatrix} \alpha_1 \\ \alpha_2 \\ \vdots \\ \alpha_k \end{bmatrix} = (A^T A)^{-1} A^T [WP_d] \qquad (3\text{-}9)$$

where A is the $24m \times k$ coefficient matrix appearing in (3-8).

Tuning the Model

Before applying the WNN method, a training phase is necessary in order to find suitable values for m and k. Generally, after using the resulting model for a certain period, prediction errors tend to increase slightly, particularly when applied to a very dynamic context, which may call for new training processes.

As will be seen below, the training phase may be relatively costly and, in any case, even if the model was trained on a daily basis, the prediction accuracy would be always limited by the intrinsic uncertainty of the time series of interest. Therefore, a compromise should be found between cost and accuracy. In this chapter, unless otherwise indicated, monthly training periods will be considered, but yearly training provides almost as accurate results.

Optimal Window Length   The number of days contained in the window that will be used to find candidate neighbors (parameter m) is determined in advance by resorting to the so-called false nearest neighbors (FNN) method [43]. This method compares the distance between a day d and a candidate neighbor i with that between $d + 1$ and $i + 1$. If the second distance is larger, as illustrated in Fig. 3-2, it is said that d and i are false neighbors, because the trajectory of the associated prices tend to diverge. Note that, according to (3-3), such distances depend on m in an implicit manner.

By trying all days contained in the training set, m can be chosen so as to minimize the number of false neighbors.

In practice, as m increases, the cost of the training process also increases and the number of candidate neighbors gets significantly reduced (in the limit, if m approached the size of the training set, a single candidate would remain). Hence, the suboptimal but cheaper scheme adopted in this work consists of choosing the minimum value of m leading to a percentage of false neighbors not exceeding a given threshold (e.g., 10%). Frequently, but not necessarily, $m = 1$ leads to forecasting errors that are very close to that of the optimal values.

Optimal Number of Nearest Neighbors   The optimal number of nearest neighbors (parameter k) is the one that minimizes the forecasting error when the

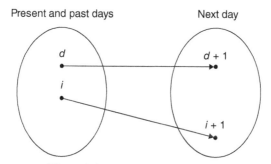

Figure 3-2   False nearest neighbors

WNN methodology is applied to the training set. Mathematically, this is equivalent to finding the value of k that minimizes the following quadratic function

$$\sum_{d \in TS} \|\hat{P}_{d+1} - P_{d+1}\|^2 \tag{3-10}$$

where $\hat{P}_{d+1}$ contains the forecasted values for day $d + 1$; according to the WNN method, P_{d+1} represents actual recorded values and TS refers to the training set. Note that, according to (3-5), $\hat{P}_{d+1}$ is an implicit function of the discrete variable k, which prevents the application of standard mathematical programming methods when searching for k. In practice, k is assigned successive integer numbers ($k = 2$, 3, ...) until a local minimum is found.

PERFORMANCE ASSESSMENT

In this chapter, one or several of the following prediction errors have been computed to assess the performance of the WNN and competing forecasting methodologies.

- Mean relative error (MRE)

$$\text{MRE} = 100 \cdot \frac{1}{n} \sum_{h=1}^{n} \frac{|p_h - \hat{p}_h|}{p_h} \tag{3-11}$$

- Mean absolute error (MAE)

$$\text{MAE} = \frac{1}{n} \sum_{h=1}^{n} |p_h - \hat{p}_h| \tag{3-12}$$

- Mean error relative to $\bar{p}_{\text{period}}$ (MMRE)

$$\text{MMRE} = \frac{1}{n} \sum_{h=1}^{n} \frac{|p_h - \hat{p}_h|}{\bar{p}_{\text{period}}} \tag{3-13}$$

where

$$\bar{p}_{\text{period}} = \frac{1}{n} \sum_{h=1}^{n} p_h \tag{3-14}$$

$\hat{p}_h$ and p_h are the predicted and actual hourly values, respectively, and n is the number of hours for the period of interest (usually a month, a week, or a day).

APPLICATION TO AGGREGATED LOAD FORECASTING

In this section, the WNN tool will be applied to the problem of forecasting the hourly energy demand served through the day-ahead for Spanish electricity market during 2005.

Description of the Data Base

The time series in this case is composed of hourly demand values corresponding to 2004, plus the values corresponding to 2005 up to the day of interest.

Figure 3-3 shows a histogram of the hourly energy that was auctioned in the Spanish daily market during 2004. The evolution of the demand during the last trimester of 2004 is represented in Fig. 3-4, where the weekly repetitive pattern is apparent.

The WNN methodology has been applied separately to the weekdays and weekends of 2005. In both cases, in order to obtain the optimal values of k and m, the 12 months preceding the current month have been used, resulting in 12 different WNN models. Then, each monthly model resorts to the most recent information to predict every day demand throughout the respective month. For instance, to predict the load on March 5, 2005, a model trained with the data base comprising from February 2004 to February 2005 is adopted. However, the model is fed subsequently with all the data up to March 4 to obtain the day-ahead prediction.

Test Results

Table 3-1 shows the resulting parameters m and k when each monthly model corresponding to 2005 is trained with the 12 preceding months. For comparison, the rightmost column collects the optimal values of k when $m = 1$ is adopted, instead of

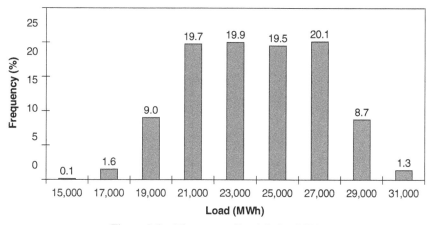

Figure 3-3   Histogram of load during 2004.

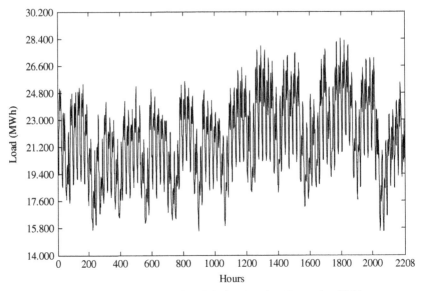

Figure 3-4 Evolution of load during October–December 2004.

the value provided by the FNN methodology. Note the larger number of neighbors involved when the sliding window comprises only the current day ($m = 1$).

Table 3-2 presents the resulting errors, along with the average monthly demand, $\bar{D}_{month}$, for the weekdays of each month in 2005. When the parameter m is determined by the FNN method ($m \geq 1$), the MRE ranges from 2.4% in July to 4.37% in March, yielding an average value of 3.19% in 2005.

Table 3-1 WNN Parameters of Monthly Trained Models to Predict the Demand in Weekdays of the Year 2005

		$m \geq 1$		$m = 1$
2005 Month	Training Set	k	M	k
Jan	Jan 2004–Dec 2004	3	13	10
Feb	Feb 2004–Jan 2005	1	10	6
Mar	Mar 2004–Feb 2005	3	1	3
Apr	Apr 2004–Mar 2005	6	1	6
May	May 2004–Apr 2005	5	5	10
Jun	Jun 2004–May 2005	1	10	9
Jul	Jul 2004–Jun 2005	1	5	8
Aug	Aug 2004–Jul 2005	1	13	10
Sep	Sep 2004–Aug 2005	8	1	8
Oct	Oct 2004–Sep 2005	6	4	8
Nov	Nov 2004–Oct 2005	3	1	3
Dec	Dec 2004–Nov 2005	5	11	5

Table 3-2   Monthly Prediction Errors Obtained with the WNN Methodology in 2005

Weekdays 2005	$m = 1$		$m \geq 1$		
	MRE (%)	MAE (MWh)	MRE (%)	MAE (MWh)	$\bar{D}_{month}$ (MWh)
Jan	3.83	1006.30	3.65	970.47	26,450.01
Feb	3.38	896.36	3.33	895.17	27,016.92
Mar	4.37	1216.73	4.37	1216.73	28,307.91
Apr	3.09	725.06	3.09	725.06	24,205.85
May	3.48	800.93	2.73	659.01	24,418.41
Jun	3.39	931.80	3.22	860.73	27,521.24
Jul	2.38	670.65	2.40	681.00	28,936.89
Aug	2.80	721.02	2.76	699.12	26,176.98
Sep	2.64	709.88	2.64	709.88	27,573.72
Oct	2.14	533.62	1.91	484.46	26,574.66
Nov	4.00	1015.11	4.00	1015.11	25,764.02
Dec	4.42	1168.24	4.17	1098.60	26,955.34
Average	3.33	866.31	3.19	834.61	26,658.50

Table 3-3 shows the standard deviation of MRE and MAE corresponding to the variable m case, which is useful to assess the variability of prediction errors.

Table 3-4 collects, for each month, the maximum and minimum values of the daily average absolute and relative prediction errors, corresponding to the variable m case. It is worth noting that the largest prediction errors arise for days immediately before and after national holidays. For instance, November 2 and May 2 are preceded by a holiday, whereas August 2 follows the day in which a majority of Spaniards start their summer break.

Table 3-3   Standard Deviation of MRE and MAE

Working Days 2005	σ_{MRE} (%)	σ_{MAE} (MWh)
Jan	1.83	500.06
Feb	1.94	525.69
Mar	2.92	837.19
Apr	1.78	401.79
May	2.16	478.93
Jun	1.28	342.11
Jul	1.96	534.38
Aug	1.91	506.48
Sep	1.48	353.04
Oct	0.81	184.71
Nov	2.38	564.25
Dec	2.58	613.45
Average	1.92	486.84

Table 3-4 Maximum and Minimum Absolute and Relative Errors, and Day that Took Place

Weekdays 2005	Maximum Daily Absolute Errors		Minimum Daily Absolute Errors		Maximum Daily Relative Errors		Minimum Daily Relative Errors	
	MWh	Day	MWh	Day	%	Day	%	Day
Jan	2495.16	Fr 28	371.36	Mo 24	9.40	Fr 28	1.43	Mo 24
Feb	2228.91	Th 24	281.13	Tu 22	8.20	Mo 14	1.02	Tu 22
Mar	2889.98	We 01	185.64	Th 10	9.36	Tu 01	0.62	Th 10
Apr	1486.13	Mo 11	164.70	Fr 22	6.90	Mo 11	0.70	Fr 22
May	2202.91	Mo 02	211.10	Fr 13	10.46	Mo 02	0.86	Fr 13
Jun	1776.50	Fr 24	349.60	Fr 17	6.20	Fr 24	1.28	Fr 17
Jul	2468.56	Mo 25	161.28	Th 21	9.16	Mo 25	0.55	Th 21
Aug	2430.66	Tu 02	138.43	Th 18	9.01	Tu 02	0.54	Th 04
Sep	1521.06	Mo 12	288.89	We 28	6.40	Mo 19	1.10	We 28
Oct	953.88	We 19	233.30	Tu 18	3.78	We 19	0.87	Tu 18
Nov	2812.46	We 02	288.17	Th 17	11.96	We 02	1.05	Th 17
Dec	2861.12	Th 29	339.46	We 21	11.75	Th 29	1.19	We 21

Table 3-5   WNN Parameters of Monthly Trained Models to Predict the Demand in Weekends of the Year 2005

| 2002 Month | Training Set | $m \geq 1$ | | $m = 1$ |
		k	m	k
Jan	Jan 2004–Dec 2004	10	4	8
Feb	Feb 2004–Jan 2005	3	10	5
Mar	Mar 2004–Feb 2005	1	7	4
Apr	Apr 2004–Mar 2005	10	13	10
May	May 2004–Apr 2005	6	8	4
Jun	Jun 2004–May 2005	1	10	1
Jul	Jul 2004–Jun 2005	8	8	9
Aug	Aug 2004–Jul 2005	9	7	4
Sep	Sep 2004–Aug 2005	1	6	4
Oct	Oct 2004–Sep 2005	6	15	1
Nov	Nov 2004–Oct 2005	10	1	10
Dec	Dec 2004–Nov 2005	3	4	9

Tables 3-5 to 3-8 are the counterparts of the above tables for the weekends of the year 2005. As can be seen, average errors for weekend predictions are, in general, larger than those of weekdays.

Table 3-9 summarizes the overall results corresponding to every month of 2005, including both weekdays and weekends. The largest monthly average error corresponds with March (4.32%) while the smallest one takes place in October (1.92%).

Table 3-6   Monthly Prediction Errors Obtained with the WNN Methodology in Weekends of the Year 2005

| Weekends 2005 | $m = 1$ | | $m \geq 1$ | | |
	MRE (%)	MAE (MWh)	MRE (%)	MAE (MWh)	$\bar{D}_{month}$ (MWh)
Jan	5.37	1273.05	4.41	1032.56	23,796.88
Feb	5.41	1301.73	3.92	937.46	24,900.97
Mar	6.84	1506.29	4.18	914.80	23,554.82
Apr	5.88	1231.17	1.97	399.35	21,482.36
May	3.60	792.72	1.54	327.07	22,085.56
Jun	5.15	1247.88	3.35	785.14	24,650.43
Jul	5.97	1530.21	2.82	698.08	25,784.64
Aug	7.69	1753.79	3.11	698.22	23,562.54
Sep	5.95	1469.99	3.63	837.51	24,504.05
Oct	4.24	998.07	1.94	433.95	22,734.48
Nov	2.74	685.82	2.74	685.82	25,645.64
Dec	3.41	885.01	2.95	752.86	26,400.15
Average	5.19	1222.98	3.05	708.57	24,091.88

Table 3-7   Standard Deviation of MRE and MAE in Weekends of the Year 2005

Weekends Days 2005	σ_{MRE} (%)	σ_{MAE} (MWh)
Jan	2.41	570.60
Feb	1.95	454.60
Mar	2.46	503.78
Apr	0.64	122.14
May	0.57	112.55
Jun	2.27	512.69
Jul	1.72	422.31
Aug	1.64	366.93
Sep	0.86	164.21
Oct	0.76	151.11
Nov	0.93	251.61
Dec	1.93	440.17
Average	1.51	339.39

Comparison with AR

The results shown earlier are compared in this section with those provided by an autoregressive (AR) model, which is applied to a time series composed of all days (i.e., no distinction is made between weekdays and weekends).

Under this approach, the demand at hour t, D_t, is estimated from the demand at hours $t-1$, $t-2$,..., etc. In order to reduce the size of the resulting model, a correlation study is performed first with D_t, D_{t-1},..., in order to determine which of the past demand values are most influential on the present demand.

Figure 3-5 presents the average correlation coefficient between the present and past demand values for the period February 2004 to December 2004. As can be clearly seen, this coefficient shows a 24-hour periodicity. The main conclusion is that the highest correlation with the demand at a given hour takes place at the same hour of the previous days. Furthermore, such a correlation decreases as the number of past hours increases.

In view of the conclusions obtained from the correlation study, the following AR model is proposed:

$$\hat{D}_t = a_0 D_{t-24} + a_1 D_{t-48} + a_2 D_{t-72} + a_3 D_{t-96} + a_4 D_{t-120} + a_5 D_{t-144} + a_6 D_{t-168}$$

$$(3\text{-}15)$$

Experimental results suggest that it is not worth including extra terms like D_{t-1}, D_{t-2}, D_{t-23} in the model, as the average forecasting error for the tested period remains unaffected.

Table 3-8 Maximum, Minimum, Absolute, and Relative Errors and Day that Took Place (Weekends 2005)

Weekends 2005	Maximum Daily Absolute Errors		Minimum Daily Absolute Errors		Maximum Daily Relative Errors		Minimum Daily Relative Errors	
	MWh	Day	MWh	Day	%	Day	%	Day
Jan	1917.82	Su 16	565.38	Sa 22	8.15	Su 30	2.39	Sa 22
Feb	1659.96	Sa 05	516.61	Su 27	6.70	Sa 05	2.19	Su 27
Mar	1573.30	Su 20	303.28	Sa 12	7.76	Su 20	1.16	Sa 12
Apr	619.77	Su 03	291.72	Su 10	3.17	Su 03	1.38	Sa16
May	549.30	Su 08	186.65	Sa 21	2.74	Su 08	0.83	Sa 21
Jun	1860.42	Su 12	329.39	Su 05	8.31	Su 12	1.46	Sa 18
Jul	1762.99	Su 17	291.82	Sa 09	7.11	Su 17	1.15	Su 09
Aug	1401.95	Sa 20	277.23	Su 14	6.26	Sa 20	1.30	Su 14
Sep	1046.93	Sa 10	478.81	Sa 03	4.50	Su 25	1.75	Sa 03
Oct	698.37	Su 23	272.57	Sa 29	3.42	Su 23	1.14	Sa 29
Nov	1055.52	Su 12	318.75	Su 06	4.07	Sa 12	1.46	Su 06
Dec	1495.53	Su 11	293.74	Su 04	6.45	Su 11	1.18	Su 04

Table 3-9 Average of the Relative Errors for All Days of Each Month in 2005

	Weekdays MRE (%)	Weekends MRE (%)	Monthly Average
Jan	3.65	4.41	3.87
Feb	3.33	3.92	3.50
Mar	4.37	4.18	4.32
Apr	3.09	1.97	2.77
May	2.73	1.54	2.39
Jun	3.22	3.35	3.26
Jul	2.40	2.82	2.52
Aug	2.76	3.11	2.86
Sep	2.64	3.63	2.92
Oct	1.91	1.94	1.92
Nov	4.00	2.74	3.64
Dec	4.17	2.95	3.82
Yearly average	3.19	3.05	3.15

The model parameters a_i are obtained from the solution of the following least squares problem:

$$\text{Min} \sum_t (D_t - \hat{D}_t)^2$$

where $\hat{D}_t$ is defined by (3-15). These parameters can be either computed only once, from the entire training set, or updated daily with the most recent time-series values.

Figure 3-6 shows the evolution of the parameters when they are updated every day for the period January–December 2005. It can be observed that the resulting coefficients remain essentially constant, which means that no new demand patterns

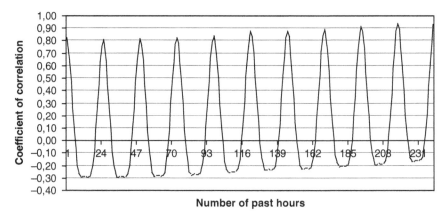

Figure 3-5 Correlation coefficients.

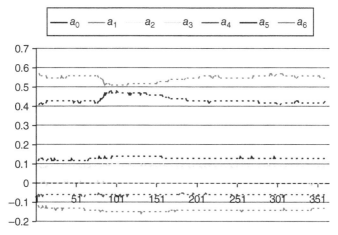

Figure 3-6   Evolution of parameters a_i during 2005.

are added to the data base each time a day is included. The average coefficient values are $a_0 = 0.44$, $a_1 = -0.14$, $a_2 = 0.09$, $a_3 = 0$, $a_4 = -0.06$, $a_5 = 0.13$, and $a_6 = 0.55$

The small values for a_2, a_3, and a_4 suggest that, when predicting D_t, the most influential values are D_{t-24} and D_{t-168}, that is, the same hour of the previous day and week.

Table 3-10 shows the resulting MRE and MAE errors arising when the AR method is applied, along with their standard deviation. The yearly average for MRE is 4.65% whereas, according to Table 3-9, that for the WNN methodology 3.15%. This increased average prediction error originates in the higher maximum errors collected in Table 3-11, a fraction of which take place in weekends.

Table 3-10   Monthly Prediction Errors and their Standard Deviation Obtained with the AR Methodology in 2005

Days 2005	MRE (%)	σ_{MRE} (%)	MAE (MWh)	σ_{MAE} (MWh)
Jan	5.59	3.16	1425.62	849.13
Feb	4.18	2.28	1095.97	597.24
Mar	7.70	6.35	1822.88	1306.89
Apr	3.78	3.12	867.56	747.39
May	3.64	2.99	851.77	749.26
Jun	4.06	2.58	1103.87	759.43
Jul	3.19	2.23	859.37	546.26
Aug	4.92	3.53	1234.59	904.65
Sep	4.25	2.87	1112.40	717.56
Oct	3.60	2.35	893.61	587.97
Nov	5.04	4.01	1245.16	884.62
Dec	5.86	3.70	1507.08	860.30
Average	4.65	3.26	1168.32	792.56

Table 3-11 Maximum, Minimum Absolute, and Relative Errors and Day That Took Place for All Days in Year 2005

Days 2005	Maximum Daily Absolute Errors		Minimum Daily Absolute Errors		Maximum Daily Relative Errors		Minimum Daily Relative Errors	
	MWh	Day	MWh	Day	%	Day	%	Day
Jan	3548.62	Th 13	280.48	Fr 28	12.26	Mo 03	1.10	V28
Feb	2479.58	Mo 21	338.08	Th 03	10.93	Su 13	1.28	J03
Mar	4665.69	Fr 25	260.87	Th 10	24.88	Fr 25	0.87	J10
Apr	3568.52	Mo 04	163.60	Fr 22	14.23	Mo 04	0.68	V22
May	3852.48	Mo 09	183.83	Fr 20	14.98	Mo 09	0.80	V20
Jun	2922.85	Mo 20	281.79	Th 16	9.72	Mo 20	1.05	J16
Jul	2174.10	Su 24	168.81	Fr 08	9.23	Su 24	0.60	V08
Aug	3971.05	Mo 22	132.43	We 10	14.44	Mo 22	0.53	X10
Sep	2654.19	Su 11	173.57	Th 29	12.14	Su 11	0.65	J29
Oct	2795.08	We 12	281.03	Tu 18	11.72	We 12	1.04	J27
Nov	4442.22	Tu 01	369.42	We 23	20.95	Tu 01	1.36	X23
Dec	4400.79	Sa 24	310.31	We 21	19.66	Sa 24	1.09	X21

Table 3-12   Comparison of Both Methods

	Mean Absolute Errors		Mean Relative Errors			
Days 2005	Average (MWh)	Average Standard Deviation (MWh)	Average (%)	Average Standard Deviation (%)	Maximum Daily Absolute Errors (MWh)	Maximum Daily Relative Errors (%)
WNN	798.60	444.71	3.15	1.80	2889.98	11.96
AR	1168.32	792.56	4.65	3.26	4665.69	20.95

Finally, Table 3-12 presents the standard deviation, the average absolute and relative forecasting errors, and the maximum daily errors obtained from the application of both the WNN method and the AR model to the entire period considered (2005). In this particular application the average figures provided by the WNN method are clearly better.

APPLICATION TO POOL ENERGY PRICE FORECASTING

This section presents and discusses the application of the WNN method to the day-ahead prediction of electric energy prices in the Spanish pool. An exhaustive comparison of results with those of several references appeared in the literature is performed.

Characterization of Spanish Electricity Market Prices

The Spanish electricity market, running since 1998, relies basically on a pool where energy is traded through an auction process [42] (bilateral transactions, although possible, have been historically negligible compared to those of other markets). Generators, retailers, and consumers submit hourly bids the day before to the market operator; containing at least the price assigned to each block of energy (more complex bids are possible). Then the resulting marginal price is used every hour to charge/remunerate all agents buying/selling energy, irrespective of their original bids. In an ideal market, submitted bids should reflect actual variable costs, but the Spanish market is still far from this perfect competitive environment, as over 70% of energy is produced by units belonging to two major generating holdings.

In this context, it is crucial for all agents to have as accurate a prediction as possible of next-day energy prices in order to develop their optimal bidding strategies.

The hourly spot market prices arising from January 2000 to August 2002 have been recorded to visually illustrate the behavior and evolution of the energy prices. As weekends and holidays constitute separate cases, only data corresponding to working days have been retained and analyzed.

Figure 3-7 shows the hourly averages and standard deviations of prices for the working days of March 2001 in cents of Euro per kWh (cE/kWh). As expected, larger

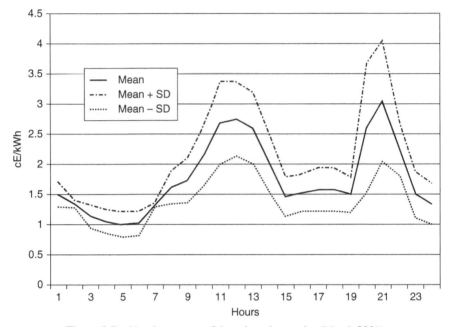

Figure 3-7   Hourly average of the prices time series (March 2001).

average spot prices occur during the morning and evening peak hours (10 a.m.–2 p.m. and 8–10 p.m., respectively). Except for a few valley hours, the standard deviation of hourly prices exceeds 20% of the mean value, reaching 40% at 8 p.m. and 9 p.m.

The way prices have historically evolved is summarized in the histograms of Fig. 3-8, corresponding to years 2000, 2001, and 2002. Apart from a clear trend for prices to spread and increase, it is evident that prices do not follow a normal distribution. Whereas over 50% of prices were lower than 1 cE/kWh during 2000, a majority of prices lie between 2 and 3 cE/kWh in 2001, reflecting the fact that the Spanish electricity market was, and still is to a certain extent, rather unstable and subject to significant price changes.

To conclude this section, Fig. 3-9 represents hourly prices during 2002 (vertical axis) versus the respective energy demand (horizontal axis), along with the resulting regression line plus two parallel lines delimiting the ±1 cE/kWh interval. When all data are considered, the correlation between price and energy is quite poor ($R = 0.63$), particularly at peak hours. However, three different patterns (lower, central, and upper) bounded by the two edge lines can be visually observed, the upper one being more diffuse. Results corresponding to each cluster of data will be separately analyzed in Test Results section. As the method presented in this chapter is intended to be of general applicability, no effort has been made to improve the prediction accuracy by taking advantage of such data clustering.

All of the preceding comments suggest that predicting energy prices in the Spanish electricity market, like in most markets, is a challenging task because of the random

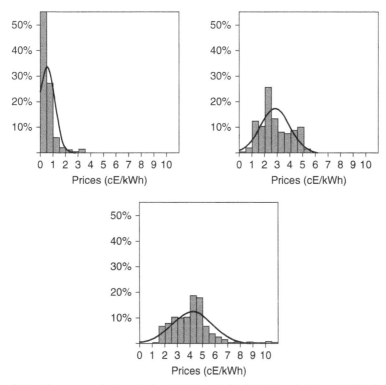

Figure 3-8   Histogram of prices during 2000 (top left), 2001 (top right), and 2002 (bottom).

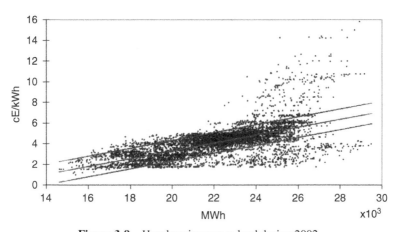

Figure 3-9   Hourly price versus load during 2002.

nature of the electricity prices, mainly caused by the uncertainty in the variables driving the price [44]. Therefore, larger prediction errors compared to those reported in the former case study are expected.

Test Results

The WNN methodology described earlier has been applied to the working days of 2002 (four atypical days of the last week of December, much closer to a holiday than to a working day, have also been excluded from the analysis). Table 3-13 shows the parameters m (sliding window size) and k (number of neighbors) that result when each monthly model corresponding to 2002 is trained with the 12 preceding months. For instance, the model corresponding to April 2002 is trained with data corresponding to the period April 2001 to March 2002. However, the data base, in which neighbors are searched for, comprises all working days from 2000 up to the day of 2002 right before the day whose prices are to be forecasted.

Note that if m is not equal to 1, then it is, or is very close to, a multiple of 5, revealing the weekly repetitive pattern of energy prices (not so regular anyway than that of the load). The fact that m sometimes equals 9 or 14, instead of the more intuitive values of 10 or 15, is explained either by some isolated holidays being removed occasionally from the data base or by the slightly different behavior of Monday valleys compared to the remaining working days.

For comparison, the rightmost column collects the values of k that result when $m = 1$ is adopted, instead of the value provided by the FNN methodology. Note the larger number of neighbors involved in general when the sliding window comprises only the current day ($m = 1$).

Table 3-14 presents the resulting prediction errors, as defined earlier, along with the monthly average price, $\bar{p}_{month}$, for each month of 2002. When the parameter m is

Table 3-13   Parameters of Monthly Trained Models to Predict Prices in 2002

| 2002 Month | Training Set | $m \geq 1$ | | $m = 1$ |
		k	m	k
Jan	Jan 2001–Dec 2001	8	5	10
Feb	Feb 2001–Jan 2002	3	9	8
Mar	Mar 2001–Feb 2002	3	14	6
Apr	Apr 2001–Mar 2002	2	15	5
May	May 2001–Apr 2002	6	15	8
Jun	Jun 2001–May 2002	4	1	4
Jul	Jul 2001–Jun 2002	4	9	9
Aug	Aug 2001–Jul 2002	3	15	10
Sep	Sep 2001–Aug 2002	3	14	5
Oct	Oct 2001–Sep 2002	9	5	9
Nov	Nov 2001–Oct 2002	3	10	3
Dec	Dec 2001–Nov 2002	5	1	5

Table 3-14   Monthly Prediction Errors Obtained with the WNN Methodology in 2002

Working Days 2002	$m = 1$ MRE (%)	$m \geq 1$ MRE (%)	MAE (cE/kWh)	MMRE (%)	$\bar{P}_{month}$(cE/kWh)
Jan	13.72	11.18	0.93	12.54	7.45
Feb	8.12	6.48	0.27	5.96	4.00
Mar	6.20	5.10	0.23	5.33	3.68
Apr	5.82	5.55	0.24	5.28	4.00
May	6.64	6.12	0.27	5.56	4.18
Jun	9.67	9.67	0.52	10.07	4.38
Jul	7.99	7.13	0.37	6.59	4.86
Aug	9.79	8.98	0.37	8.73	3.37
Sep	11.54	9.70	0.35	8.75	4.00
Oct	8.93	8.75	0.29	7.83	3.76
Nov	16.69	13.34	0.41	12.78	3.18
Dec	12.05	12.05	0.32	12.02	2.68
Average	9.76	8.67	0.38	8.45	4.07

determined by the FNN ($m \geq 1$), the MRE ranges from 5.1% in March to 13.3% in November, yielding an average value of 8.7% for 2002. In a majority of cases, the MMRE is smaller than the MRE, which is logical considering that, unlike $\bar{p}_{month}$, p_h is sometimes relatively close to zero. Note that the MRE corresponding to $m = 1$, also included for comparison in the second column, is larger than the one provided by the value of m determined through the FNN method.

Apart from December, which is a quite atypical month in Spain, unusually large prediction errors can be observed in January, June, and November. Unlike the demand, whose behavior is rather predictable, there are no definite clues about prices behaving so irregularly in those months (to the authors' knowledge, no exhaustive analysis has been published to date about this issue). Nevertheless, the market and system operators' monthly reports reveal that:

- In January, extremely low temperatures gave rise to sustained demand peaks. In addition, important blackouts that had recently affected Madrid and other large cities, along with reduced levels of hydraulic production, led to expensive fuel-fired units to be resorted to for security of supply reasons. This probably explains the fact that the average price in January was 197% higher than that of the same month in 2001, and also significantly higher than those of the remaining months in 2002.

- After several dry years, hydro-plant generation beat minimum historic records in June. Furthermore, a general strike on June 20 (Thursday) significantly affected the demand and prices for several consecutive days, as many people took advantage of the resulting long weekend for a break. For example, the average price on June 20 was 30% lower than that of the day before, yielding an average prediction error of about 31.2% for that day and 23% for the next (the peak error on Thursday exceeded 80% and took place at 8 a.m.). In spite of that, the monthly average price was 15% larger than that of June 2001, clearly showing that June was also a rather special month.

- Somewhat the opposite happened late in 2002. Owing to the unusually mild temperatures and, for the first time in the year, hydro-production well over the average, prices in November were abnormally lower than expected (20.1% lower in average than those of the same month in 2001). For unknown reasons, the demand forecasted by the system operator on Wednesday 6 was about 10% lower than the actual demand, yielding an exceptionally low daily average price of 1.4 cE/kWh (in fact, the price was 0.0 cE/kWh for 4 consecutive hours). Note, however, that the absolute prediction error in November is not far from the average which corresponds to 2002.
- In December, mild temperatures and rain continued, leading to low demand levels and twice as much hydro-plant production as that of November.

In summary, extreme weather factors (very cold and dry season early in 2002, rather mild and rainy fall) along with the peculiarities of the generation mix in Spain, where the hydro-plant production in 2002 ranged from virtually null to nearly 27% in December, may partly explain the irregular price behavior and associated prediction errors. Furthermore, market agents are very sensitive to energy-related government policies, among which the so-called "transition-to-competition costs" were quite relevant those years. Such costs, intended to compensate existing GENCOs for potential profit losses originated by the regulatory change, were somewhat conditioned to the "reasonable" evolution of the energy pool prices, constituting in this way an economic signal whose real influence on prices is very difficult to analyze.

Table 3-15 shows the standard deviation of the MRE and MAE corresponding to the variable m case, which is useful to assess the variability of prediction errors.

Table 3-15 Standard Deviation of MRE and MAE

Working Days 2002	σ_{MRE} (%)	σ_{MAE} (cE/kWh)
Jan	6.28	0.87
Feb	4.98	0.17
Mar	2.64	0.11
Apr	3.51	0.15
May	2.36	0.08
Jun	7.47	0.45
Jul	3.99	0.19
Aug	4.77	0.19
Sep	5.86	0.18
Oct	7.61	0.17
Nov	6.26	0.19
Dec	7.83	0.18
Average	5.30	0.24

Table 3-16   Maximum and Minimum Mean Daily Absolute Errors

Working Days 2002	Maximum Daily Error (cE/kWh)	Minimum Daily Error (cE/kWh)
Jan	3.21	0.25
Feb	0.83	0.07
Mar	0.45	0.08
Apr	0.65	0.08
May	0.43	0.11
Jun	1.81	0.11
Jul	0.82	0.12
Aug	0.80	0.08
Sep	0.81	0.15
Oct	0.92	0.11
Nov	0.82	0.13
Dec	0.74	0.15

Table 3-16 collects, for each month, the maximum and minimum daily average prediction errors, in absolute value, corresponding to the variable m case. Interestingly enough, the largest prediction errors corresponding to July, August, and September took place respectively on Monday 15, Monday 5, and Monday 16, typical days in which a majority of Spaniards start their summer holiday season.

Figures 3-10 and 3-11 show the evolution of actual and forecasted prices corresponding to the best and worst day of the third week of May (Monday 20 to Friday 24) respectively, in terms of prediction accuracy. The resulting MMRE error for those days is 1.5% (May 23) and 5.6 % (May 24), respectively.

To conclude this section, prediction errors corresponding to 2002 will be separately analyzed for the three clusters visually identified in Fig. 3-9. Table 3-17 presents the absolute and relative mean prediction errors, along with the number of hours comprising each cluster. Figures in the rightmost column, referring to the hourly averages for the entire period considered, slightly differ from those in the last row of Table 3-14, because in that case the average is computed by months.

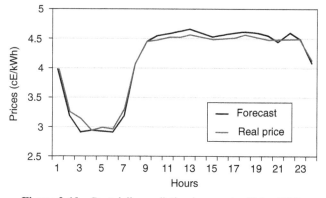

Figure 3-10   Best daily prediction in a week of May 2002.

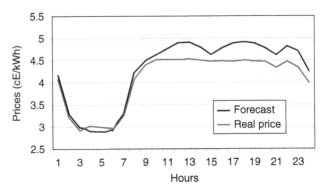

Figure 3-11   Worst daily prediction in a week of May 2002.

The performance of the WNN method is better for the central cluster, where the correlation between price and demand is larger. As expected, the largest absolute prediction errors arise in the upper cluster. It is also interesting to split the data in Table 3-17 by months, as in Table 3-18. A careful examination of this table shows that a majority of hours corresponding to November and December lie in the lower cluster, where prices are statistically small for the associated demand, while somewhat the opposite happens in January. Such findings support the explanations provided earlier regarding the anomalous behavior of prices for certain months in 2002.

For simplicity, holidays and weekends have been removed from the records to obtain the preceding results, but they could be gathered in a separate data base to be processed in the same way as labor days.

Comparison with Other Techniques

In this section, the performance of the WNN method is compared, whenever possible, with that of other published approaches, namely the ANN approach described in Reference [35], the neuro-fuzzy system described in Reference [40], the GARCH model described in Reference [32], and the proposal of Reference [45] combining the Wavelet transform with an ARIMA model.

Comparison with a Plain ANN   The ANN described in Reference [35] has been used for comparison purposes, as both the ANN and the WNN approach resort to nonlinear models.

Table 3-17   Absolute and Relative Prediction Errors by Clusters of Data

	Lower	Central	Upper	Total
MRE (%)	13.57	7.7	8.96	8.6
MAE (cE/kWh)	0.27	0.29	0.75	0.33
No. hours	767	4473	592	5832

Table 3-18   Monthly Relative Prediction Errors by Clusters of Data

	Lower		Central		Upper	
	No. hours	MRE (%)	No. hours	MRE (%)	No. hours	MRE (%)
Jan	4	24.56	218	9.66	234	12.30
Feb	29	15.69	448	5.86	3	10.11
Mar	27	2.81	423	5.38	6	15.94
Apr	0	0.00	440	5.31	88	6.73
May	0	0.00	437	5.99	91	6.73
Jun	0	0.00	444	9.80	36	8.19
Jul	8	25.68	430	7.93	114	4.28
Aug	25	19.57	474	8.31	5	19.90
Sep	22	19.29	476	9.75	6	8.87
Oct	41	16.67	509	8.10	2	12.65
Nov	289	14.16	160	11.62	7	19.24
Dec	322	12.07	14	11.59	0	0.00

For this application, the ANN is fed with a shifting window of prices comprising 24 hours, which means that the input layer is composed of 24 perceptrons. As far as the number of output perceptrons is concerned, 24 outputs corresponding to the prices of a whole day are used, whose values are thus determined by those of the previous day. This implies that the window is shifted 24 hours each time. Finally, the intermediate layer is also composed of 24 perceptrons (this number is determined in an optimal way after several tests). The period used in Reference [35] to train and tune the ANN is January–February 2001, while the period March–August is devoted to check the forecasting errors.

Table 3-19 presents the resulting errors when market clearing prices of working days are forecasted using both the ANN tested in Reference [35] and the WNN approach, along with $\bar{p}_{month}$, for each tested month. As can be observed, the WNN method clearly outperforms this particular ANN implementation.

Table 3-19   Monthly Prediction Errors Provided by the ANN and WNN Methodologies in 2001

	ANN			WNN			
Working Days 2001	MRE (%)	MAE (cE/kWh)	MMRE (%)	MRE (%)	MAE (cE/kWh)	MMRE (%)	$\bar{p}_{month}$ (cE/kWh)
Mar	21.45	0.37	20.75	14.24	0.25	14.23	1.77
Apr	16.92	0.35	16.34	11.06	0.22	10.33	2.15
May	10.71	0.32	11.29	6.08	0.18	6.36	2.84
Jun	10.82	0.42	10.99	5.54	0.21	5.44	3.84
Jul	9.25	0.34	9.05	7.11	0.25	6.64	3.74
Aug	18.57	0.52	16.97	10.06	0.30	9.78	3.08

Table 3-20 MRE Provided by the WNN Methodology on the Ontario Case Studies

	Case A	Case B
Jun	19.99	17.08
Jul	23.82	22.42
Aug	22.25	20.85
Sep	35.17	31.48
Average	24.58	22.34

Comparison with a Neuro-Fuzzy System   In order to compare the WNN method with a more sophisticated neural network arrangement, the work reported in Reference [40] has been considered, for which Ontario market data have been downloaded from Reference [46]. For this comparison, four WNN monthly models corresponding to the period June–September 2002 have been developed and tested, additional data corresponding to May being included in the data base (otherwise, there would be no price patterns to compare with during the first days of June). Like in Reference [40], the MRE has been obtained and tabulated (note, however, that this error is termed MAPE (Mean Absolute Percentage Error) in Reference [40]). Table 3-20 shows the prediction errors provided by the WNN method when all days are considered (case A) and when several extreme days (June 11, July 2, August 13, and September 3, according to Reference [40]) are omitted (case B). The WNN average errors for the considered period are 24.58 % and 22.34 % for cases A and B, respectively.

Such results could be significantly improved if other critical days, like September 15 for which the prediction error reaches 115%, were omitted in addition to those already ignored in case B.

Several prediction results provided by an ANN are presented in Reference [40], according to different variables used as inputs by the model (demand, capacity short-fall, and both), number of neurons at the intermediate layer, and learning techniques (LM and momentum algorithms). Table 3-21 summarizes the best results given in Reference [40] for cases A and B. Note that, in spite of its simplicity and reduced size of the data base, the average results provided by the WNN method are better than those of Table 3-21.

A more complex model, based on a neuro-fuzzy system, is also reported in Reference [40]. In addition to those cited above, forecasted energy imports and

Table 3-21 MRE Provided by the ANN on the Ontario Case Studies

Input	Case A	Case B
Demand	25.88	23.30
Capacity shortfall	27.37	24.28
Both	27.04	24.03

Table 3-22   MRE Provided by the Neuro-Fuzzy System on the Ontario Case Studies

Input	Case A	Case B
Demand	23.03	20.17
Capacity shortfall	27.11	22.96
Import	30.52	28.09
Demand + Capacity shortfall	24.48	21.86
Demand + Outages	21.85	19.83
Demand + Import	21.67	20.4

expected generator outages are used as inputs to the model, several membership functions being tested. The best results obtained by this approach are collected in Table 3-22. Compared with Table 3-20, it can be concluded that this sophisticated model, that needs to be periodically trained, outperforms the WNN method only when the predicted demand is considered.

Comparison with GARCH Model   Following Reference [32], the data set in this case comprises the period ranging from January 1999 to November 2000 (both included), and the resulting monthly models are only used to predict the last week of every month in 2000 (except for December, whose last week refers to 1999).

For each of the 12 weeks under study, the MRE error is computed. Table 3-23 compares the results provided by the WNN methodology with those taken from Reference [32]. Both the plain GARCH model and the GARCH model including the forecasted load as additional input are included.

Table 3-23   Comparison of Mean Weekly Prediction Errors, MRE (%), Provided by WNN and GARCH Methods

	WNN	GARCH	GARCH with Load
Jan	6.00	9.25	8.62
Feb	4.49	7.24	6.64
Mar	9.34	9.94	9.75
Apr	9.70	12	11.91
May	5.58	5.19	4.62
Jun	9.70	8.92	8.67
Jul	7.60	8.49	8.23
Aug	7.52	7.28	7.2
Sep	5.55	9.46	9.08
Oct	9.26	8.99	8.83
Nov	9.28	10.92	10.24
Dec	18.89	16.96	15.41
Average	8.57	9.55	9.10
Average (without Dec)	7.64	8.88	8.53

Table 3-24   Comparison of the MMRE Provided by WNN and ARIMA Methods for the 4 Weeks Analyzed

	WNN (Weekly Model)					
Selected Week of	Working Days	Weekend	Average	ARIMA	Wavelet and ARIMA	Naïve
Feb	5.25	4.89	5.15	6.32	4.78	7.68
May	3.98	5.24	4.34	6.36	5.69	7.27
Aug	9.88	13.41	10.89	13.39	10.7	27.30
Nov	11.27	13.21	11.83	13.78	11.27	19.98
Average	7.60	9.19	8.05	9.96	8.11	15.56

Note that, in average, the WNN method outperforms both GARCH models. The average MRE is 8.57% for the WNN method and 9.55% for the plain GARCH model, neither of them using information about the expected load (better indices are obtained when December is excluded).

Comparison with a Hybrid Wavelet-ARIMA Model   According to Reference [45], the data set in this case comprises the working days of years 2001 and 2002, and the following 4 weeks of 2002 are selected to perform the experiments: February 18 to 22, May 20 to 24, August 19 to 23, and November 18 to 22. For comparison, in this case the WNN approach has been tuned specifically for each week, separate models being developed for working days and weekends.

Table 3-24 compares, for each week, the error provided by the WNN method with those corresponding to the plain ARIMA and the Wavelet-ARIMA models, both taken from Reference [45]. Results provided by the "naive" approach, in which the prices of the present week are directly taken as estimates for the next week, are also included.

In spite of its simplicity, the WNN method outperforms in all cases the plain ARIMA model and, in average, also the enhanced Wavelet-ARIMA model. The average MMRE for the whole test period is 8.05% when the WNN method is applied, whereas the Wavelet-ARIMA model yields an average of 8.11% for the same period.

APPLICATION TO CUSTOMER-LEVEL FORECASTING

This section discusses the application of the WNN technique to predict the demand of a mid-size single customer, which is served from the medium-voltage distribution level. In Spain, consumers in this sector are being given increased incentives to get their electricity either from the pool or from bilateral contracts, for which an accurate prediction of the short- and medium-term demand is crucial. Government or institutional buildings constitute good examples of such type of customers. The time series used below refers specifically to the electricity consumed by the Engineering School building at the University of Seville, in Spain.

Characterization and Processing of Data Records

Unlike the aggregated demand of a whole country or region, that follows a rather repetitive daily and weekly pattern, the demand of an academic institution, like that of a university faculty, may experience a large variation from one day to another, depending not only on weather conditions but also on many other circumstances such as if the day is a weekday or not, if it is a holiday period for students, if there is any special event, etc. Owing to this increased demand variability, it is not an easy task to find as accurate predictions as in the former applications discussed in this chapter.

This kind of customers has four different ways to go through a power-supply contract: official electricity tariffs inherited from the past that will be transiently available, rates freely offered by electricity trading companies, bilateral contracts, and the pool market. However, according to Spanish regulations, the old system based on annually revisited electricity rates is going to disappear for mid-size consumers on July 1, 2008, which makes it mandatory for these customers to choose any of the other possibilities. Apart from other management and technical issues, preparing reasonable buying bids to pool markets, or reaching acceptable agreements with energy brokers, requires that suitable demand-prediction tools be developed. The uncertainty of the resulting load estimation is directly related to the risk of engaging in a bilateral contract or the amount paid for deviation penalties in the pool.

In order to implement any prediction tool, it is first necessary to duly analyze existing records of energy consumption. The following steps explain the process that has been carried out in order to get a solid data base for this study:

1. Obtaining records: The hourly energy consumption has been measured with the help of power system analyzers, the data sets being saved in spreadsheets.

2. Data filtering: Some measurements were missing for a period of time (sometimes a whole day), or the assigned time stamp was doubtful. These irregularities, possibly due to errors in the data loggers or in the communication system, called for a pre-filtering stage aimed at getting a regular time series, suitable for the prediction study.

3. Restoring: Records for days with less than five consecutives missing hours were restored. The missing data were estimated as the mean of the three nearest complete days within the data base.

Proceeding as explained, 68 days of the database were fully restored, out of 168 incomplete days. Figures 3-12 to 3-14 show a restoration example of the demand curve for March 1, 2005.

The graph in Fig. 3-12 shows the records initially available for that day, in which some missing values can be observed. In Fig. 3-13, the three nearest days from which the missing data are going to be estimated are superimposed. The restored curve is finally shown in Fig. 3-14.

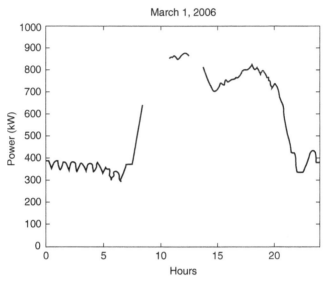

Figure 3-12   Initial available data.

The variability of the demand as a function of the building occupancy will be illustrated below. The pictures in Fig. 3-15 provide an example of the daily demand for a typical week during the teaching season in Spain (February 13–19, 2005).

It can be noted how the demand drops during the weekend compared to the weekdays. On a typical weekday the power demand curve is made up of different

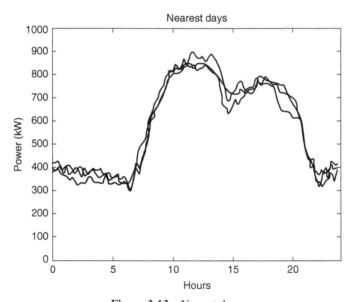

Figure 3-13   Nearest days.

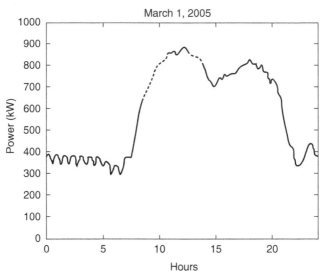

Figure 3-14   Restored curve.

parts: two valley periods, approximately covering the hours from midnight to 8:00 a.m. and from 9:00 p.m. to midnight; and a high-demand time during the rest of the day. The demand peaks at around noon and then it reaches a lower maximum at around 4 p.m. It is worth noting that classes start at 8:30 a.m. and finish at 9:00 p.m.

Another week (August 28 to September 3, 2005), representing the end of the summer break (August) and the beginning of the exam period (September) is shown in Fig. 3-16. During the weekdays of August the demand is lower than usual (compared to the Fig. 3-15), as the occupancy of the building is very low. Then, on September

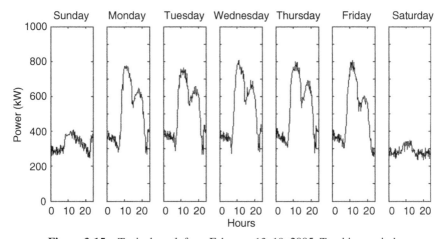

Figure 3-15   Typical week from February 13–19, 2005. Teaching period.

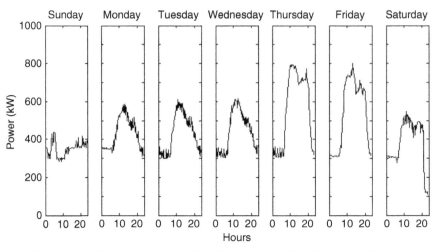

Figure 3-16   Week from August 28 to September 3, 2005. Exam period begins.

1, the examination period begins, the building occupancy increases and so does the demand. It can also be noted how on that particular Saturday the demand is higher than usual, which happens because some of the examinations take place on Saturdays.

Figure 3-17 shows an example of a holiday in the middle of the week, in this case, Tuesday, November 1, 2005. It can be seen how the demand for that day behaves like the one corresponding to a weekend day.

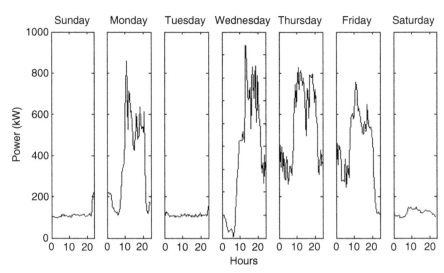

Figure 3-17   Week from October 30 to November 5, 2005. Special holiday on November 1.

Test Results

The WNN method will be applied next to predict the daily demand throughout the year 2005. Since the actual demand for those days is known, it will be possible to evaluate the prediction error.

For this application, both the least squares (LS) and the IPD approaches for computing the WNN weighting coefficients have been tested and compared. Furthermore, unlike in the applications described in earlier sections, no special effort has been made in this case to optimize the number of selected neighbors or the size of the window data. The chosen window data length is 24 hours, as experience shows it to be the most appropriate value for load forecasting, while three neighbors have been selected in all experiments. In this particular application, demand measurements are available every 15 minutes, which means that each vector of data contains 96 rather than 24 points.

An important issue refers also to the convenience of previously classifying the entire set of days contained in the data base into several characteristic clusters, so that nearest neighbors are only searched for in the right subset. As will be seen below, this classification may significantly affect the resulting prediction accuracy.

The tested classification methods are explained next:

Without any classification: The day before the prediction day (current day) is compared to the entire data base in order to choose its nearest neighbors.

Day of the week: The present day is only compared to the days of the data base corresponding to the same day of the week.

Weekday or weekend (W&W): If the prediction day is Saturday or Sunday, the previous day is only compared to Fridays and Saturdays, respectively. If the prediction day is any day from Monday to Friday, the day before is compared to all days from Sunday to Thursday.

Business day or holiday (B&H): In this method, each day has been assigned an identification flag with two letters. The first letter refers to the previous day, being a "B" if it is a business day or an "H" if it is a holiday, a Saturday or a Sunday. The second letter, with the same meaning, refers to the day to which the flag belongs. Therefore, four kinds of days result:
- BB: business day preceded by a business day.
- HB: business day preceded by a holiday.
- BH: holiday preceded by a business day.
- HH: holiday preceded by a holiday.

Then, the day preceding the prediction day is only compared to days of the database which are followed by a day with the same flag as the prediction day.

The advantage of this method is that the flag of any day can be modified well in advance if it is known to be a special day such as, for example, a holiday in between the weekdays. Note that if a flag is modified, the first flag letter of the following day must also be modified.

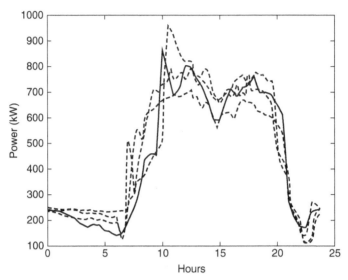

Figure 3-18 Day before the prediction day (current day) and its three nearest neighbors using the WNN method with LS coefficients and "W&W" classification.

Figures 3-18 to 3-20 show each step of the WNN prediction process. This result is obtained for October 5, 2005, by applying the LS method of computing coefficients and the "W&W" classification scheme for the data base.

Figure 3-18 shows the demand for the day that precedes the day of interest (solid line), as well as its three nearest neighbors (dotted lines). Figure 3-19 shows the

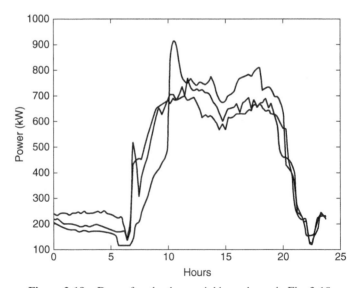

Figure 3-19 Days after the three neighbors shown in Fig. 3-18.

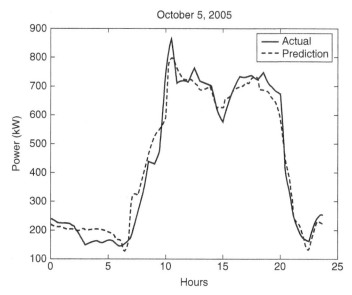

Figure 3-20   Actual and predicted demands for October 5, 2005.

3 days following the three neighbors represented in Fig. 3-18. In Fig. 3-20, the actual and the predicted demands are shown in solid and dotted lines, respectively.

The graph in Fig. 3-21 represents the prediction for the same day without applying any classification to the database.

Compared with Fig. 3-20, it can be clearly observed that a much less accurate prediction is obtained (28.4% relative error versus 7.6% for the total daily energy).

The bar diagram of Fig. 3-22 represents the daily energy relative errors that result by applying the four database classification schemes and the IPD coefficient computation method for 2 weeks of July 2005.

As expected, the worst results are obtained when no classification of days is attempted (black bar). In this example, both the "W&W" and "B&H" schemes provide the same results (grey bar), which are always the best except for July 5 and 6.

Figure 3-23 represents the actual and predicted demands for the same 2 weeks when the best classification method ("W&W" or "B&H") is applied.

Although in the last case the "W&W" and "B&H" classification schemes provide the same results, in general, the "B&H" approach performs better than all of the other methods, as will be shown through the following example.

Figure 3-24 illustrates the case of a week with a special holiday arising on a Tuesday (November 1, All-Saint's Day in Spain). The black and white bars represent the errors for the "W&W" and "B&H" methods, respectively, when the LS method of computing coefficients is adopted. The reason why the "B&H" classification method performs much better in this case lies in the fact that the demand for November 1 behaves like that of a weekend day, as shown in Fig. 3-25, where a comparison is made with a regular labor Tuesday. This also happens when long weekends take place.

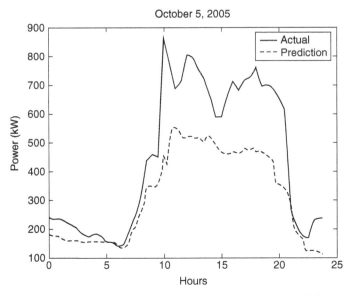

Figure 3-21   Actual demand and prediction. WNN method with LS coefficients and without any data base classification.

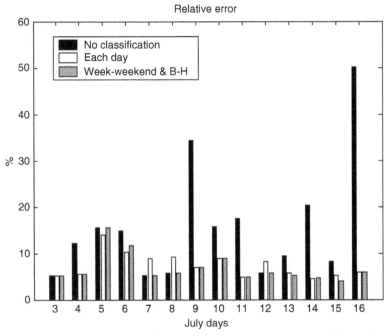

Figure 3-22   Relative errors for different data base classifications and IPD coefficients.

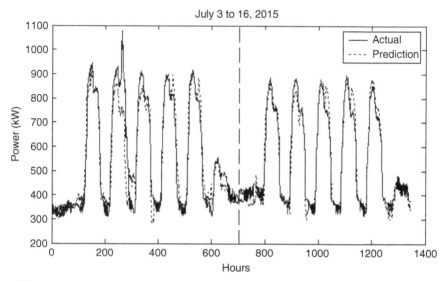

Figure 3-23   Actual and predicted demands for the 2-week period, July 3 to 16, 2005.

Comparison with an AR Model

An autoregressive (AR) model has been developed in order to compare the results with those obtained with the WNN method. For simplicity, the AR model has been applied to predict just weekdays by creating a new data base without weekends. This way, past

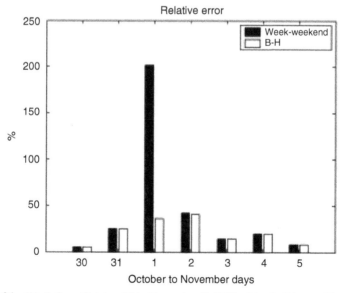

Figure 3-24   Week from October 30 to November 5, 2005 with a holiday on November 1.

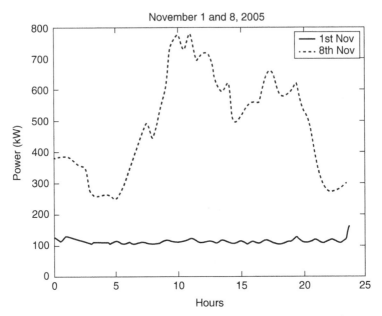

Figure 3-25   Demands corresponding to November 1 and 8, Tuesdays.

data up to the same day of the week as the prediction day can be considered, giving rise to a lower order AR model. With this simplification, considering the entire past week to predict the next day, the resulting order is 480 (96 records per day, captured at 15-minute intervals, times 5 days). Resorting to a longer past period would lead to a higher order model, but the higher the order the longer the computation time.

Figure 3-26 shows the relative errors of predicting, by means of the AR and the WNN methods, the weekdays from July 11 to 15, 2005. The WNN method makes use of the "B&H" classification method and the IPD computation of weighting coefficients. As can be seen in Fig. 3-23, the demand is very regular for those days, all the weekdays following the same pattern. For this specific week, the AR method gives overall a more accurate prediction than the WNN method does (except for the first day).

However, it does not always happen that way. As explained before, the demand for this kind of buildings on weekdays depends on many factors, not all of them fully predictable, and it is seldom as uniform as in the previously tested week. Figure 3-27 shows the relative prediction errors of both methods for the weekdays represented in Fig. 3-24, that is, from October 31 to November 4, 2005.

It can be observed for this week that the WNN method, when combined with the "B&H" data base classification, adapts much better to sudden demand changes from the usual patterns, providing better predictions in overall.

Table 3-25 shows the average weekly errors obtained by applying the WNN method for the weekdays July 11 to 15 and October 31 to November 4, 2005.

Table 3-26 is the counterpart of Table 3-25 for the AR method and the same 2 weeks.

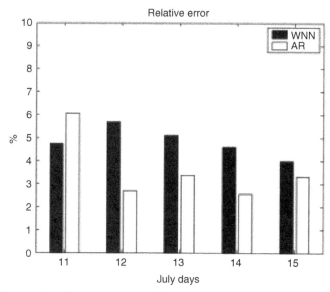

Figure 3-26   Relative errors for WNN and AR methods. Regular week.

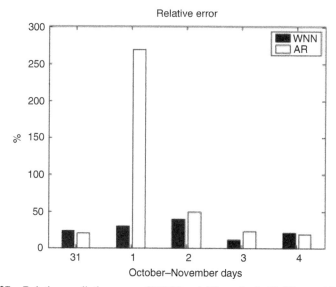

Figure 3-27   Relative prediction error of WNN and AR methods. Holiday on November 1.

Table 3-25   Weekly errors. WNN Method, IPD Coefficients, "B&H" Data Base Classification

Weekly Errors:	July 11–15	October 31–November 4
Average relative error (%)	4.4	23.7
Average absolute error (kW)	26.3	85.6
Average maximum error (kW)	341.6	1024.5

Table 3-26   Weekly Errors. AR Method.

Weekly Errors	July 11–15, 2005	October 31–November 4, 2005
Average relative error (%)	3.4	75.2
Average absolute error (kW)	20.3	157.5
Average maximum error (kW)	204.5	1464.5

From the two preceding tables it can be concluded that the AR method gives more accurate results for the weekdays of a regular week, whereas for an atypical week the WNN method is more accurate (and also for the weekends). However, the difference between the prediction accuracy of both methods in the regular week case is much less noticeable than in the irregular one. Therefore, taking also into account its simplicity and small computational effort, the WNN method can be deemed the preferable choice for this application as well.

CONCLUSIONS

In this chapter, a forecasting methodology, based on the WNN technique, is described in detail, including several ways of tuning the model parameters for a particular application. Such parameters have to do with the window length of the time series, the number of neighbors chosen for the prediction, and the way the weighting coefficients are computed.

This technique provides a very simple approach to forecast power system magnitudes characterized by daily and weekly repetitive patterns, such as energy demand and prices. Three case studies are used in this chapter to illustrate the potential of the WNN method.

The first application refers to the energy demand auctioned every hour in the day-ahead Spanish electricity market. When applied to the weekdays of year 2005 this provides a monthly relative prediction error ranging from 2.4% to 4.37%. The average yearly error of the WNN forecasts (3.15%) compares well with that provided by an AR model (4.65%).

The second application is focused on a more irregular time series, namely the hourly marginal prices of the day-ahead Spanish electricity market. In order to allow comparisons with some previously published methods, forecasting results corresponding to the market of mainland Spain for the entire year 2002 are reported, yielding an average monthly error which is close to 8%. An exhaustive analysis of the simulation results, for the weeks and scenarios reported in the literature, shows that the WNN prediction accuracy is generally better, on average, than that of other more sophisticated techniques such as ANN, GARCH, and ARIMA (with and without applying the Wavelet transform). This can be considered a satisfactory performance, particularly when the uncertainty of prices for the Spanish system is taken into account.

The third case study involves predicting the hourly demand of a faculty building owned by the University of Seville during 2005. This application is becoming of paramount importance for presently deregulated systems, in which medium and large customers are being forced to buy electricity from the pool or from bilateral contracts and, therefore, have to perform their own demand prediction in order to devise the best energy procurement strategy. In this case, the time series is much more volatile, as there are many ill-defined factors affecting the electricity consumption of an engineering faculty, which means that larger prediction errors are to be expected. Average weekly errors ranging from 4.4% to over 20% are provided by the WNN approach, depending on whether the weekly demand follows a regular pattern or not. This compares well with an AR model, yielding weekly errors from 3.4% to 75% for the same test periods.

The good results provided by the WNN method are somewhat unexpected, considering that exogenous variables, such as weather conditions or equipment availability, are not explicitly considered, unlike in other competing methods. Future efforts should be directed to devise hybrid techniques, taking advantage of the capability of conventional procedures to predict very repetitive or slowly changing time series and the potential shown by the WNN and other data mining techniques to react more quickly to sudden variations of the usual patterns.

REFERENCES

1. T. Cover and P. Hart, "Nearest neighbor pattern classification," *IEEE Transactions on Information Theory*, vol. 13, no. 1, pp. 21–27, 1967.
2. T. Cover, "Estimation by nearest neighbor rule," *IEEE Transactions on Information Theory*, vol. 14, pp. 50–55, 1968.
3. S. Dudani, "The distance-weighted k-nearest-neighbor rule," *IEEE Transactions on Systems, Man and Cybernetics, SMC-6*, vol. 4, no. 1, pp. 325–327, 1975.
4. R.D. Short and K. Fukunaga, "The optimal distance measure for nearest neighbour classification," *IEEE Transactions on Information Theory*, vol. 27, no. 5, pp. 622–627, 1981.
5. K. Fukunaga and T.E. Flick, "An optimal global nearest neighbour metric," *IEEE Transaction on Pattern Analysis and Machine Intelligence*, vol. 6, no. 3, pp. 314–318, 1984.
6. B.V. Dasarathy, *Nearest Neighbour (NN) Norms: NN Pattern Classification Techniques*, IEEE Computer Society Press, 1991.
7. D. Bunn and E.D. Farmer, "Economical and operational context of electric load prediction," in *Comparative Models for Electrical Load Forecasting*, D.W. Bunn and E.D. Farmer (eds), John Wiley & Sons, New York, 1985, pp. 13–30.
8. J.S. Griffith, S.R. Erwin, J.T. Wood, K.D. Le, J.T. Day, and C.K. Yin, "Using Cost Reconstruction Tool to Analyze Historical System Operation," Proceedings of the 16th Power Industry Computer Applications International Conference, PICA'1989, Seattle, WA, pp. 311–317, May 1989.
9. D.K. Ranaweera, G.G. Karady, and R.G. Farmer, "Economic impact analysis of load forecasting," *IEEE Transactions on Power Systems*, vol. 12, no. 3, pp. 1388–1392, August 1997.
10. O. Mohammed, D. Park, D.R. Merchant, T. Dinh, C. Tong, A. Azeem, J. Farah, and C. Drake, "Practical experiences with an adaptive neural network short-term load forecasting

system," *IEEE Transactions on Power Systems*, vol. 10, no. 1, pp. 254–265, February 1995.

11. P. Douglas, A.M. Breipohl, F.N. Lee, and R. Adapa, "The impacts of temperature forecast uncertainty on Bayesian load forecasting," *IEEE Transactions on Power Systems*, vol. 13, no. 4, pp. 1507–1513, November 1998.

12. I. Drezga and S. Rahman, "Input variable selection for ANN-based short-term load forecasting," *IEEE Transactions on Power Systems*, vol. 13, no. 4, pp. 1238–1244, November 1998.

13. G. Gross and F.D. Galiana, "Short-term load forecasting," *Proceedings of the IEEE*, vol. 75, no. 12, pp. 1558–1573, December 1987.

14. A. Papalexopoulos and T.C. Hesterberg, "A regression-based approach to short-term load forecasting," *IEEE Transactions on Power Systems*, vol. 5, no. 4, pp. 1535–1550, November 1990.

15. N. Mbamalu and M.E. El-Hawary, "Load forecasting via suboptimal autoregressive models and iteratively reweighted least squares estimation," *IEEE Transactions on Power Systems*, vol. 8, no. 1, pp. 343–348, February 1993.

16. J. Toyoda, M. Chen, and Y. Inoue, "An application of state estimation to short-term load forecasting, Part I and II," *IEEE Transactions on Power Apparatus and Systems*, vol. PAS-89, pp. 1678–1688, September 1970.

17. M.T. Hagan and S.M. Behr, "The time series approach to short-term load forecasting," *IEEE Transactions on Power Apparatus and Systems*, vol. PAS-2, no. 3, pp. 785–791, August 1987.

18. S. Rahman and O. Hazim, "A generalized knowledge-based short-term load-forecasting technique," *IEEE Transactions on Power Systems*, vol. 8, no. 2, pp. 508–514, May 1993.

19. H. Mori and H. Kobayashi, "Optimal fuzzy inference for short-term load forecasting," *IEEE Transactions on Power Systems*, vol. 11, pp. 390–396, February 1996.

20. A. Khotanzad, R. Afkhami-Rohani, and D. Maratukulam, "ANNSTLF-Artificial neural network short-term load forecaster-generation tree," *IEEE Transactions on Power Systems*, vol. 13, no. 4, pp. 1413–1421, November 1998.

21. H. Mori and S. Urano, "Short-term Load Forecasting with Chaos Time Series Analysis," Proceedings of the IEEE ISAP '96, Orlando, FL, pp. 133–137, February 1996.

22. D.G.C. Smith, "Combination of forecasts in electricity demand prediction," *Journal of Forecasting*, vol. 8, pp. 349–356, 1989.

23. A. Khotanzad, R. Afkhami-Rohani, and D. Maratukulam, "ANNSTLF-Artificial neural network short-term load forecaster-generation tree," *IEEE Transactions on Power Systems*, vol. 13, no. 4, pp. 1413–1421, November 1998.

24. K.H. Kim, J.K. Park, K.J. Hwang, and S.H. Kim, "Implementation of hybrid short-term load forecasting system using artificial neural networks and fuzzy expert systems," *IEEE Transactions on Power Systems*, vol. 10, pp. 1534–1539, August 1995.

25. S.J. Huang and K.R. Shih, "Short-term load forecasting via ARMA model identification including non-Gaussian process considerations," *IEEE Transactions on Power Systems*, vol. 18, no. 2, pp. 673–679, May 2003.

26. H. Yang and C. Huang, "A new short-term load forecasting approach using self-organizing fuzzy ARMAX models," *IEEE Transactions on Power Systems*, vol. 13, no. 1, pp. 217–225, February 1998.

27. S. Tomonobu, P. Mandal, K. Uezato, and T. Funabashi, "Next day load curve forecasting using hybrid correction method," *IEEE Transactions on Power Systems*, vol. 20, no. 1, pp. 102–109, February 2005.

28. A. Troncoso, J.M. Riquelme, J.C. Riquelme, A. Gómez, and J.L. Martínez, "Time-series prediction: Application to the short-term electric energy demand," *Lecture Notes in Artificial Intelligence*, vol. 3040, pp. 577–586, 2004.

29. A. Troncoso, J.C. Riquelme, J.L. Martínez, J.M. Riquelme, and A. Gómez, "Influence of kNN-based load forecasting errors on optimal energy production," *Lecture Notes in Artificial Intelligence*, vol. 2902, pp. 189–203, 2003.

30. A.M. González, A.M.S Roque, and J. García-González, "Modeling and forecasting electricity prices with input/output hidden Markov models," *IEEE Transactions on Power System*, vol. 20, no. 1, pp. 13–24, February 2005.

31. A.J. Conejo, J. Contreras, J.M. Arroyo, and S. de la Torre, "Optimal response of an oligopolistic generating company to a competitive pool-based electric power market," *IEEE Transactions on Power System*, vol. 17, no. 2, pp. 404–410, 2002.

32. R.C. García, J. Contreras, M. van Akkeren, and J.B.C. García, "A GARCH forecasting model to predict day-ahead electricity prices," *IEEE Transactions on Power System*, vol. 20, no. 2, pp. 867–874, 2005.

33. J. Contreras, R. Espínola, F.J. Nogales, and A.J. Conejo, "ARIMA models to predict next-day electricity prices," *IEEE Transactions on Power System*, vol. 18, no. 3, pp. 1014–1020, 2003.

34. F.J Nogales, J. Contreras, A.J. Conejo, and R. Espinola, "Forecasting next-day electricity prices by time series models," *IEEE Transactions on Power System*, vol. 17, pp. 342–348, May 2002.

35. J.L. Martínez-Ramos, A. Gómez-Expósito, J.M. Riquelme-Santos, A. Troncoso, and A.R. Marulanda Guerra, "Influence of ANN-based Market Price Forecasting Uncertainty on Optimal Bidding," PSCC Power System Computation Conference, Seville, Spain, November 2002.

36. B.R. Szkuta, L.A. Sanabria, and T.S. Dillon, "Electricity price short-term forecasting using artificial neural networks," *IEEE Transactions on Power System*, vol. 14, no. 3, pp. 851–857, 1999.

37. B. Ramsay and A.J. Wang, "A neural network based estimator for electricity spot-pricing with particular reference to weekend and public holidays," *Neurocomputing*, vol. 23, pp. 47–57, 1998.

38. L. Zhang, P.B. Luh, and K. Kasiviswanathan, "Energy clearing price prediction and confidence interval estimation with cascaded neural networks," *IEEE Transactions on Power System*, vol. 18, no. 1, pp. 99–105, 2003.

39. A. Troncoso, J. Riquelme, J. Riquelme, J. L. Martínez, and A. Gómez, "Electricity market price forecasting: Neural networks versus weighted-distance k nearest neighbours," *Lecture Notes in Computer Science*, vol. 2453, pp. 321–330, 2002.

40. C.P. Rodriguez and G.J. Anders, "Energy price forecasting in the Ontario competitive power system market," *IEEE Transactions on Power System*, vol. 19, no. 1, pp. 366–374, February 2004.

41. A. Troncoso, J.M. Riquelme, A. Gómez, J.L. Martínez, and J.C. Riquelme, "Electricity market price forecasting based on weighted nearest neighbors techniques," *IEEE Transactions on Power System*, vol. 22, no. 3, pp. 1294–1301, August 2007.

42. Compañía Operadora del Mercado Español de Electricidad (Spanish Market Operator), Available at http://www.omel.es

43. M.B. Kennel, R. Brown, and H.D.I. Abarbanel, "Determining embedding dimension for phase-space reconstruction using a geometrical construction," *Physical Review A, vol.* 45, no. 6, pp. 3403–3411, 1992.

44. A. Canoyra, C. Illán, A. Landa, J.M. Moreno, J.I. Pérez Arriaga, C. Sallé, and C. Solé, "*The Hierarchical Market Approach to the Economic and Secure Operation of the Spanish Power System,*" Bulk Power System Dynamic and Control IV, Santorini, Greece, August 24–28, 1998.

45. A.J. Conejo, M.A. Plazas, R. Espínola, and A.B. Molina, "Day-ahead electricity price forecasting using the wavelet transform and ARIMA models," *IEEE Transactions on Power System*, vol. 20, no. 2, pp. 1035–1042, May 2005.

46. IESO. Power to Ontario. On Demand, Available at http://www.ieso.ca/

Chapter 4

Electricity Prices as a Stochastic Process

Yunhe Hou, Chen-Ching Liu, and Harold Salazar

INTRODUCTION

The transition of the power industry from a traditional utility company to a competitive electricity market creates the need for accurate models of electricity prices [1–4], both for suppliers and consumers. Participants in the power markets have to meet the challenge of managing revenues driven by volatile electricity prices. In a competitive market, good insights into the dynamics of electricity price processes are important [5]. However, modeling the price behavior of electricity is a challenging task due to the distinguishing characteristics of electricity. The purpose of this chapter is to provide a survey of the stochastic process models for electricity prices.

Electricity prices vary with time. Stochastic processes are functions of time whose values at different times are random variables. They have been used to model prices [6]. Stochastic models of electricity prices are fundamentally important for electricity market participants. The participants' activities include price forecasting [4, 7, 8], risk management of project investment, risk assessment of physical and financial contracts [9], bidding strategy development [10], and evaluation of operating policies for generation and transmission assets [5, 11]. Risk management and asset valuation are essential issues for investors and operators of electricity markets. The two issues require in-depth understanding and detailed modeling of electricity prices [12, 13]. The reason is simple—risk is associated with uncertainty. When there is no uncertainty, there is no risk. If one knows for sure what is going to happen next, good or bad, it will be possible to take necessary precautions. It is the uncertainty that causes anxiety about risk; therefore, models of stochastic prices are needed [14].

Advances in Electric Power and Energy Systems: Load and Price Forecasting, First Edition.
Edited by Mohamed E. El-Hawary.
© 2017 by The Institute of Electrical and Electronics Engineers, Inc. Published 2017 by John Wiley & Sons, Inc.

There have been studies on modeling of spot prices of electricity. Generally, electricity price models are classified into four groups [15]: production costs models, continuous time stochastic models, time series models, and game theory models.

Fundamental models consider not only production costing but also the impact of agents' strategic behavior on the market price. Schweppe [1] is the pioneer of this group. Based on the concept of nodal price, Chen et al. [16] present a method to provide a detailed description of each nodal price. In the work by Boogert and Dupont [17,18], the relationship between supply and demand is considered. Based on the supply–demand framework, the hourly day-ahead price of electricity is modeled. The level and probability of a spike in the spot price can be forecasted. Since real-time price elasticity of electricity contains important information on the demand response to the volatility of peak prices. Lijesen [19] provides a quantification of the real-time relationship between total peak demand and spot market prices. A low value for the real-time price elasticity is found. Kirschen et al. [20] present an analysis of the effect that the market structure can have on the elasticity of demand for electricity. The relationship between electricity price, volatility, and electricity market structure of Singapore's electricity market is studied in Reference [21]. Reference [22] examines extensive hourly or half-hourly power price data from 14 deregulated power markets. It analyzes average diurnal patterns, relationship of system load, volatility, and consistency over time. To model the mid-term spot price of the Nordic market, Vehvilainen and Pyykkonen [23] model the fundamentals affecting the spot price and propose a market equilibrium model to combine them and form the spot price. The impact of temperature [24], water temperatures [25] ,and information [26] is also studied in this framework.

Continuous time stochastic models of commodity prices led to the success of geometric Brownian motion (GBM) in modeling stock prices (see References [27] and [28]). Early studies in this area typically assume that commodity prices are governed by GBM. More recently, a number of researchers, such as Alvarado and Rajaraman [29], Deng [30], Lucia and Schwartz [31], considered the use of mean reversion price models. The basic idea is that deviations of the price from its equilibrium level are corrected and subjected to random perturbations. Schwartz [32] proposes a two-factor mean reversion model. The underlying idea is that the short-term deviations correspond to temporary changes in prices that are expected to revert toward zero. Changes in the equilibrium price level reflect fundamental longer-term changes that are expected to persist, such as expectations of the exhaustion of existing supply, improving technology for production, as well as political and regulatory effects [24]. References [2] and [33] indicate the inadequacy of GBM and mean reversion processes for modeling of electricity prices. To describe the spikes of electricity price, Deng [30] examines three types of mean reversion jump-diffusion electricity price models: mean reversion jump-diffusion process with deterministic volatility, mean reversion jump-diffusion process with regime switching, and mean reversion jump-diffusion process with stochastic volatility. The modified jump-diffusion process is also employed in References [30, 34–43]. Considering the spot price dynamics of electricity which includes seasonality of the prices and spikes, Benth et al. [44] describe the dynamics of electricity price as a sum of non-Gaussian Ornstein–Uhlenbeck processes with jump processes, the normal variations and spike behavior of the prices are involved as well. Comparing with the mean reversion process, Levy process has a more general form. In the work presented in References [15] and [45], a set of stochastic mean reversion models for electricity prices with Levy-process-driven Ornstein–Uhlenbeck (OU) processes is proposed. Simulation results show that this approach effectively captures not only the anomalous tail behavior but also the correlation structure present in the electricity price series. Each market state is characterized by a particular density function, which represents the relationship between explanatory variables such as load; hydro, thermal, and nuclear resources available; and the electricity spot price through a dynamic regression model. Hence, their model can be seen as a switching model in which the system evolves through different states, where a particular dynamic regression model is adjusted in each one. Based on this idea, some references, for example, [34–36, 46–48] are focused on the regime-switching characteristics of electricity prices.

Time series models [49, 50] study price evolution from a statistical point of view. The well-known generalized autoregressive conditional heteroscedastic (GARCH) models [51], are representative in this group. GARCH [52] is used to describe electricity price volatility and the marginal cost of congestion on the NYISO market, and to examine properties of spot prices and price volatility among the five regional electricity markets in the Australian National Electricity Market [53]. One of the essential characteristics of price is the fat tail of its distribution. To deal with the fat tail property, that is, electricity prices exhibit extremely large skewness or kurtosis, an exponential GARCH with extreme values is adopted in Reference [54]. A similar method is employed in Reference [55]. Vector autoregression (VAR) [56] is used to describe price dynamics among 11 US electricity spot markets. VAR allows regularities in the data to be studied without imposing as many prior restrictions as structural models impose. Based on the model of real-time balancing power market, Olsson and Soder [57] employ the combined seasonal autoregressive integrated moving average (SARIMA) and discrete Markov processes to model prices of a power market. Considering the mean-reverting characteristics of price, an ARMAX model is selected. To improve the performance of time series method, Swider and Weber [58] use the normality hypothesis of electricity prices and apply an autoregressive moving average (ARMA) model combined with GARCH, Gaussian-mixture, and switching-regime approaches to study electricity prices of the spot and two reserve markets in Germany. It is shown that the proposed extended models lead to significantly improved representations of stochastic price processes. As an improvement, a general form, known as autoregressive integrated moving average (ARIMA) [59], is used for price forecasting. Considering different trading periods, Guthrie and Videbeck [60] use a standard autoregression, known as a periodic autoregression (PAR), to model spot price data. Comparing with AR models, PAR takes different values in different trading periods. Misiorek et al. [61], assess the short-term forecasting power of different time series models in the electricity spot market. They calibrate AR/ARX ("X" stands for exogenous/fundamental variable—system load), AR/ARX-GARCH, TAR/TARX, and Markov regime-switching models with California Power Exchange (CalPX) system spot prices. They also use these models for out-of-sample point and interval forecasting in normal and extremely volatile periods preceding the market crash in winter 2000/2001.

Artificial neural network (ANN) is an approach to curve approximation based on series of prices. Combining with different time series models, the performance for price forecasting may be improved. Many references, for example, [59, 62–73] focus on these approaches. Pino et al. [66] use a multilayer perceptron (MLP) to train ART-type neural network, which is used to forecast next-day price of electricity in the Spanish energy market. Vahidinasab et al. [73] use Levenberg–Marquardt modification to the Gauss–Newton method to improve the efficiency of ANN training. Zhao et al. [74] employ a support vector machine (SVM) to forecast the prices. SVM is also used in Reference [75]. Wavelet transform is an important empirical study approach based on the series of prices. References [71] and [76] employ this method to approximate electricity prices. All the above time series models are based on state space models. Pedregal and Trapero [77] presented a univariate dynamic harmonic regression model set up in a state space framework for price forecasting. In this work, a fast automatic identification and estimation procedure is proposed based on the frequency domain. Good forecasting performance and rapid adaptability of the model are illustrated with actual prices from the PJM interconnection and for the Spanish market for the year 2002.

Another group of models aim to obtain reasonable price estimations and to analyze market power based on game theory. In Reference [78], a production-based approach is introduced to take into account different attitudes and liabilities of market participants in the equilibrium analysis of day-ahead prices. Considering the availability of units and demand, three oligopoly models—Bertrand, Cournot, and supply function equilibrium (SFE), were employed to analyze the mean and variance of electricity prices [79]. Genc and Sen [80] use games with probabilistic scenarios to predict Nash equilibrium of capital investment and price trajectories of the Ontario electricity market. The implications of two policies are compared. Borenstein and Bushnell [81] took a different approach to analyze the existence of market power in California's electricity market. Using a Cournot simulation approach, they found that the potential for strategic behavior existed in the California electricity market. Furthermore, their analysis suggests that the relevant geographic market for California spot electricity, in the absence of transmission constraints, should encompass member utilities of the WSCC. The price equilibrium is investigated in References [82] and [83] by different game theory models.

This chapter is organized in the following manner. The Characteristics of Electricity Prices describes some significant features of electricity prices, such as mean reversion, seasonality, volatility and spikes. Based on the characteristics of electricity prices, Stochastic Process Models for Electricity Prices discusses continuous time stochastic models. A variety of time series models for electricity prices are discussed in Numerical Examples.

CHARACTERISTICS OF ELECTRICITY PRICES

The electricity industry has been transformed from a primarily technical business to one in which products are treated in much the same way as other commodities. In a commodity market, trading and risk management are key tools for success. However, electricity is unique in that essentially once produced it cannot be stored. This poses a challenge for modeling of the price behavior. Moreover, the aggregate electricity supply and demand have to be balanced constantly so as to maintain the quality of power supply and system reliability. Since the supply and demand shocks cannot be smoothed by inventories, electricity spot prices are more volatile than those of other commodities [15, 30].

This section addresses the challenge of capturing the behavior of electricity prices. The behavior of electricity prices of several markets is reviewed. The comparison serves to highlight the important characteristics of electricity prices. Those characteristics illustrate the economic and physical fundamentals of electricity markets.

The data used in this study consist of hourly electricity prices. Several markets around the world are examined. Figures 4-1 to 4-4 show, respectively, hourly prices of (1) PJM market from April 2, 1998 to September 30, 2003, (2) England and Wales market from February 4, 2001 to March 3, 2004, (3) California market from April 1, 1998 to March 2, 2000, and (4) Ontario market from May 1, 2002 to April 30, 2005.

The figures illustrate the cyclical and volatile nature of electricity prices in each market. Some properties of electricity are shown. First, electricity spot prices show a strong trend of mean reversion. This is a common feature in prices of other traded commodities [84]. The reason is that when the price of electricity is high, its supply tends to increase, putting a downward pressure on the price. Similarly, when the price is low, the supply of electricity tends to decrease, providing an upward lift to the price.

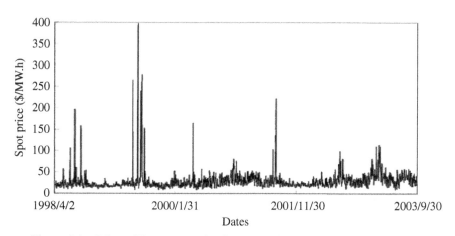

Figure 4-1   Prices of PJM spot market from April 2, 1998 to September 30, 2003.

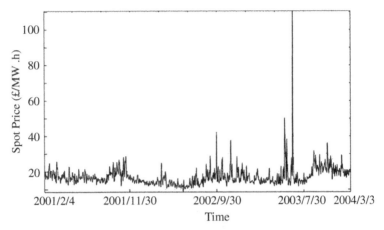

Figure 4-2   Prices of England and Wales market from February 4, 2001 to March 3, 2004.

The next salient feature of electricity prices is seasonality. The seasonal property of electricity spot prices is a direct consequence of the corresponding fluctuations in demand. It arises from the changing conditions, that is, temperature and daylight. Figure 4-5 illustrates a sample of hourly price and quantity data for the time period from August 1, 1998 to August 22, 1998 in California. It is clear that price follows demand. Figure 4-6 plots the Nord pool market prices from January 1, 1997 to April 25, 2000. The annual cycle can be well approximated by a sinusoid with a linear trend [4, 36].

Electricity price curves demonstrate strong and stochastic volatility or spikes [85]. Figure 4-7 presents the histogram for the 2 months of PJM spot prices in summer 2000. Existence of the fat tails suggests that the probability of rare events is much

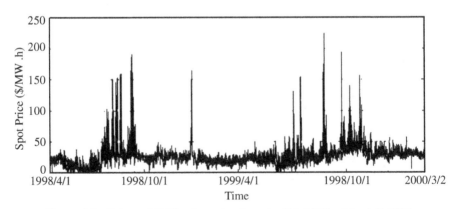

Figure 4-3   Prices of California market from April 1, 1998 to March 2, 2000.

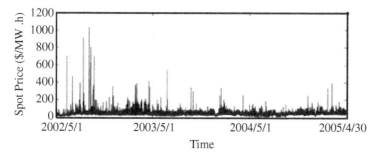

Figure 4-4 Prices of Ontario market from May 1, 2002 to April 30, 2005.

higher than what is predicted by a Gaussian distribution. The implication is that the spikes in electricity data cannot be captured by simple Gaussian shocks.

The jump behavior of electricity spot prices is primarily due to the fact that a typical regional aggregate supply cost curve for electricity almost always has a kink at a certain capacity level and a steep upward slope beyond that capacity level. A forced outage of a major power plant or a sudden surge in demand will either shift the supply curve left or lift the demand curve up so that the regional electricity demand curve crosses the regional supply curve at its steep-rise portion, causing a jump in the

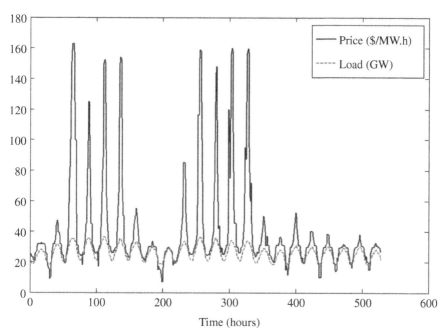

Figure 4-5 Sample hourly electricity prices and load from August 1, 1998 to August 22, 1998, in California.

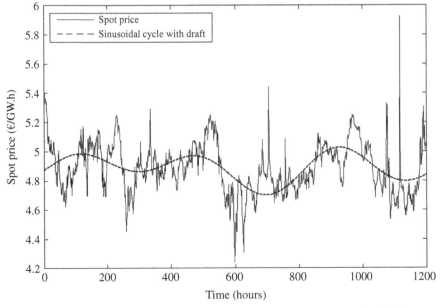

Figure 4-6   Nord pool market prices from January 1, 1997 to April 25, 2000.

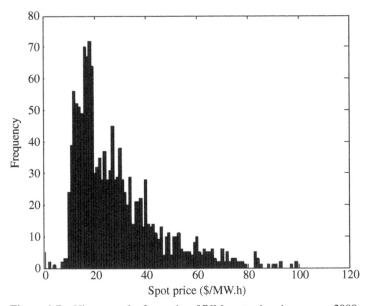

Figure 4-7   Histogram for 2 months of PJM spot prices in summer 2000.

price. When the contingency disappears in the short term, the high price will drop to its normal range, forming a spike.

To summarize, in order to construct a model for simulation of the electricity price process in the spot market, the following characteristics need to be considered [3, 33]:

(i) Strongly seasonal nature of prices;

(ii) Mean reversion nature;

(iii) Price-dependent volatilities;

(iv) Correlation between electricity price and load.

STOCHASTIC PROCESS MODELS FOR ELECTRICITY PRICES

Electricity prices vary over time due to the constant change in demand. Random events, such as load variations, contingencies, network congestions, and market power, cause prices to fluctuate. Stochastic processes have been widely used to model such random behavior. In this section, stochastic processes that are suitable for modeling electricity prices are discussed. This section starts with a building block of stochastic processes, that is, the Brownian motion. More advanced stochastic processes are constructed based on this fundamental model. GBM and mean reversion process are also discussed. Those models can be applied to electricity prices over different time horizons. For instance, weekly and hourly prices can be modeled by a geometric mean reversion (GMR) process with a time-varying mean.

From Random Walk to Brownian Motion

Brownian motion, sometimes called the Wiener process, is an essential stochastic process. It serves as a building block of other more elaborate models. Brownian motion has made a profound impact on physical as well as social sciences. Consider the simplest discrete case, in which the stochastic process is defined as a sequence of random variables. Denote by $\tilde{Z}(t)$ a random variable at time t, and assume that the difference between two consecutive time values t_i and t_{i+1} is constant and equal to Δt. A random walk is a stochastic process defined by

$$\tilde{Z}(t_k) = \tilde{Z}(t_{k-1}) + \tilde{\varepsilon}_k \sqrt{\Delta t}; \tilde{\varepsilon}_k \sim N(0, 1) \tag{4-1}$$

where $\tilde{\varepsilon}_k$ is a normal distribution with zero mean and one standard deviation, that is, $N(0, 1)$. A random walk can also be represented as the difference between two consecutive values of $\tilde{Z}(t)$, that is,

$$\Delta \tilde{Z}(t_k) = \tilde{\varepsilon}_k \sqrt{\Delta t} \tag{4-2}$$

where $\Delta \tilde{Z}(t_k) = \tilde{Z}(t_k) - \tilde{Z}(t_{k-1})$.

To distinguish between random and regular variables, the symbol $(\sim)$ over a variable, for example, $\tilde{z}$, will be used. z is the value (a real number) that the random variable assumes.

The following properties characterize a random walk [6]:

1. $E\{\tilde{Z}(t_k)\} = 0$
2. $\text{Var}\{\tilde{Z}(t_k)\} = t_k$
3. $E\{\tilde{Z}(t_k), \tilde{Z}(t_m)\} = 0$ if $k \neq m$

where $E\{\bullet\}$ and $\text{Var}\{\bullet\}$ denote expectation and variance, respectively. By property 2, the variance of random walk increases proportionally over time and property 3 implies that the walk is uncorrelated at different time values. Note that due to the normal distribution of small increments $\Delta\tilde{Z}(t_k)$, a random walk may take negative values and, therefore, violate the non-negative characteristic of electricity prices. Property 3 deserves special attention since prices might be correlated. This model, therefore, needs to be modified in order to accurately reproduce the behavior of electricity prices. More elaborate models are constructed by relaxing some properties and addition of more terms to 4-1.

At the limit $(\Delta t \to 0)$, the random walk is represented by

$$d\tilde{Z}(t) = \tilde{\varepsilon}(t)\sqrt{dt} \qquad (4\text{-}3)$$

which implies that the variance of infinitesimal increments is equal to dt. A formal definition of a Brownian motion is as follows:

A stochastic process $\tilde{Z}(t)$ defined in the interval $0 \leq t \leq T$, that is, $\{\tilde{Z}(t), 0 \leq t < T\}$ is a *Brownian motion process* in $[0, T)$ if

(i) $\tilde{Z}(0) = 0$.
(ii) $\tilde{Z}(t)$ has stationary independent increments, that is, for any finite discrete sample of time $0 \leq t_1 \leq t_2 \leq \ldots \leq T$, the random variables $\tilde{Z}(t_2) - \tilde{Z}(t_1), \tilde{Z}(t_3) - \tilde{Z}(t_2), \ldots, \tilde{Z}(t_n) - \tilde{Z}(t_{n-1})$ are independent.
(iii) The difference or increment $\tilde{Z}(t) - \tilde{Z}(s)$ from time t to s for $0 \leq s \leq t < T$ has a normal distribution with zero mean and variance $t - s$.

The last property implies that the increment from s to t does not depend on the history of the process up to time s (memoryless), but only on the time interval $t - s$. A stochastic process that is memoryless is called a Markov process. Hence, Brownian motion is, based on property 3, a Markov process.

Brownian motion can be simulated by sampling a standard normal distribution. Figure shows different realizations (sample paths) of a Brownian motion. For a fixed value of Δt, the sample paths are obtained based on the following simple routine:

(i) Set $Z(t_0) = 0$.

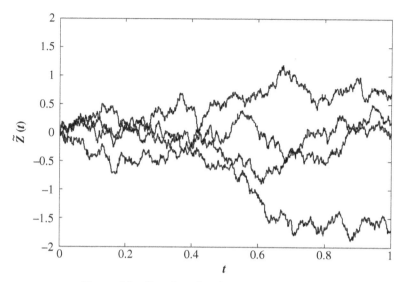

Figure 4-8   Sample paths of a Brownian motion.

(ii) Obtain $Z(t_i) = Z(t_{i-1}) + \varepsilon_i \sqrt{\Delta t}$, where $\varepsilon_1 \dots \varepsilon_i \dots \varepsilon_n$ are samples from a normal distribution. Note that the absence of a tilde above of Z denotes a realization of the random variable $\tilde{Z}(t)$.

Numerical simulation of various stochastic processes is useful for validation of different models. For instance, Figure 4-8 confirms that the Brownian motion can take negative values, which is a violation of the non-negativity assumption on electricity prices. There are, however, pathological cases in which electricity prices are indeed negative.

The probability density function (pdf) of a Brownian motion is illustrated in Figure 4-9. The horizontal plane represents time t and values of $\tilde{Z}(t)$. The pdf of $\tilde{Z}(t)$ is shown on the vertical axis. The dashed lines on the horizon plane trace the variance of the pdf of $\tilde{Z}(t)$, which increases linearly over time. Consequently, uncertainty grows as time goes on.

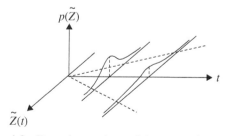

Figure 4-9   Brownian motion and the range of uncertainty.

Brownian Motion with Drift

Starting with Brownian motion as a building block, the following random walk model can be constructed with two additional terms.

$$\tilde{x}(t_k) = \tilde{x}(t_{k-1}) + v\Delta t + \sigma \tilde{\varepsilon}_k \sqrt{\Delta t} \tag{4-4}$$

where v and σ are constant parameters called drift and variance, respectively. In the limit $(\Delta t \rightarrow 0)$, the *Brownian motion with drift* is given by

$$d\tilde{x} = vdt + \sigma d\tilde{z} \tag{4-5}$$

where $d\tilde{z} = \tilde{\varepsilon}\sqrt{dt}$ is the Brownian motion defined in From Random Walk to Brownian Motion. Let $\bar{x}(t) = E\{\tilde{x}(t)\}$ be the expected value of $\tilde{x}(t)$. The expectation of $\tilde{x}(t)$, considering that $E\{\tilde{\varepsilon}_k\} = 0$, is given by

$$E\{\tilde{x}(t)\} = \bar{x}(t) = x(0) + vt \tag{4-6}$$

Note that the expectation, according to (4-4), increases linearly over time. The drift, defined by v, establishes the trend of the process. Using the integral of (4-5) and the expected value (4-6), the Brownian motion with drift can be represented as

$$\tilde{x}(t) = \bar{x}(t) + \sigma \int_0^t d\tilde{z}(\tau) \tag{4-7}$$

The variance of Brownian motion can be calculated as

$$\mathrm{Var}\{\tilde{x}(t)\} = E\{[\tilde{x}(t) - \bar{x}(t)]^2\} = \sigma^2 \int_0^t \int_0^t E\{d\tilde{z}(\tau_1)d\tilde{z}(\tau_2)\} \tag{4-8}$$

and, from the properties of Brownian motion

$$E\{d\tilde{z}(\tau_1)d\tilde{z}(\tau_2)\} = \begin{cases} d\tau & \text{if } \tau_1 = \tau_2 = \tau \\ 0 & \text{if } \tau_1 \neq \tau_2 \end{cases} \tag{4-9}$$

the variance is, therefore, given by

$$\mathrm{Var}\{\tilde{x}(t)\} = \sigma^2 \int_0^t d\tau = \sigma^2 t \tag{4-10}$$

Note that increments $\Delta\tilde{x}$ of Brownian motion over any time interval Δt follow a normal distribution with $E\{\Delta\tilde{x}\} = v\Delta t$ and $\mathrm{Var}\{\Delta\tilde{x}\} = \sigma^2 \Delta t$. Figure 4-10 shows a realization

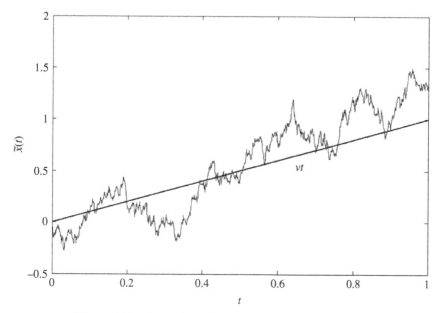

Figure 4-10   A sample path of Brownian motion with drift.

of the Brownian motion with drift that emphasizes the fact that the expected value increases linearly over time.

Ito Process

Brownian Motion with Drift shows that the Brownian motion can be modified in such a way that a different behavior can be fitted well. The nonconstant expected value of the Brownian motion with drift is an example. In order to model electricity prices, consider first the generalization of Brownian Motion to an Ito process.

$$d\tilde{x} = a(\tilde{x}, t)dt + b(\tilde{x}, t)d\tilde{z} \qquad (4\text{-}11)$$

In (4-11), the drift and the variance are nonrandom functions that depend on the current state of the process and time. A special case of the Ito process is considered in which $a(\tilde{x}, t)$ and $b(\tilde{x}, t)$ are linear functions as follows

$$a(\tilde{x}, t) = a_1(t)\tilde{x} + a_2(t) \qquad (4\text{-}12)$$
$$b(\tilde{x}, t) = b_1(t)\tilde{x} + b_2(t) \qquad (4\text{-}13)$$

Therefore, Eq. (4-11) can be written as a continuous time equation as

$$d\tilde{x} = [a_1(t)\tilde{x} + a_2(t)]dt + [b_1(t)\tilde{x} + b_2(t)]d\tilde{z} \qquad (4\text{-}14)$$

To derive a solution of Eq. (4-14), and in general to differentiate and integrate functions of the Ito process, a mathematical tool, Ito's lemma [86], is needed. Using Ito's lemma, the solution of (4-12) is given by [87]

$$\tilde{x} = \rho_{t_0}(t) \left\{ x(t_0) + \int_{t_0}^{t} [a_2(s) - b_1(s)b_2(s)]\rho_{t_0}^{-1}(s)ds + \int_{t_0}^{t} b_2(s)\rho_{t_0}^{-1}(s)d\tilde{z} \right\} \quad (4\text{-}15)$$

where $\rho_{t_0}(t) = \exp \left\{ \int_{t_0}^{t} \left[a_1(s) - \frac{b_1^2(s)}{2} \right] ds + \int_{t_0}^{t} b_1(s)d\tilde{z} \right\}$

Considering a special case in which the variance of (4-11) is constant,

$$d\tilde{x} = [a\tilde{x} + b(t)]dt + \sigma d\tilde{z} \quad (4\text{-}16)$$

The solution of (4-16) is given as follows with the assumption that the initial condition $x(0) = x_0$.

$$\tilde{x}(t) = e^{at}x_0 + \int_{0}^{t} b(\tau)e^{a(t-\tau)}d\tau + \sigma \int_{0}^{t} e^{a(t-\tau)}d\tilde{z} \quad (4\text{-}17)$$

Note that (4-17) has three terms. The first term is the effect of an initial condition. The second term is the effect of $b(t)$. The last term is a stochastic integral corresponding to the limit of linear combination of normal distributions. Hence, $\tilde{x}(t)$ follows a normal distribution. The expectation of (4-17) is given by

$$E[\tilde{x}(t)] = E\left[e^{at}x_0 + \int_{0}^{t} b(\tau)e^{a(t-\tau)}d\tau \right] + E\left[\sigma \int_{0}^{t} e^{a(t-\tau)}d\tilde{z} \right] = e^{at}x_0$$

$$+ \int_{0}^{t} b(\tau)e^{a(t-\tau)}d\tau \quad (4\text{-}18)$$

Equation (4-18) can be interpreted as the solution of the differential equation.

$$\frac{d\bar{x}}{dt} = a\bar{x} + b(t) \quad (4\text{-}19)$$

The variance of this process is given by $\text{Var}\{\tilde{x}(t)\} = E\{[\tilde{x}(t) - \bar{x}(t)]^2\}$. Considering $E\{\tilde{x}(t)\tilde{x}(s)\}$ for $s \geq t$

$$E\{\tilde{x}(t)\tilde{x}(s)\} = \bar{x}(t)\bar{x}(s) + \sigma^2 \int_{0}^{t}\int_{0}^{t} e^{a(t-\tau_1)}e^{a(t-\tau_2)}E\{d\tilde{z}(\tau_1)d\tilde{z}(\tau_2)\}, \quad s \geq t \quad (4\text{-}20)$$

Considering property of Brownian motion defined by (4-9), it can be seen that

$$\int_0^t \int_0^t e^{a(t-\tau_1)} e^{a(t-\tau_2)} E\{d\tilde{z}(\tau_1) d\tilde{z}(\tau_2)\} = \int_0^t e^{2a(t-\tau)} d\tau = \frac{1}{-2a}(1 - e^{2at}) \quad (4\text{-}21)$$

Therefore,

$$\text{Var}\{\tilde{x}(t)\} = \frac{\sigma^2}{-2a}(1 - e^{2at}) \quad (4\text{-}22)$$

According to (4-22), the variance of a linear Ito process depends on the sign of a. If a is positive, the variance grows exponentially over time, faster than the linear growth of a generalized Brownian motion. On the other hand, if a is negative, the variance of a linear Ito process approaches the limit as time goes to infinity.

Geometric Brownian Motion

GBM has been widely used to model financial assets. It constitutes the first model that is often used for electricity prices because of the non-negative characteristic of the actual electricity prices. Before introducing GBM, there are two other models (additive and multiplicative model) that help to provide a better understanding of GBM.

Additive and Multiplicative Models Consider the following random walk:

$$\tilde{x}(t + \Delta t) = \tilde{x}(t) + v\Delta t + \tilde{\varepsilon}(t) \quad (4\text{-}23)$$

where $\tilde{x}(t)$ is the variable of interest, that is, the price, v is the growth rate per unit time, and $\tilde{\varepsilon}(t)$ represents random disturbances. The previous model is known as additive model and it is the first approach to model electricity prices. Assuming that the random disturbance follows a normal distribution with zero mean and variance σ^2, that is, $\tilde{\varepsilon}(t) \sim N(0, \sigma^2)$.

The drawback of the additive model is that the random disturbance can take negative values and, therefore, it is theoretically possible that a large disturbance can cause a negative value of $\tilde{x}(t + \Delta t)$ which violates the non-negative assumption of electricity prices.

The multiplicative model is an alternative that eliminates the difficulty of the additive model. Consider the following adjustment that intends to guarantee positive values:

$$S(t + \Delta t) = S(t)e^{v\Delta t} \quad (4\text{-}24)$$

For a small increment, the exponential function $e^{v\Delta t}$ can be approximated by $e^{v\Delta t} \approx 1 + v\Delta t$, $\Delta t \to 0$, and hence

$$S(t + \Delta t) = S(t)e^{v\Delta t} = S(t)(1 + v\Delta t) \tag{4-25}$$

or

$$\frac{S(t + \Delta t) - S(t)}{S(t)} = \frac{\Delta S(t)}{S(t)} = v\Delta t \tag{4-26}$$

The factor v, which is different from the v in the additive model, is interpreted as the percentage growth rate. Note that increments of $S(t)$, $\Delta S(t)$, are positive if the growth rate is positive.

Considering the same random disturbance $\tilde{\varepsilon}(t) \sim N(0, \sigma^2)$ used in the additive model, to ensure that the random disturbance is non-negative, the following transformation is applied to the model.

$$\tilde{\varepsilon}(t) = e^{\tilde{w}(t)} \tag{4-27}$$

The multiplicative model, taking all transformation into account, is therefore given by

$$\tilde{S}(t + \Delta t) = \tilde{S}(t)e^{v\Delta t}e^{\tilde{w}(t)} \tag{4-28}$$

The model has important implications in electricity price modeling. For instance, it only allows positive values.

Geometric Brownian Motion   A special case of the multiplicative model is called the GBM with drift. This model is widely used for modeling financial assets. A GBM model is described by the following stochastic differential equation:

$$d\tilde{S} = \mu\tilde{S}dt + \sigma\tilde{S}d\tilde{z} \tag{4-29}$$

where μ represents the percentage rate of return, or the growth rate, of electricity price $\tilde{S}$, and σ is the volatility. The random component, $d\tilde{z}z$, is a Brownian motion. GBM is widely used to model stocks prices, commodities, etc.

Based on (4-15), the solution of (4-29) is given by

$$\tilde{S}(t) = S(0) \cdot \exp\left[\left(\mu - \frac{1}{2}\sigma^2\right)t\right] \cdot \exp[\sigma\tilde{z}(t)] \tag{4-30}$$

The most important characteristics of GBM [3], are

(i) $\tilde{S}(t)$ is positive for all t, which is a consequence of the multiplicative model.

(ii) Returns are independent, that is, $(S(t_3) - S(t_2))/S(t_2)$ and $(S(t_3) - S(t_2))/S(t_2)$ for $t_1 < t_2 < t_3$ are independent.

(iii) The logarithm of the price $\tilde{S}(t)$ is normally distributed with mean $\ln S(0) + (\mu - \frac{1}{2}\sigma^2)t$, and variance $\sigma^2 t$.

(iv) The system (4-29) has the following implications:

If $\mu < \dfrac{\sigma^2}{2}$, then $\tilde{S}(t) \to 0$.

If $\mu > \dfrac{\sigma^2}{2}$, then $\tilde{S}(t) \to \infty$.

If $\mu = \dfrac{\sigma^2}{2}$, $\tilde{S}(t)$ is called exponentially Brownian motion.

The path of price via Monte Carlo simulations can be obtained as follows. Let $0 \le t_1 < t_2 < \ldots < t_n$ be the points of time and $\Delta t_i = t_i - t_{i-1}$, a path is generated by the following steps:

Step 1: Generate standard normally distributed random numbers $\tilde{\varepsilon}_1, \tilde{\varepsilon}_2, \ldots, \tilde{\varepsilon}_n$

Step 2: Starting with $S(t_0)$ calculate iteratively

$$\tilde{S}(t_i) = \tilde{S}(t_{i-1}) \exp\left[\left(\mu - \frac{1}{2}\sigma^2\right)\Delta t_i + \sigma\sqrt{\Delta t_i}\,\varepsilon_i\right] \qquad (4\text{-}31)$$

Figure 4-11 illustrates several sample paths of GBM and their mean.

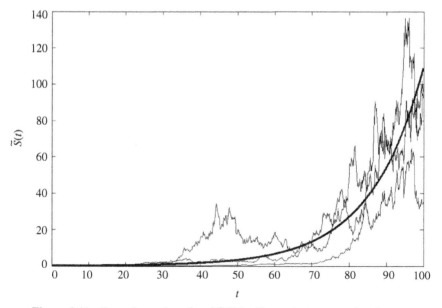

Figure 4-11   Several sample paths of GBM with constant mean and variance.

Geometric Brownian Motion with Seasonal Behavior   A drawback of GBM with constant drift and variance is that the model does not represent the seasonal behavior of some commodities. The model can be modified in such a way that the mean and variance vary over time or season. Consider the following generalization of the GBM

$$d\tilde{S} = \mu(t)\tilde{S}dt + \sigma(t)\tilde{S}d\tilde{z} \tag{4-32}$$

A closed form solution of (4-32) based on (4-15) is given by

$$\tilde{S}(t) = S(0)\exp\left\{\int_0^t\left[\mu(\tau) - \frac{\sigma^2(\tau)}{2}\right]d\tau + \int_0^t\sigma(\tau)d\tilde{z}\right\} \tag{4-33}$$

The logarithmic term $\ln\tilde{S}(t)$ is normally distributed with mean and variance given by $\int_0^t\left[\mu(\tau) - \frac{\sigma^2(\tau)}{2}\right]d\tau$ and $\int_0^t\sigma^2(\tau)d\tau$, respectively. Note that seasonality implies that $\int_0^t\left[\mu(\tau) - \frac{\sigma^2(\tau)}{2}\right]d\tau$ is periodic. In other words,

$$0 = E[\ln\tilde{S}(t+T)] - E[\ln\tilde{S}(t)]$$

$$= \int_0^{t+T}\left[\mu(\tau) - \frac{\sigma^2(\tau)}{2}\right]d\tau - \int_0^t\left[\mu(\tau) - \frac{\sigma^2(\tau)}{2}\right]d\tau$$

$$= \int_t^{t+T}\left[\mu(\tau) - \frac{\sigma^2(\tau)}{2}\right]d\tau$$

Hence, the condition required for a seasonal process is given by

$$\int_t^{t+T}\left[\mu(\tau) - \frac{\sigma^2(\tau)}{2}\right]d\tau = 0 \tag{4-34}$$

where T corresponds to the period of the season.

Mean Reversion Process

Additive Mean Reversion Models

Constant Mean Reversion Process   Empirical observations show that prices of certain securities tend to revert over time to a fixed value. Consider the following random walk

$$\tilde{x}(t + \Delta t) - \tilde{x}(t) = k[m - \tilde{x}(t)] + \sigma\tilde{\varepsilon}(t) \tag{4-35}$$

where k, σ are positive factors, and m is a value over which the process fluctuates. The first term on the right-hand side of (4-35) represents the difference between price and the "mean" value. The last term is the random disturbance. Statistically, when the current price is less than m, the price, in general, tends to increase. Similarly, when m is lower than the current value, the price tends to decrease. Note that this characteristic cannot be modeled by (4-23). The limiting case of this model can be written as

$$d\tilde{x}(t) = k[m - \tilde{x}(t)]dt + \sigma d\tilde{z} \tag{4-36}$$

Alternatively, the mean reversion can be described as a combination of two terms as

$$\tilde{x}(t) = m + \tilde{x}_r(t) \tag{4-37}$$

where $d\tilde{x}_r(t) = -k\tilde{x}_r dt + \sigma d\tilde{z}$. Equation (4-36) is called a constant mean reversion process, and it has a deterministic and a stochastic term. The process is expected to revert to a constant m. The stochastic term is characterized by the following expected value and variance.

$$E[\tilde{x}(t)] = e^{-kt} x_r(0) + m \tag{4-38}$$

$$\text{Var}\{\tilde{x}(t)\} = \frac{\sigma^2}{2k}(1 - e^{-2kt}) \tag{4-39}$$

Time-Varying Mean Reversion Process   Consider the following stochastic process with a time-varying deterministic term

$$\tilde{x}(t) = x_d(t) + \tilde{x}_r(t) \tag{4-40}$$

where

$$d\tilde{x}_r(t) = -k\tilde{x}_r dt + \sigma d\tilde{z} \tag{4-41}$$

Combining (4-40) and, (4-41). the mean reversion process becomes

$$d\tilde{x} = k[g(t) - \tilde{x}]dt + \sigma d\tilde{z} \tag{4-42}$$

where

$$g(t) = x_d(t) + \frac{1}{k}\frac{d}{dt}x_d(t) \tag{4-43}$$

Equation (4-42) is called *time-varying mean reversion process*, where the mean and variance are given by

$$E[\tilde{x}(t)] = e^{-kt} x_r(0) + x_d(t) \tag{4-44}$$

$$\text{Var}\{\tilde{x}(t)\} = \frac{\sigma^2}{2k}(1 - e^{-2kt}) \tag{4-45}$$

Note that the mean (4-44) approaches $x_d(t)$ in the limit, and that the process does not revert to $g(t)$. It reverts, instead, to $x_d(t)$.

Comparison of Additive Mean Reversion Process and Brownian Motion with Drift   For the following Brownian motion with drift

$$d\tilde{x}_1 = vdt + \sigma d\tilde{z} \tag{4-46}$$

as discussed in Brownian Motion with Drift

$$E\{\tilde{x}_1(t)\} = x_1(0) + vt \tag{4-47}$$

To match the mean of Brownian motion with drift by mean reversion process, $\tilde{x}_2(t)$, one can set the deterministic component of the mean reversion process

$$\tilde{x}_2(t) = \bar{x}_1(t) + \tilde{x}_r(t) \tag{4-48}$$
$$d\tilde{x}_r = -k\tilde{x}_r dt + \sigma d\tilde{z} \tag{4-49}$$

Combining the two expressions, the following is obtained:

$$d\tilde{x}_2 = k\left[\bar{x}_1(t) + \frac{v}{k} - \tilde{x}_2\right] dt + \sigma d\tilde{z} \tag{4-50}$$

Substituting $x_2(0) = x_1(0) + x_r(0)$, the mean of stochastic process $\tilde{x}_2(t)$ satisfies

$$E\{\tilde{x}_2(t)\} = e^{-kt}x_r(0) + x_1(0) + vt \tag{4-51}$$

Under the usual assumption that $x_r(0) = 0$, the two processes have the same mean function, that is, $x_1(t) = x_2(t)$. The two processes fluctuate around this mean function due to the stochastic component.

The behavior of the stochastic component is somewhat different. Comparing the variances of two processes, the variance of Brownian motion with drift is

$$\text{Var}\{\tilde{x}_1(t)\} = \sigma^2 t \tag{4-52}$$

whereas the variance of the mean reversion process is

$$\text{Var}\{\tilde{x}_2(t)\} = \frac{\sigma^2}{2k}(1 - e^{-2kt}) \tag{4-53}$$

Figure 4-12 compares the variance (volatility) of both processes.

Multiplicative Mean Reversion Model   For a multiplicative model of the price $\tilde{S}(t)$,

$$\tilde{S}(t) = S_0 G(t)\tilde{e}_1 \tag{4-54}$$

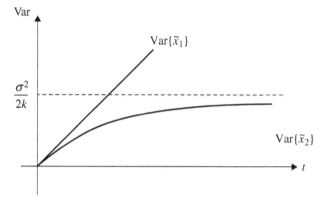

Figure 4-12   Variances of mean reversion process and Brownian motion with drift.

where $G(t)$ is the deterministic component and $\tilde{\varepsilon}_1$ is the stochastic component of $\breve{S}(t)$. By taking the expected value, $E\{\breve{S}\} = S_0 G(t) E\{\tilde{\varepsilon}_1\}$, it is obvious that for $G(t)$ to represent the deterministic part of $\breve{S}(t)$, it is required that

$$E\{\tilde{\varepsilon}_1\} = 1 \qquad (4\text{-}55)$$

It is commonly assumed that in the multiplicative model the stochastic variable $\tilde{\varepsilon}_1$ has a lognormal distribution. Taking the logarithm of (4-54), it is obtained that

$$\ln\left(\frac{\breve{S}(t)}{S_0}\right) = \ln G(t) + \ln \tilde{\varepsilon}_1 \qquad (4\text{-}56)$$

Define $\tilde{x} = \ln \breve{S}(t) - \ln S_0$ and $\tilde{x}_1 = \ln \tilde{\varepsilon}_1$, then (4-56) can be written as

$$\tilde{x} = \ln G(t) + \tilde{x}_1 \qquad (4\text{-}57)$$

and the requirement (4-55) becomes

$$E\{e^{\tilde{x}_1}\} = 1 \qquad (4\text{-}58)$$

Note that a mean reversion process with zero mean, given by

$$d\tilde{x}_r = -k\tilde{x}_r dt + \sigma d\tilde{z} \qquad (4\text{-}59)$$

has the property that $E\{\tilde{x}_r\} = 0$ if $x_r(0) = 0$, otherwise $E\{\tilde{x}_r\} = e^{-kt}x_r(0)$, and $\mathrm{Var}\{\tilde{x}_r\} = \frac{\sigma^2}{2k}(1 - e^{-2kt})$. Due to the property of lognormal distribution,

$$E\{e^{\tilde{x}_r}\} = e^{\frac{\sigma^2}{4k}(1-e^{-2kt})} \qquad (4\text{-}60)$$

which is not equal to one and does not satisfy the requirement (4-58). It is therefore necessary to compensate for the shift of the mean in lognormal distribution. For this purpose, define

$$\tilde{x}_1 = x_m(t) + \tilde{x}_r \tag{4-61}$$

where

$$x_m(t) = -\frac{\sigma^2}{4k}(1 - e^{-2kt}) \tag{4-62}$$

Taking the differential of (4-61), one obtains

$$d\tilde{x}_1 = x'_m(t) \cdot dt + d\tilde{x}_r \tag{4-63}$$

Substituting (4-59) and (4-61) into (4-63), it is seen that

$$d\tilde{x}_1 = k\left[x_m(t) + \frac{1}{k} \cdot x'_m(t) - \tilde{x}_1\right]dt + \sigma d\tilde{z} \tag{4-64}$$

Taking the expected values of both sides of (4-61), one obtains

$$E\{\tilde{x}_1\} = x_m(t) = -\frac{\sigma^2}{4k}(1 - e^{-2kt}) \tag{4-65}$$

From (4-61), the variance of $\tilde{x}_1$ is the same as that of $\tilde{x}_1$, $\mathrm{Var}\{\tilde{x}_1\} = \frac{\sigma^2}{2k}(1 - e^{-2kt})$. Therefore, by the property of lognormal distribution, $\tilde{x}_1$ satisfies the requirement (4-58), that is, $E\{e^{\tilde{x}_1}\} = 1$.

Now consider (4-57) and substitute (4-61) into it,

$$\tilde{x} = \ln G(t) + \tilde{x}_1 = \ln G(t) + x_m(t) + \tilde{x}_r \tag{4-66}$$

Taking the differential of (4-66),

$$d\tilde{x} = \left[\frac{G'(t)}{G(t)} + x'_m(t)\right]dt + d\tilde{x}_r \tag{4-67}$$

Substituting (4-59), (4-61), and (4-66) into (4-67), it is seen that

$$d\tilde{x} = k[g_1(t) - \tilde{x}]dt + \sigma d\tilde{z} \tag{4-68}$$

where

$$g_1(t) = \ln G(t) + \frac{1}{k}\frac{G'(t)}{G(t)} + x_m(t) + \frac{1}{k}x'_m(t) \tag{4-69}$$

By the application of Ito's lemma, it is concluded that

$$d\tilde{S} = \left\{ k[g_1(t) - \ln \tilde{S}]\tilde{S} + \frac{1}{2}\sigma^2\tilde{S} \right\} dt + \sigma\tilde{S}d\tilde{z} \tag{4-70}$$

or

$$\frac{d\tilde{S}}{\tilde{S}} = k[g(t) - \ln \tilde{S}]dt + \sigma d\tilde{z} \tag{4-71}$$

where

$$g(t) = g_1(t) + \frac{1}{2k}\sigma^2 \tag{4-72}$$

or

$$g(t) = \ln G(t) + \frac{1}{k}\frac{G'(t)}{G(t)} + \frac{\sigma^2}{4k}(1 - e^{-2kt}) \tag{4-73}$$

Therefore, the multiplicative mean reversion process with time-varying mean can be represented either in the S-domain by (4-71) and (4-73) or in the x-domain by (4-68) and (4-69).

Mean Reversion Process with Seasonal Behavior   If it is intended to construct a model to simulate the electricity price process in a spot market, the characteristics of multiple seasonality of price need to be considered. This section discusses how to model the price with a mean reversion process.

Considering electricity price in the mean reversion process use x-domain, that is, (4-68) and (4-69), in the following form

$$d\tilde{x}(t) = k\left[\sum_{i=1}^{n} \alpha_i(t) - \tilde{x}(t) \right] dt + \sigma d\tilde{z} \tag{4-74}$$

Denote

$$E[\tilde{x}(t)] = \pi(t) \tag{4-75}$$

$E[\tilde{x}(t)]$ is the solution of the following ordinary differential equation

$$\frac{d\pi(t)}{dt} = k\left[\sum_{i=1}^{n} \alpha_i(t) - \pi(t) \right] \tag{4-76}$$

Furthermore, $\pi(t)$ can be expressed as

$$\pi(t) = \sum_{i=1}^{n} \pi_i(t) \tag{4-77}$$

where $\pi_i(t)$ satisfies,

$$\frac{d\pi_i(t)}{dt} = k[\alpha_i(t) - \pi_i(t)] \tag{4-78}$$

To describe the price with multiple seasonality characteristics, note that the solution of (4-78) is periodic. It can be shown that, for the system (4-78), if $\alpha_i(t)$ is periodic, then $\pi_i(t)$ converges to a unique periodic oscillation.
First, if $k > 0$, the solution of system (4-78) is unique.
Then, the solution of system (4-78) converges to a periodic oscillation.
Let $\eta(t)$ be a solution of system (4-78). That is

$$\frac{d\eta(t)}{dt} = k[\alpha_i(t) - \eta(t)] \tag{4-79}$$

For time $t+T$, it is seen that

$$\eta'(t + T) = k[\alpha_i(t + T) - \eta(t + T)] = k[\alpha_i(t) - \eta(t + T)] \tag{4-80}$$

where T is the period of $\alpha_i(t)$.
Then $\eta(t+T)$ is a solution of system (4-78). Since the solution of system (4-78) is unique, it follows that: $\forall \varepsilon > 0, \exists t_1 > 0$, when $t > t_1$

$$|\eta(t + \omega) - \eta(t)| < \varepsilon \tag{4-81}$$

This means that the solution of system (4-78) converges to a unique periodic oscillation.

Comparison of Geometric Brownian Motion and Geometric Mean Reversion Process
It can be shown that the price of electricity, modeled either as a GBM or a mean reversion process, at time t, $\tilde{S}(t)$, is a random variable with a lognormal distribution. The mean $\bar{x}(t)$ and variance $\sigma_x^2(t)$ of the random variable $\tilde{x} = \ln \tilde{S}$ are given in Table 4-1 [88].

Properties of a Lognormal Distribution

CDF and PDF of a Lognormal Distribution
According to the properties of GBM and the mean reversion process, for a fixed time, the distribution of price is a lognormal distribution. This section explains the properties of a lognormal distribution.

Table 4-1 Summary of GBM Process and GMR Process

Model	GBM	GMR
Distribution of $\tilde{S}$ equation	Lognormal $d\tilde{S} = \mu\tilde{S}dt + \sigma\tilde{S}d\tilde{z}$ $d\tilde{x} = vdt + \sigma d\tilde{z}$	Lognormal $d\tilde{S} = k[g(t) - \ln\tilde{S}]\tilde{S}dt + \sigma\tilde{S}d\tilde{z}$ $d\tilde{x} = k[g_1(t) - \tilde{x}]dt + \sigma d\tilde{z}$
Equation in $\tilde{x} = \ln\tilde{S}$	$v = \mu - \dfrac{1}{2}\sigma^2$	$g_1(t) = g(t) - \dfrac{1}{2k}\sigma^2$
Mean $\bar{x}(t)$	$x(0) + vt$	$e^{-kt}x(0) + \int_0^t e^{-k(t-\tau)}kg_1(\tau)d\tau$
Variance $\sigma_x^2(t)$	$\sigma^2 t$	$\dfrac{\sigma^2}{2k}(1 - e^{-2kt})$

A random variable $\tilde{u}$ is said to be a lognormal distribution if its logarithm, $\ln\tilde{u}$, is a normal distribution, that is,

$$\tilde{u} = e^{\tilde{w}} \tag{4-82}$$

where $\tilde{w}$ is normally distributed, and the density is given by

$$f_{\tilde{w}}(w) = \frac{1}{\sigma\sqrt{2\pi}}\exp\left[-\frac{(w-\mu)^2}{2\sigma^2}\right] \tag{4-83}$$

Mathematically, the cumulative distribution function (cdf) and the pdf of a lognormal random variable $\tilde{u}$ can be derived [89]. Since the exponential function is a monotonically increasing function, the cdf of a lognormal distribution $\tilde{u}$, $F_{\tilde{u}}(u)$, can be defined via its logarithmic function

$$F_{\tilde{u}}(u) = P(\tilde{u} \le u) = P(\tilde{w} \le \ln u) = \int_{-\infty}^{\ln u}\frac{1}{\sigma\sqrt{2\pi}}\exp\left[-\frac{(w-\mu)^2}{2\sigma^2}\right]dw = N\left(\frac{\ln u - \mu}{\sigma}\right) \tag{4-84}$$

where $N(\bullet)$ is a cumulative normal distribution function of the standard normal distribution. The pdf of a lognormal distribution $\tilde{u}$, $f_{\tilde{u}}(u)$, can be obtained by differentiation of the cumulative normal distribution function.

$$f_{\tilde{u}}(u) = \frac{d}{du}F_{\tilde{u}}(u) = \frac{d}{du}F_{\tilde{w}}(u) = f_{\tilde{w}}(\ln u)\frac{d(\ln u)}{du} = \frac{1}{u\sigma\sqrt{2\pi}}\exp\left[-\frac{(\ln u - \mu)^2}{2\sigma^2}\right] \tag{4-85}$$

Moments of a Lognormal Distribution A lognormal distribution $\tilde{u}$ is characterized by the mean and variance of the distribution of its logarithm $\tilde{w} = \ln\tilde{u}$, which is normal. It is no longer characterized by its own mean and variance. Once the pdf is available, one can calculate the mean, variance, and higher-order moments of any random variable. The mean, variance, and skewness can be expressed in terms of the moments as follows [90].

(i) Mean: $E(\bar{u}) = \exp\left(\mu + \dfrac{\sigma^2}{2}\right)$ $\hspace{3cm}$ (4-86)

(ii) Variance: $\text{Var}(\tilde{x}) = e^{2\mu+\sigma^2}(e^{\sigma^2} - 1)$ $\hspace{2.5cm}$ (4-87)

(iii) Skewness: $C_S(\tilde{x}) = (e^{\sigma^2} - 1)^{\frac{1}{2}}(e^{\sigma^2} + 2)$ $\hspace{2cm}$ (4-88)

VaR and CVaR of a Lognormal Distribution   Although variances and standard deviations are useful measures, they do not relate directly to what concerns the financial sector–huge loss. A more direct measure of the risk is the probability of a huge loss. For that purpose, one needs to specify what is huge, which is difficult to standardize. An alternative is by fixing a probability level first and finding the corresponding amount of loss. That is, an acceptable probability level is specified first and then the potential loss relative to that confidence level is determined. For example, it is desirable to have a confidence level of, say $\beta = 99\%$, regarding possible losses. Is it possible that one can be at least 99% sure that the loss will not exceed α dollars. There may be more than one value of α. Since α corresponds to a loss in dollars, it is better to be lower. Value-at-risk (VaR) is defined as the smallest value of α for which the probability that the loss $\tilde{\rho}$ is less than α is equal to or higher than β. VaR is defined as:

$$\text{VaR} = \min[\alpha : P\{\tilde{\rho} < \alpha\} \geq \beta] \hspace{2cm} (4\text{-}89)$$

The VaR of a lognormal distribution variable is given by

$$\text{VaR} = \exp[\sigma \cdot N^{-1}(1 - \beta) + \mu] \hspace{2cm} (4\text{-}90)$$

where $N^{-1}(\bullet)$ is the inverse function of a cumulative normal distribution function.

When there is a more than 99% chance that the loss will not exceed α dollars (VaR), it implies that there is, nevertheless, less than 1% chance that the loss might exceed α dollars. For that less than 1% chance, what could be the loss? This question is not answered by the VaR. The consequence of having loss exceeding VaR could be devastating depending on the situation. How bad could it be? How much loss is expected in such a case? This question is answered by another measure of risk called conditional value-at-risk (CVaR).

CVaR is defined as the expected loss when the loss exceeds VaR.

$$\text{CVaR}(\tilde{\rho}) = E\{\tilde{\rho}|\tilde{\rho} > \text{VaR}(\tilde{\rho})\} \hspace{2cm} (4\text{-}91)$$

The above definition of CVaR as a conditional expectation can be expressed in terms of the pdf, $f_{\tilde{\rho}}(y)$, of the loss. Noting that $P\{\tilde{\rho} > \text{VaR}(\tilde{\rho})\} = 1 - \beta$,

$$\text{CVaR}(\tilde{\rho}) = \frac{1}{1 - \beta} \int_{\text{VaR}(\tilde{\rho})}^{\infty} y f_{\tilde{\rho}}(y) dy \tag{4-92}$$

CVaR answers the question: what could be the expected loss if the loss exceeds VaR. The expected value is often associated with the average value in layman's terms. CVaR can be interpreted as the average value of the possible losses when the loss exceeds VaR.

CVaR of a lognormal distribution variable $\tilde{u}$ is given by

$$\text{CVaR} = \frac{e^{\mu + \frac{\sigma^2}{2}}}{(1 - \beta)\sqrt{2\pi}} \{1 - N[N^{-1}(1 - \beta) - \sigma]\} \tag{4-93}$$

where $N(\bullet)$ is a cumulative normal distribution function [90].

Properties of Measures of Risk   Usually, there are three measures of risk: standard deviation (SD), VaR, and CVaR. Mathematically, a measure of risk is a real-valued function of loss, $m(\tilde{\rho})$. This section discusses several desirable properties of the risk measure, starting from intuitive notions and then translating them into mathematical properties. The first one expresses the idea that the risk measure should be scalable in the sense that if the loss becomes α times larger, then the corresponding measure of risk should be magnified by the same ratio α. This is called *homogeneity* property

(i) Homogeneity: For $a \geq 0$,

$$m(a\tilde{\rho}) = am(\tilde{\rho}) \tag{4-94}$$

The next desirable property relates a combined risk to the risks of its constituent parts. It states that the combined risk should be lower than the sum of risks of its constituent parts. This property is called *subadditivity*.

(ii) Subadditivity: For all $\tilde{\rho}_1$ and $\tilde{\rho}_2$,

$$m(\tilde{\rho}_1 + \tilde{\rho}_2) \leq m(\tilde{\rho}_1) + m(\tilde{\rho}_2) \tag{4-95}$$

The subadditivity property is a natural requirement for the measure of risk in most situations, especially finance. It si'mply confirms the intuitive notion that merger, by virtue of diversification, does not create extra risk.

A useful result is that homogeneity and subadditivity imply *convexity*.

(iii) Convexity: For any $\tilde{\rho}_1$, $\tilde{\rho}_2$, and $0 \leq p \leq 1$

$$m(p\tilde{\rho}_1 + (1 - p)\tilde{\rho}_2) \leq (p)m(\tilde{\rho}_1) + (1 - p)m(\tilde{\rho}_2) \tag{4-96}$$

Table 4-2 Properties of Risk Measures

	Homogeneity	Subadditivity	Convexity
SD	√	√	√
VaR	√	×	×
CVaR	√	√	√

The implication is significant. Convexity is an extremely important property in optimization. It leads to a host of desirable mathematical properties, such as existence of solution, uniqueness of solution, and guaranteeing a local optimum to be globally optimum. Mathematical characterization of optimality is usually expressed in statements that only guarantee local optimality. Computational methods to find optimal solutions are derived accordingly. Convexity ensures that such methods will find the optimal solution.

It can be shown that both SD and CVaR satisfy the homogeneity and subadditivity properties; hence, they are convex functions, whereas VaR is not subadditive, nor convex [90]. These properties are summarized in Table 4-2.

The fact that VaR is not a convex function while CVar is, has significant implications in the selection of risk measures in the formulation of the optimization problem. The objective of an investment is usually to maximize the profit and, at the same time, minimize the risk. Classical portfolio optimization in investment science is based on the mean-variance approach in which the profit is measured in terms of the expected profit and risk is measured in terms of the standard deviation (or variance), resulting in a convex optimization problem. Attempts have been made in using VaR as an alternative measure of risk, which is physically more meaningful and direct, in the investment problem. Because of nonconvexity, optimization involving VaR often encounters computational problems. Advances has been made recently in using CVaR to formulate portfolio optimization problem in finance [91].

Risk-Adjusted Price

A financial instrument whose value is a function of (or derived from) the value of an underlying asset is called a derivative. Examples of derivatives include forward, futures, and option contracts. The value of a derivative, strictly speaking, is specified only at the time of maturity of the contract, that is, a future time, by its payoff function. The value of a derivative at any time prior to that is really the present value of the expected payoff. The price of a derivative, theoretically, should be equal to its value.

Due to its nonstorable nature, electricity is different from most other tradable commodities. The price of electricity is stochastic. A derivative of electricity price is defined as any physical variable, not necessarily a financial instrument, whose value is a function of the electricity price.

Considering a random variable $\tilde{S}$ whose time function satisfies the following differential equation

$$d\tilde{S} = \mu(\tilde{S}, t)\tilde{S}dt + \sigma_1(\tilde{S}, t)\tilde{S}d\tilde{z} \qquad (4\text{-}97)$$

where $d\tilde{z}$ is a Brownian motion. This model includes GBM and GMR process as special cases. In this work, $\tilde{S}$ is a random variable and S is a deterministic quantity.

Assume that there are two *derivatives* of $\tilde{S}$, whose values are denoted by $\tilde{f}_1(\tilde{S}, t)$ and $\tilde{f}_2(\tilde{S}, t)$, and they satisfy the following differential equations, respectively,

$$d\tilde{f}_1 = \mu_1(\tilde{f}_1, t)\tilde{f}_1 dt + \sigma_1(\tilde{f}_1, t)\tilde{f}_1 d\tilde{z} \qquad (4\text{-}98)$$
$$d\tilde{f}_2 = \mu_2(\tilde{f}_2, t)\tilde{f}_2 dt + \sigma_2(\tilde{f}_2, t)\tilde{f}_2 d\tilde{z} \qquad (4\text{-}99)$$

Note that the values or prices of both derivatives $\tilde{f}_1(\tilde{S}, t)$ and $\tilde{f}_2(\tilde{S}, t)$ are derived from $\tilde{S}$ and hence they inherit the same uncertainty $d\tilde{z}$.

Consider the interval between t and $t + dt$. At time t, S is known, so, by definition, are $f_1(S, t)$ and $f_2(S, t)$, and hence they are deterministic. Due to the presence of uncertainty $d\tilde{z}$, $d\tilde{f}_1$ and $d\tilde{f}_2$ are stochastic. To eliminate $d\tilde{z}$ from (4-98) and (4-99), start with

$$d\tilde{\pi} = (\sigma_2 f_2)d\tilde{f}_1 - (\sigma_1 f_1)d\tilde{f}_2 \qquad (4\text{-}100)$$

The $d\tilde{\pi}$ in (4-100) can be obtained from

$$\pi(S, t) = (\sigma_2 f_2)f_1 - (\sigma_1 f_1)f_2 \qquad (4\text{-}101)$$

at t. The increment $d\tilde{\pi} = (\sigma_2 f_2)d\tilde{f}_1 - (\sigma_1 f_1)d\tilde{f}_2$ should be stochastic, as a result of $d\tilde{f}_1$ and $d\tilde{f}_2$ being stochastic. After substituting (4-98) and (4-99) into (4-100), it is obtained

$$d\tilde{\pi} = (\mu_1 \sigma_2 f_1 f_2 - \mu_2 \sigma_1 f_1 f_2)dt \qquad (4\text{-}102)$$

(4-102) is independent of $d\tilde{z}$ and is deterministic. One can write it as a deterministic quantity

$$d\pi = (\mu_1 \sigma_2 f_1 f_2 - \mu_2 \sigma_1 f_1 f_2)dt \qquad (4\text{-}103)$$

The two derivatives $f_1(S, t)$ and $f_2(S, t)$ may be combined (to form a portfolio) to eliminate the uncertainty, $d\tilde{z}$, called delta hedging [28] in the finance literature. Since the return on investment of the portfolio is deterministic, if $d\pi$ earns more than the riskless return ($r\pi dt$), one could borrow money at the riskless rate r and make a sure profit. In finance, a fundamental assumption is that there is no arbitrage opportunity in the market, that is, it is not possible to make profit without risk by buying and selling financial instruments in a market. This is called *no-arbitrage* assumption. Now if $d\pi$

earns more than the riskless return, it would contradict the no-arbitrage assumption. It is concluded that $d\pi$ could only earn no more and no less than the riskless return

$$d\pi = r\pi dt \qquad (4\text{-}104)$$

Combining (4-103), (4-104), and (4-101), it is seen that

$$(\mu_1\sigma_2 f_1 f_2 - \mu_2\sigma_1 f_1 f_2) = r(\sigma_2 f_2 f_1 - \sigma_1 f_1 f_2)$$

After rearrangement of terms, the *Ross–Cox equation* [28] is obtained

$$\frac{\mu_1 - r}{\sigma_1} = \frac{\mu_2 - r}{\sigma_2} \qquad (4\text{-}105)$$

This ratio is called *the market price of risk* and is denoted by λ.

$$\lambda = \frac{\mu_1 - r}{\sigma_1} = \frac{\mu_2 - r}{\sigma_2} \qquad (4\text{-}106)$$

Since $\lambda = (\mu_1 - r)/\sigma_1$, λ must be independent of f_2. By the same token, $\lambda = (\mu_2 - r)/\sigma_2$ must be independent of f_1. So λ is independent of both derivatives f_1 and f_2. However, in general, μ_1, μ_2, σ_1, and σ_2 may be functions of S and t, hence the market price of risk λ may also be a function of S and t. It is reasonable that the market price of risk λ is independent of any specific derivative but is dependent on the value of the underlying variable S of all derivatives. This is because all derivatives share the same source of uncertainty and they are exchangeable as assets in the financial market. Therefore, a quantity is available here that depends on the common source of uncertainty S.

For the prices described by Brownian motion or mean reversion process, the risk-adjusted forms are summarized in Table 4-3.

Electricity Price with Spikes

Mean Reversion Jump-Diffusion Model for Electricity Price Process with Spikes

Constant Parameters for Spikes One of the important problems in simulating the electricity price process is to take into account the price spikes in the spot market. As discussed earlier, the electricity price can be described as a mean reversion process. The physical characteristics of electricity and power system operation are considered in modeling. The price spikes that occur with a much lower probability can be caused by the short-term generation shortage. This kind of generation shortage can be described by a unit forced outage rate (FOR). When there is a generator failure, backup units are committed immediately with a higher marginal cost, which shows a sudden price spike. The generator failure probability is described by the unit FOR,

Table 4-3 Summary of Risk-Adjusted Price

	Brownian Motion		Mean Reversion	
	Brownian Motion	Geometric Brownian Motion	Mean Reversion	Geometric Mean Reversion
Model	$d\tilde{S} = \mu(t)dt$ $+ \sigma(t)d\tilde{z}$	$d\tilde{S} = \mu(t)\tilde{S}dt$ $+ \sigma(t)\tilde{S}d\tilde{z}$	$d\tilde{S} = k[g(t) - \tilde{S}]dt$ $+ \sigma(t)d\tilde{z}$	$d\tilde{S} = k[g(t) - \ln\tilde{S}]\tilde{S}dt$ $+ \sigma(t)\tilde{S}d\tilde{z}$
Risk-adjusted form	$d\tilde{S} = [\mu(t) - \lambda\sigma(t)]dt$ $+ \sigma(t)d\tilde{z}$	$d\tilde{S} = [\mu(t) - \lambda\sigma(t)]\tilde{S}dt$ $+ \sigma(t)\tilde{S}d\tilde{z}$	$d\tilde{S} = \left[\mu(t) - \dfrac{\lambda\sigma(t)}{k} - \tilde{S}\right]dt$ $+ \sigma(t)d\tilde{z}$	$d\tilde{S} = \left[\mu(t) - \dfrac{\lambda\sigma(t)}{k} - \ln\tilde{S}\right]\tilde{S}dt$ $+ \sigma(t)\tilde{S}d\tilde{z}$

and the fix time for the units can be denoted as mean time to repair (MTTR). When the generator is fixed and recommitted to the system, the spot price will drop down to the normal level. Assume there is no elasticity for the demand curve (this is true for most conditions), if the supply curve of the system is given, the uplift of the price spike can be obtained.

Consider the mean reversion model for the logarithm of electricity prices. The mean reversion price model can be rewritten with the additional random jumps that follow the probability distribution of a Poisson process. Let $d\tilde{q}$ denote the Poisson process for jump occurrence, Φ denote the non-negative logarithm jump size, the mean reversion stochastic price process with jumps can be described as

$$d\tilde{x} = \kappa[g(t) - \tilde{x}]dt + \sigma d\tilde{z} + \Phi d\tilde{q} \tag{4-107}$$

where the logarithm jump size Φ is related to the outage generator capacity and system supply curve, and dq follows a Poisson process that most of the time is zero and sometimes jumps of logarithm size Φ occur with arrival rate ρ. The parameter ρ is related to the probability distribution of FOR for the units.

$$d\tilde{q} = \begin{cases} 0 & \text{with probability } (1 - \rho)dt \\ 1 & \text{with probability } \rho dt \end{cases} \tag{4-108}$$

The jump process $d\tilde{q}$ is assumed independent of $d\tilde{z}$.

According to the physical characteristics of system operations, the non-negative logarithm jump size Φ is related to the capacity of the outage unit and the system supply curve. Therefore, the uplift of the price spike caused by generator failure can be calculated. Assume the capacity of generator i is $cap(i)$, if a failure occurs, there will be an instant shortage of capacity

$$\Delta q = cap(i) \tag{4-109}$$

In order to satisfy the demand of consumers, the backup units should be committed to the system. Comparing with the outage generator, the marginal cost of backup peak units is much higher, which will pull up the price as a spike. If one assumes there is no elasticity on the consumer demand curve, the jump size ϕ_i for unit i outage can be described as

$$\phi_i = a \times \Delta q_i \tag{4-110}$$

where a is the slope of linear system cumulative supply curve, that is,

$$P = aq + b \tag{4-111}$$

where P is the electricity spot price.

Based on Eqs. (4-110) and (4-111), the price spike jump size ϕ can be described as follows for the whole system

$$\phi = \sum_i \phi_i \qquad (4\text{-}112)$$

and the logarithm jump size Φ for X in the mean reversion stochastic process is

$$\Phi = \ln(\phi/P_0) \qquad (4\text{-}113)$$

where P_0 denotes the spot price if no failure occurs.

Since the jump is defined as non-negative, when the outage unit is fixed, the price will drop to the normal level. The fix duration can be defined as MTTR in system operation. From the derivation given above it is difficult to solve (4-107) analytically to achieve a price process if detailed physical characteristics are considered. Within the simulation framework, since the jump process is independent of the process dz, there is no necessity to transfer the price jumps to a logarithm format and then convert it back after solving the differential equation. Thus price jumps caused by generation adequacy can be rewritten as

$$
\begin{aligned}
X &= \ln S_0 \\
\tilde{x} &= \kappa[g(t) - \tilde{x}]dt + \sigma d\tilde{z} \\
S &= S_0 + \phi \\
\phi &= \sum_i \rho_i \times \phi_i^k = \sum_i \rho_i \times a \times \Delta q_i, k = 1, \ldots, i
\end{aligned}
\qquad (4\text{-}114)
$$

where a is the slope of the system integrated supply curve, and ρ_i is the FOR of unit i.

Stochastic Parameters for Spikes   In this model, a popular extension of the standard diffusion process—the jump-diffusion process, is employed. The price process is specified by appending an additional term to (4-68), yielding

$$d\tilde{x} = k[g_1(t) - \tilde{x}]dt + \sigma d\tilde{z} + \tilde{\theta}d\tilde{q} \qquad (4\text{-}115)$$

where $q(t)$ is a Poisson process with intensity ρ, $\tilde{\theta}$ is a draw from a normal distribution with mean μ_θ and standard deviation σ_θ. It is assumed that the Brownian motion, Poisson process, and jump size are mutually independent.

From (4-108), during the observation interval dt, the probability for no jump is $(1 - \rho)dt$. This is equivalent to drawing the price at time t from a normal distribution with mean $e^{-kt}x(0) + \int_0^t e^{-k(t-\tau)}kg(\tau)d\tau$, and variance $\frac{\sigma^2}{2k}(1 - e^{-2kt})$. The probability for a jump is ρdt. Now the price is drawn from a normal distribution with mean $e^{-kt}x(0) + \int_0^t e^{-k(t-\tau)}kg(\tau)d\tau + \mu_\theta$, and variance $\frac{\sigma^2}{2k}(1 - e^{-2kt}) + \sigma_\theta^2$. Note that while the mean may rise or fall when a jump occurs, the variance always increases.

The model specification is refined by allowing the jump intensity parameter to vary. There are several reasons for this including the fact that jumps are more likely to occur when transmission lines become congested. This suggests that during high demand periods a jump in prices is more likely. Thus, one allows the jump intensity to vary by the time of day and season. That is

$$\rho(t) = \rho_0 + \rho_{\text{peak}} \text{ Peak}_t + \rho_{\text{weekend}} \text{ Weekend}_t + \rho_{\text{Fall}} \text{ Fall}_t + \rho_{\text{Win}} \text{ Winter}_t$$
$$+ \rho_{\text{Spr}} \text{ Spring}_t + \rho_{\text{Sum}} \text{ Summer}_t \tag{4-116}$$

In both pre-crisis and crisis samples, the probability of a jump increases during peak hours and decreases during the spring and winter months. In addition, a significant weekend effect in the jump intensity can be seen.

Structural Model Based on Supply and Demand   Four characteristics are often used to describe electricity prices: mean reversion, seasonality, stochastic volatility, and spikes. Of these, spikes are the most important characteristic for risk management in electricity markets. Thus, many models have been developed to describe the spikes. Among these models, regime switching between a nonspike regime (where prices are unlikely to spike) and a spike regime is a method. Most regime-switching models assume constant transition probabilities from one regime to another, for example, from a nonspike to a spike regime. This simplification allows models to incorporate the magnitude and frequency of price spikes. For example, Figure 4-13 shows scatter plots between demand (i.e., supply) levels and prices in

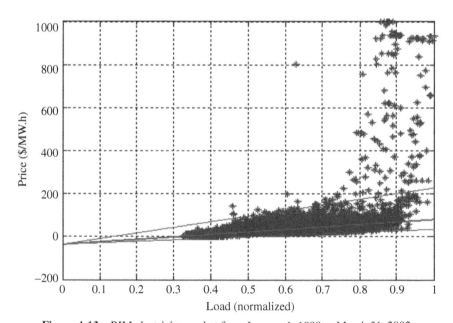

Figure 4-13   PJM electricity market from January 1, 1999 to March 31, 2002.

the PJM electricity market from January 1, 1999 to March 31, 2002. Clearly, price spikes occur only when demand is high. In a competitive electricity market, the supply curve has an upward slope, reflecting the increasing marginal cost of power generation. When demand is low, the corresponding supply level is low so that the marginal cost of generation, or the price, is low. As demand increases, the marginal cost of supply increases, and as it approaches the limit of supply capacity, the marginal cost of generation rapidly increases, causing sudden and large price increases.

In this case, prices are unlikely to spike when demand is low compared to supply, but as the demand increases and approaches the supply capacity, prices are more likely to spike. Then, the transition probability from a nonspike to a spike regime cannot be constant. It should depend on the underlying supply and demand relationship.

As an approximation, a hockey-stick-shaped curve is shown in Figure 4-14. The supply curve has this shape because of the increasing marginal cost, the limited capacity of electricity supply, and the FOR. In the short run, available power-generating facilities in the whole market are almost fixed. As demand increases, power companies increase their supply with facilities at higher marginal costs, but they cannot produce more electricity than the available capacities. As a result, near the limit of supply capacity, prices suddenly increase so that the slope of supply curve suddenly becomes steeper. Furthermore, the forced outage of generator increases this tendency.

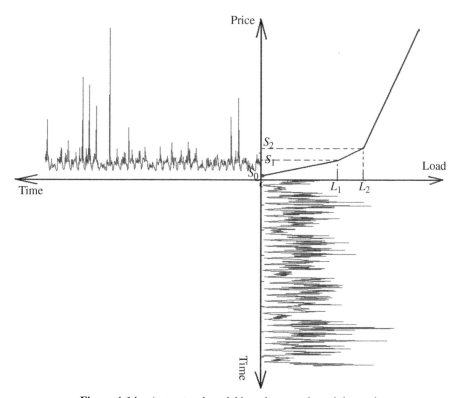

Figure 4-14   A structural model based on supply and demand.

The supply curve is modeled by two segments, that is, base segment and spiking segment, as shown in Figure 4-14. To simplify the problem, slopes of the segments are deterministic. Load is a stochastic process, that is, a mean reversion process. The intersections of segments and the random demand settle the spot prices. Mathematically, this model can be written as

$$d\check{L} = \eta[f(t) - \ln \check{L}]\check{L}dt + v\check{L}d\check{z} \tag{4-117}$$

$$\check{S}(t) = \alpha_1\check{L}(t) + b_1 \quad 0 < L \le L_1 \tag{4-118}$$

$$\check{S}(t) = \alpha_2\check{L}(t) + b_2 \quad L_1 < L \tag{4-119}$$

where a_1, a_2, b_1, and b_2 are constants, η is a parameter, $f(t)$ is the deterministic component of the load, and v is its volatility.

According to this model, by Monte Carlo simulations, sample paths of electricity price can be generated. $\check{L}(t)$ is the classical GBM, and sample paths of (4-117) can be generated. Based on piecewise linear functions (4-118) and (4-119), the sample paths of electricity price are derived.

TIME SERIES METHODS FOR ELECTRICITY PRICES

A time series is a sequence of observations arranged according to the time of occurrence. Clearly, electricity price is a time series. Time series analysis utilizes methods to understand the underlying context of data points, or to make predictions. Several most common time series models that have been used in modeling electricity prices are presented in this section.

Autoregressive (AR) Model

Basic Model This section starts with the simplest *autoregressive* model, and progresses to its extensions. The building block for time series models is the white noise denoted as $\tilde{\varepsilon}_t$. In the general case

$$\tilde{\varepsilon}_t \sim \text{i.i.d. } N(0, \sigma_\varepsilon^2)$$

Notice three implications of this assumption:

(i) $E\{\tilde{\varepsilon}_t\} = E\{\tilde{\varepsilon}_t | \tilde{\varepsilon}_{t-1}, \tilde{\varepsilon}_{t-2}, \ldots\} = E\{\tilde{\varepsilon}_t | \tilde{\varepsilon}_{t-1}\} = 0$

(ii) $E\{\tilde{\varepsilon}_t\tilde{\varepsilon}_{t-j}\} = \text{cov}\{\tilde{\varepsilon}_t\tilde{\varepsilon}_{t-j}\} = 0$

(iii) $\text{var}\{\tilde{\varepsilon}_t\} = \text{var}\{\tilde{\varepsilon}_t | \tilde{\varepsilon}_{t-1}, \tilde{\varepsilon}_{t-2}, \ldots\} = \text{var}\{\tilde{\varepsilon}_t | \tilde{\varepsilon}_{t-1}\} = \sigma_\varepsilon^2$

The first and second properties are the absence of any *serial correlation*. The third property is *conditional homoskedasticity*.

A real-valued stochastic process, $\tilde{x}_t$, is said to be an autoregressive process of order p, denoted by AR(p) if there exist $c, \alpha_i \in R, i = 1, 2, \dots, p$ with $\alpha_i \neq 0$, and white noise ($\tilde{\varepsilon}_t$) such that

$$\tilde{x}_t = c + \sum_{i=1}^{p} \alpha_i \tilde{x}_{t-i} + \tilde{\varepsilon}_t \qquad (4\text{-}120)$$

The value of an AR(p) at time t is, therefore, regressed on its own past p values plus a random shock. The constant term, c, is omitted for simplicity. By this, (4.1) may be written as

$$\tilde{x}_t = \sum_{i=1}^{p} \alpha_i \tilde{x}_{t-i} + \tilde{\varepsilon}_t \qquad (4\text{-}121)$$

The following result [50] provides a sufficient condition on the constants $\alpha_i \in R, i = 1, 2, \dots, p$ implying the existence of a uniquely determined stationary solution ($\tilde{x}_k$) of (4-120). The result is that AR(p) (4-120) with the given constants $\alpha_i, i = 1, 2, \dots, p$ and white noise ($\tilde{\varepsilon}_t$) has a stationary solution if all p roots of equation $1 - \sum_{i=1}^{p} \alpha_i z^i$ fall outside of the unit circle.

Properties   The simplest form of AR model is AR(1), which is given by

$$\tilde{x}_t = c + \alpha \tilde{x}_{t-1} + \tilde{\varepsilon}_t, \quad 0 < \alpha < 1 \qquad (4\text{-}122)$$

where $\tilde{\varepsilon}_t$ is a Brownian motion with zero mean and variance σ^2. Given x_0, it is obtained that

$$\begin{aligned}
\tilde{x}_t &= \alpha^t x_0 + c(1 + \alpha + \cdots + \alpha^{t-1}) + \alpha^{t-1}\tilde{\varepsilon}_1 + \alpha^{t-2}\tilde{\varepsilon}_2 + \cdots + \tilde{\varepsilon}_t \\
&= \alpha^t x_0 + c\left(\frac{1 - \alpha^t}{1 - \alpha}\right) + \alpha^{t-1}\tilde{\varepsilon}_1 + \alpha^{t-2}\tilde{\varepsilon}_2 + \cdots + \tilde{\varepsilon}_t
\end{aligned} \qquad (4\text{-}123)$$

$$E\{\tilde{x}_t\} = \alpha^t x_0 + c\left(\frac{1 - \alpha^t}{1 - \alpha}\right) \qquad (4\text{-}124)$$

$$\text{Var}\{\tilde{x}_t\} = \frac{1 - \alpha^{2t}}{1 - \alpha^2}\sigma^2 \qquad (4\text{-}125)$$

when $t \to \infty$, one obtain

$$E\{\tilde{x}_t\} = \frac{c}{1 - \alpha} \qquad (4\text{-}126)$$

$$\text{Var}\{\tilde{x}_t\} = \frac{\sigma^2}{1 - \alpha^2} \qquad (4\text{-}127)$$

From (4-126), it is noted that AR(1) assumes: as time t moves on, the mean of prices approaches a constant $E\{\tilde{x}_t\} = \dfrac{c}{1-\alpha}$. Furthermore, the autocovariance is given by

$$B_n = E(\tilde{x}_{t+n}\tilde{x}_t) - E(\tilde{x}_t)^2 = \frac{\sigma^2}{1-a^2}a^{|n|} \qquad (4\text{-}128)$$

The spectral density function is the Fourier transform of the autocovariance function. In discrete terms this will be the discrete-time Fourier transform

$$\Phi(\omega) = \frac{1}{\sqrt{2\pi}}\sum_{-\infty}^{+\infty} B_n e^{-i\omega n} = \frac{1}{\sqrt{2\pi}}\left(\frac{\sigma^2}{1+a^2 - 2a\cos(\omega)}\right) \qquad (4\text{-}129)$$

If it is assumed that the sampling time ($\Delta t = 1$) is much smaller than the decay time (τ), then one can use a continuum approximation to B_n

$$B(t) \approx \frac{\sigma^2}{1-a^2}a^{|t|} \qquad (4\text{-}130)$$

For the p order autoregressive model, that is, AR(p) model, more lagged prices are considered

$$\tilde{x}_t = c + (\alpha_1 \tilde{x}_{t-1} + \alpha_2 \tilde{x}_{t-2} + \cdots + \alpha_p \tilde{x}_{t-p}) + \tilde{\varepsilon}_t \qquad (4\text{-}131)$$

That is

$$\tilde{x}_t - \alpha_1 \tilde{x}_{t-1} - \alpha_2 \tilde{x}_{t-2} - \cdots - \alpha_p \tilde{x}_{t-p} = c + \tilde{\varepsilon}_t \qquad (4\text{-}132)$$

A compact form of this system can be written as

$$a(L)\tilde{x}_t = c + \tilde{\varepsilon}_t \qquad (4\text{-}133)$$

where, $a(L) = 1 - \alpha_1 L^{-1} - \alpha_2 L^{-2} - \cdots - \alpha_p L^{-p}$

To find the mean of AR(p), take the expectation of both sides of (4-133),

$$E\{\tilde{x}_t - \alpha_1 \tilde{x}_{t-1} - \alpha_2 \tilde{x}_{t-2} - \cdots - \alpha_p \tilde{x}_{t-p}\} = E\{c + \tilde{\varepsilon}_t\}$$
$$E\{\tilde{x}_t\} - \alpha_1 E\{\tilde{x}_{t-1}\} - \alpha_2 E\{\tilde{x}_{t-2}\} - \cdots - \alpha_p E\{\tilde{x}_{t-p}\} = c + E\{\tilde{\varepsilon}_t\}$$
$$E\{\tilde{x}_t\} - \alpha_1 E\{\tilde{x}_t\} - \alpha_2 E\{\tilde{x}_t\} - \cdots - \alpha_p E\{\tilde{x}_t\} = c$$
$$E\{\tilde{x}_t\}(1 - \alpha_1 - \alpha_2 - \cdots - \alpha_p) = c$$

The following is obtained:

$$E\{\tilde{x}_t\} = \frac{c}{1-\alpha_1 - \alpha_2 - \cdots - \alpha_p} \qquad (4\text{-}134)$$

AR models have been used for decades in economics, signal processing, as well as electricity load and price forecasting. AR models often serve as benchmarks for more sophisticated approaches. In order to model the daily shape, intra-week pattern and yearly seasonality of electricity prices, it is necessary to add a deterministic function for modeling the time-varying mean of electricity prices.

AR with Time-Varying Mean   AR with time-varying mean (ARV) models assume that prices are a sum of a deterministic time-varying function and an AR process. Specifics of AR process are mostly determined in calibration, while the deterministic function requires more elaboration. Assume a deterministic function $f(t)$, which describes the time-varying mean of electricity prices. For electricity price $\tilde{S}_t$, it can be written as

$$\tilde{S}_t = f(t) + \tilde{x}_t \qquad (4\text{-}135)$$

where $\tilde{x}_t$ follows a simple AR(1) process with constant c set as 0, that is,

$$\tilde{x}_t = \alpha \tilde{x}_{t-1} + \tilde{\varepsilon}_t \qquad (4\text{-}136)$$

Using the form of AR(1), $\tilde{S}_t$ can be written as

$$\begin{aligned}\tilde{S}_t &= [f(t) - \alpha f(t-1)] + \alpha \tilde{S}_{t-1} + \tilde{\varepsilon}_t \\ &= c_t + \alpha \tilde{S}_{t-1} + \tilde{\varepsilon}_t \end{aligned} \qquad (4\text{-}137)$$

where, $c_t = f(t) - \alpha f(t-1)$.

A more general AR with time-varying mean and higher-order lagged prices is

$$\tilde{S}_t = c_t + (\alpha_1 \tilde{S}_{t-1} + \alpha_2 \tilde{S}_{t-2} + \cdots + \alpha_m \tilde{S}_{t-m}) + \tilde{\varepsilon}_t \qquad (4\text{-}138)$$

The specifics of c_t depend on the problem under study.

For the short-term electricity price model, that is, next-day prices, c_t is to model the intra-day price shape, the weekly pattern, and even the yearly seasonality. Whether to model the yearly seasonality depends on the data used for calibration. If the data covers several months, only the deterministic levels for these months should be modeled. If the data cover several years, however, a deterministic function modeling the yearly seasonality should be included.

For next-week daily prices forecasting, because the intra-day shape has been averaged out, c_t only needs to describe the weekly pattern and yearly seasonality.

The deterministic time-varying function c_t may be forged with binary variables or sinusoidal functions. Intra-day shape and weekly pattern can be described by binary variables. There are two options for modeling the yearly seasonality. The first one is monthly binary variables; and the second one is sinusoidal functions.

In Reference [92], ARV is applied to model hourly prices from Leipzig Power Exchange. In this work,

$$\tilde{S}_t = c_t + a\tilde{S}_{t-1} + \tilde{\varepsilon}_t \qquad (4\text{-}139)$$

where $\tilde{S}_t = a_0 t + \sum_{i=1}^{24} a_{1,i} I_{\text{Hour}(i)} + \sum_{j=1}^{7} a_{2,j} I_{\text{Day}(j)} + \sum_{l=1}^{12} a_{3,l} I_{\text{Month}(l)}$, $I(\bullet)$ is an indicator function taking value one if the argument is true, and zero otherwise.

In Reference [31], two deterministic mean functions are provided. In the first one, the function takes the following form:

$$f(t) = \alpha + \beta D_t + \sum_{i=2}^{12} \beta_i M_{it} \qquad (4\text{-}140)$$

where

$$D_t = \begin{cases} 1 & \text{if data } t \text{ is holiday or weekend} \\ 0 & \text{otherwise} \end{cases}$$

$$M_{it} = \begin{cases} 1 & \text{if data } t \text{ belongs to the } i\text{th calendar month} \\ 0 & \text{otherwise} \end{cases}, \text{ for } i = 2, 3, \ldots, 12$$

and α, β, β_i for $i = 2, 3, \ldots, 12$ are all constant parameters.

In this case, the parameters β_i try to capture the changes for holidays and weekends, and for the different months of the year, respectively.

The second model takes the following form

$$f(t) = \alpha + \beta D_t + \gamma \cos\left[(t + \tau)\frac{2\pi}{365}\right] \qquad (4\text{-}141)$$

where

$$D_t = \begin{cases} 1 & \text{if data } t \text{ is holiday or weekend} \\ 0 & \text{otherwise} \end{cases}$$

$\cos(\bullet)$ is the cosine function measured in radian, and α, β, γ, and τ are constant parameters. Here, β tries to capture the changes for weekends and holidays. The cosine function is expected to reflect the seasonal pattern in evolution of the relevant variable throughout the year; hence it has an annual periodicity.

In a related study, Reference [93] proposes an AR model with an optimum threshold stratification algorithm, which determines the minimum number of parameters required to represent the random component, improving the forecast accuracy. In Reference [4], a two-level seasonal autoregressive model is described. This model consists of 24 separated AR models, each describing an hour of a day; and a sinusoidal annual component and dummy variables for each day of the week.

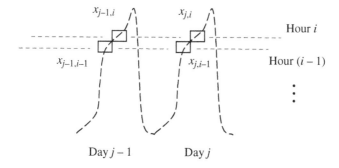

Figure 4-15   PAR model.

Periodic Autoregressive Model   The purpose of periodic autoregressive model (PARM) is to model not only the time-varying mean, but also the time-varying mean-reverting feature. The key concept of PAR is named separate hours, that is, each hour in consecutive days is treated separately. For N days, there are $24N$ hourly prices. Instead of considering $24N$ prices as a whole, they are grouped according to the hour index. For example, take the first hour of each day, order them by Day 1, Day 2, etc. A new segment series can be obtained consisting of $x_{1,1}, x_{2,1}, \ldots$ Continue the same procedure for all 24 hour indices. It results in 24 segment time series. As illustrated in Fig. 4-15.

A separate hour AR model uses a distinct AR to describe each segment time series. Recall AR(1),

$$\tilde{x}_t = c + \alpha \tilde{x}_{t-1} + \tilde{\varepsilon}_t \tag{4-142}$$

a separate hour AR(1) model for hour i can be written as

$$\tilde{x}_{j,i} = c_i + \alpha_i \tilde{x}_{j-1,i} + \tilde{\varepsilon}_{j,i} \quad i = 1, \ldots 24 \tag{4-143}$$

The disadvantage of separate hour AR model is that it dismantled the original time series. Thus it is unable to model the effect of prices in the immediate previous hours. For example, a separate hour model is not able to model the effect of $\tilde{x}_{j,i-1}$ on $\tilde{x}_{j,i}$. Including the price in the immediately previous hour to the separate hour AR model, a simplest PAR Model is obtained as

$$\tilde{x}_{j,i} = c_i + \alpha_i \tilde{x}_{j-1,i} + \beta_i \tilde{x}_{j,i-1} + \tilde{\varepsilon}_{j,i}, \quad i = 1, \ldots 24 \tag{4-144}$$

A more general PAR model could possess more price lags. The key concept of a PAR model is that prices with different hour indices are modeled by different AR models. PAR is designed to model the distinct behavior of prices with different hour indices, in terms of the mean and mean-reverting features.

The setting of hourly prices is to facilitate explanation of the PAR model. This idea also applies to daily prices which possess weekly patterns. PAR model may serve as a good option for modeling short-term prices for two reasons. Firstly, it differentiates prices during different periods. Then, it reduces seasonality by one order. For hourly prices, it is observed and could be explained by market fundamentals. Since prices during low demand behave distinctively from those during peak demand, it is reasonable to describe low and peak prices with different AR models. Hourly prices possess three orders of seasonality: intra-day shape, intra-week pattern, and intra-year seasonality. A PAR model describes each hour separately and avoids modeling the intra-day shape. For daily average prices, it is observed that prices during weekdays behave differently from these on weekends. It is reasonable to model weekday and weekend prices with different AR models. A PAR model treats each weekday separately and avoids modeling the intra-week pattern of prices.

Periodic models were studied in References [60], [92], and [46]. Reference [60] studied prices from New Zealand. They separated half-hourly price series into 48 subseries, and found that these series featured by different mean and standard deviations. With this observation, they propose a PAR model with periodicity of 48. According to the autocorrelation of intra-day prices, they find that the 48 trading periods falling into five groups: overnight, morning peak, day time, evening peak, and late evening. To model these featured groups, they use state space methods. Reference [46] studied the day-ahead hourly prices from the Dutch, German, to French markets. The conclusion is that day-ahead hourly prices should not be treated as normal time series, but as cross-sections, because day-ahead prices are not generated consecutively but by cross-sections of 24 hours in day-ahead market clearing.

AR with Exogenous Variable   The other idea to model the time-varying function is to describe it with exogenous variables. Models of this type are called ARX models, dynamic regression model, or AR with distributed lag.

An AR model with distributed lags of electricity demand d_t is

$$\tilde{S}_t = (c + \gamma_0 d_t + \gamma_1 d_{t-1} + \cdots + \gamma_n d_{t-n}) + (\alpha_1 \tilde{S}_{t-1} + \alpha_2 \tilde{S}_{t-2} + \cdots + \alpha_m \tilde{S}_{t-m}) + \tilde{\varepsilon}_t$$

(4-145)

where c, γ_i for $i = 0, 1, \ldots, n$, and α_i for $i = 1, \ldots, m$ are all constant parameters, $\tilde{\varepsilon}_t$ is white noise. Moving the price terms to the left, one obtains

$$\tilde{S}_t - \alpha_1 \tilde{S}_{t-1} - \alpha_2 \tilde{S}_{t-2} - \cdots - \alpha_m \tilde{S}_{t-m} = c + \gamma_0 d_t + \gamma_1 d_{t-1} + \cdots + \gamma_n d_{t-n} + \tilde{\varepsilon}_t$$

(4-146)

A compact form of (4-146) can be written as

$$\left(1 - \alpha_1 L^{-1} - \alpha_2 L^{-2} - \cdots - \alpha_m L^{-m}\right) \tilde{S}_t = c + \left(\gamma_0 + \gamma_1 L^{-1} + \cdots + \gamma_n L^n\right) d_t + \tilde{\varepsilon}_t$$

(4-147)

That is

$$a(L)\tilde{S}_t = c + \beta(L)d_t + \tilde{\varepsilon}_t \tag{4-148}$$

where

$$a(L) = 1 - \alpha_1 L^{-1} - \alpha_2 L^{-2} - \cdots - \alpha_m L^{-m} \tag{4-149}$$

$$\beta(L) = \gamma_0 + \gamma_1 L^{-1} + \cdots + \gamma_n L^n \tag{4-150}$$

ARX is a good option for modeling short-term electricity prices because (1) it is convenient to model the autocorrelation of prices, (2) the effect of demand on prices, and, (3) the model can be kept simple. The ARX model [94] has been used to forecast the next day's 24-hour prices in Spain and California markets. In this work, ARX predict prices accurately, with the average errors within 5%.

Smooth Transition AR Model   Until now, all models which were described in this part are linear. A natural question is whether a linear model is an adequate representation for the data-generating process. It is suggested that the structure and history of an electricity pool may lead to a nonlinear representation of the data. Therefore, nonlinear time series models are necessary. A smooth transition AR (STAR) model was used to model prices from England and Wales markets by Robinson [95]. A STAR model can be interpreted as an autoregressive model in which local dynamics are dependent on a past value of the series. This method allows the price to switch between several regimes which represent several phases of behavior. Moreover, the transition between these regimes can be smooth, so that there can be a continuum of states between these regimes.

In this area, Robinson's work on the STAR model is motivated by observations of nonlinearity of the electricity prices. Low prices are tranquil while high prices are volatile and less mean reverting. The nonlinearity in prices from England and Wales market is due to the capacity payment—the higher the demand, the higher is the capacity payment.

Mathematically, a STAR model of order p takes the following form:

$$\tilde{S}_t = \alpha_0 + \alpha_1 \tilde{w}_t + (\beta_0 + \beta_1 \tilde{w}_t)F(\tilde{S}_{t-d}) + \tilde{\varepsilon}_t \tag{4-151}$$

where $\tilde{S}_t$ is the electricity price at time t, $\tilde{\varepsilon}_t$ is error term, $\tilde{w}_t = (\tilde{S}_{t-1}, \tilde{S}_{t-2}, \tilde{S}_{t-p})^T$, and $F(\bullet)$ is a transition function which is bounded between one and zero. In that work, two types of transition function are considered. The first one is the logistic function

$$F(\tilde{S}_{t-d}) = \frac{1}{1 + \exp[-\gamma(\tilde{S}_{t-d} - c)]} \tag{4-152}$$

and the second one is the exponential function

$$F(\tilde{S}_{t-d}) = 1 - \exp[-\gamma(\tilde{S}_{t-d} - c)^2] \tag{4-153}$$

where γ and c are parameters.

The STAR family of models can represent a variety of price behaviors. As $\gamma \to \infty$ in (4-152), $F(\tilde{S}_{t-d})$ becomes a Heaviside function. That is,

$$F(\tilde{S}_{t-d}) = \begin{cases} 0, & \text{if } \tilde{S}_{t-d} < c \\ 1, & \text{if } \tilde{S}_{t-d} > c \end{cases} \qquad (4\text{-}154)$$

Hence (4-151) and (4-152) form a threshold autoregressive (TAR) model of order p. As $\gamma \to 0$, (4-151) with (4-153) becomes a linear AR(p).

Equations and represent an exponential STAR (ESTAR) model in which dynamics of the series change symmetrically around c with $\tilde{S}_{t-d}$. As $\gamma \to \infty$, the ESTAR model becomes linear. This is the case if $\gamma \to 0$ since one regime has probability one and the other probability zero on the boundary. An ESTAR model can describe price behavior where reductions in price from a "high" level to a "normal" level behave in a similar way as a return to a normal price from a low price.

Autoregressive Moving Average (ARMA) Model

In statistics, ARMA models, sometimes called Box–Jenkins models are used to fit time series data. Given a stochastic time series of data $\tilde{x}_t$, the ARMA model is a tool for understanding and predicting future values in this series. The model consists of two parts, an *autoregressive* (AR) part and a *moving average* (MA) part. The model is referred to as the ARMA(p,q) model, where p is the order of the autoregressive part and q is the order of the moving average part.

AR models have been described in Autoregressive (AR) Model. Moving average (MA) model will be described as follows. The notation MA(q) refers to the moving average model of order q is

$$\tilde{x}_t = c + \tilde{\varepsilon}_t + \sum_{i=1}^{q} \beta_i \tilde{\varepsilon}_{t-i} \qquad (4\text{-}155)$$

where $\beta_1, \ldots, \beta_q$ are parameters of the model, $\tilde{\varepsilon}_t, \tilde{\varepsilon}_{t-1}, \ldots$ are the error terms, and c is a constant.

The moving average model is essentially a finite impulse response filter. A compact form of this system can be written as

$$\tilde{x}_t = c + (\beta_0 \tilde{\varepsilon}_t + \beta_1 \tilde{\varepsilon}_{t-1} + \cdots + \beta_q \tilde{\varepsilon}_{t-q}) = c + b(L)\tilde{\varepsilon}_t \qquad (4\text{-}156)$$

where

$$b(L) = \beta_0 + \beta_1 L + \cdots + \beta_q L^q \qquad (4\text{-}157)$$

The simplest moving average model, that is, MA(1) model can be written as

$$\tilde{x}_t = c + \beta_0 \tilde{\varepsilon}_t + \beta_1 \tilde{\varepsilon}_{t-1} \tag{4-158}$$

where x_t equals a constant plus the weighed sum of past errors.

The simplest ARMA model is a combination of AR(1) and MA(1), namely, ARMA(1,1). It can be written as

$$(1 - \alpha_1 L)\tilde{S}_t = c + (\beta_0 - \beta_1 L)\tilde{\varepsilon}_t \tag{4-159}$$

The purpose of extending AR model to ARMA is to model the autocorrelation of residuals. Generally, an ARMA model is

$$a(L)\tilde{S}_t = c + b(L)\tilde{\varepsilon}_t \tag{4-160}$$

The mean of ARMA(p,q) is the same as AR(p)

$$E(\tilde{x}_t) = \frac{c}{1 + \alpha_1 + \alpha_2 + \cdots + \alpha_p} \tag{4-161}$$

ARMA with Time-Varying Mean (ARMAV)   To model the time-varying mean of electricity prices, the constant term c is generalized to a time-varying c_t. The resulting model is

$$a(L)\tilde{S}_t = c_t + b(L)\tilde{\varepsilon}_t \tag{4-162}$$

In Reference [92], variants of AR(1) and general ARMA processes are applied to short-term price forecasting of the German market. They conclude that specifications where each hour of the day is modeled separately present better forecasting properties than specifications for the entire time series. Furthermore, the inclusion of simple probabilistic processes for the arrival of jumps leads to improvement in the forecasting capabilities of electricity price models.

ARMAX and Transfer Function   In the same manner as the extension from AR to ARX, the constant term c is extended to a polynomial function of lagged demand d_t as

$$c_t = \gamma_0 d_t + \gamma_1 d_{t-1} + \cdots + \gamma_n d_{t-n} \tag{4-163}$$

Therefore, a compact form of this system can be written as

$$a(L)\tilde{S}_t = c + c(L)d_t + b(L)\tilde{\varepsilon}_t \tag{4-164}$$

Divided $a(L)$ from both sides, ARMAX can be described in the form of a *transform function*

$$\tilde{S}_t = \frac{c}{a(L)} + \frac{c(L)}{a(L)}d_t + \frac{b(L)}{a(L)}\tilde{\varepsilon}_t \qquad (4\text{-}165)$$

Nogales et al. [94] utilize ARMAX and ARX models to predict hourly prices in Spain and California markets. Both models fit the data well. The mean weekly error, for the first week of April 2000 in California is below 3%, and 5% for the third weeks of August and November in Spain. The results are significantly better than those obtained by ARIMA and ARIMA with load as an exogenous variable.

To forecast the CASIO's day-ahead prices, Nowicka-Zagrajek and Weron [96] apply the rolling volatility technique and modeled the deseasonalized loads from California market by using standard and adaptive ARMA processes with hyperbolic noise.

ARIMA and Its Extensions   In statistics, an ARIMA model is a generalization of an ARMA model. This model is fitted to time series either to better understand the data or to predict future points in the series. The model is generally referred to as an ARIMA(p,d,q) model where p, d, and q are integers greater than or equal to zero and refer to the order of the autoregressive, integrated, and moving average parts of the model, respectively.

An ARIMA(p,d,q) process is obtained by integrating an ARMA(p,q) process. That is,

$$\left(1 - \sum_{i=1}^{p} \alpha_i L^i\right)(1-L)^d \tilde{x}_k = \left(1 + \sum_{i=1}^{q} \beta_i L^i\right)\tilde{\varepsilon}_k \qquad (4\text{-}166)$$

where L is the lag operator, α_i are the parameters of the autoregressive part of the model, β_i are the parameters of the moving average part, and $\tilde{\varepsilon}_k$ is white noise. Here d is a positive integer that controls the level of differencing (or, if $d = 0$, this model is equivalent to an ARMA model). Conversely, applying term-by-term differencing d times to an ARMA(p,q) process gives an ARIMA(p,d,q) process.

The simplest ARIMA model is an ARIMA(0,1,0), which is a random walk, that is,

$$\tilde{x}_k - \tilde{x}_{k-1} = \Delta\tilde{x}_k = \tilde{\varepsilon}_k \qquad (4\text{-}167)$$

Physically, it describes a process whose change from time $k - 1$ to k is random and the increase is characterized by a standard normal distribution.

The following references deal with dynamics of the electricity prices: Contreras [97], Zhou [98,99], and Conejo [59]. In Reference [59], several different methods of short-term price forecasting, such as three time series specifications (including ARIMA), a wavelet multivariate regression technique, and a multilayer ANN model, are compared. Based on the PJM data in 2002, the ARIMA model performs better than the ANN but worse than time series models with exogenous variables. A related work [100] proposes a wavelet-ARIMA algorithm. A discrete wavelet transform is used to decompose price series first, and then approximate series with ARIMA to obtain 24 hourly predicated values. The inverse wavelet transform is applied to yield the forecasted prices for the next 24 hours. The performance of wavelet ARIMA is generally better than that of a standard ARIMA process.

Volatility of Prices

Constant Volatility   Volatility is a feature of the stochastic process. Price volatility describes how much the price varies over small time intervals. If one plots a histogram of price returns, the width of distribution would be directly related to the volatility of price return. A higher volatility results in a greater width of the distribution. Volatility, which exhibits a combination of deterministic and random behavior, exhibits different characteristics in the long term when compared with the short term [85].

Volatility, σ, is the price returns' standard deviation normalized by time, which can be written as

$$\sigma = \sqrt{\frac{\text{Var}(d\tilde{S}/S)}{dt}} \tag{4-168}$$

An intuitive measure of price randomness is described by volatility. For example, a Brownian motion, $d\tilde{z}$, is normally distributed, with a mean value of zero and a standard deviation of $\sqrt{dt}$, that is,

$$d\tilde{z} \sim N(0, dt) \tag{4-169}$$

For the price return, $\sigma d\tilde{z}$, the expectation of this stochastic term is zero, and the expected value of the stochastic term squared is $\sigma^2 dt$. This means that the standard deviation of price return is proportional to both volatility and square root of the time period between price observations. For a constant volatility, the longer the time period between observations, the greater is the standard deviation of price returns.

The variance of an entire path can be generalized for cases where volatility is not the same for different steps, that is, σ_i, between time t_i and t_{i+1}. Hence, the path's variance, $\bar{\sigma}$, is given by

$$\bar{\sigma}^2 t_N = \sum_{i=1}^{N-1} \left[\sigma_i^2(t_{i+1} - t_i)\right] \tag{4-170}$$

and, as a result, volatility of the path is

$$\bar{\sigma} = \sqrt{\frac{\sum_{i=1}^{N-1} \left[\sigma_i^2 (t_{i+1} - t_i) \right]}{t_N}} \tag{4-171}$$

For the continuous case, volatilities of historical data of spot prices can be observed. Volatility of the price can be estimated by

$$\sigma^2 = \frac{E[(d\tilde{S}/\tilde{S})^2]}{dt} \tag{4-172}$$

For a data set of N price returns, it is therefore concluded that

$$\sigma = \sqrt{\frac{1}{N} \frac{E[(d\tilde{S}/\tilde{S})^2]}{dt}} \tag{4-173}$$

Time-Varying Volatility   Volatility measures the degree of uncertainty. Uncertainty of electricity prices arises from both supply and demand. On the demand side, uncertainty depends on the extent of load fluctuations. On the supply side, it depends on the FOR of generators, number of generators, and shape of the supply stack bid into the electricity pool.

Over the 24 hours in a day, on the demand side, the electricity load during late night and early morning is usually low. While during day time, load rises to its peak, stays for a few hours and then falls to its night low. There are more factors that affect load during day time than those during night time. Thus, the day time load is more uncertain than that during night time. On the supply side, during the night time, fewer generators are operating. In the day time, more generators have to be started to support the increasing load and kept running until the load comes down. Due to the larger number of generators, the extent of uncertainty during day time is higher than that in the night.

In the horizon of 1 week, on the demand side, electricity load during weekends is lower and more stable than that during weekdays. On the supply side, accordingly, there is a better chance to suffer unpredicted events during weekdays than in weekends.

Over the horizon of a year, there are more random events during high electricity load seasons than during low load seasons. Thus, electricity volatility is usually higher in high demand seasons. Time-varying volatility across various horizons is by nature an important trait of electricity prices.

Deterministic Models   In deterministic methods, σ_t as a deterministic function $v()$, is augmented by time t, electricity load level, or other exogenous variables. It can be written as

$$\sigma = v(\bullet) \tag{4-174}$$

Empirical evidence suggests that price volatility increases with demand. One interpretation relates to the technical risks of the more expensive, flexible plants, which gradually start their operation at this stage and dominate pricing. These stations require a premium for the implicit operational risk, which increases with the intensity of plant utilization. As demand increases, technical risk becomes a more influential determinant of bidding or collusion becomes easier. Based on this idea, the following models are proposed in Reference [101].

$$\sigma_t^2 = \left(a + v_t^b\right)^2 \tag{4-175}$$

and

$$\sigma^2 = c v_t^b \tag{4-176}$$

where a, b, and c are parameters, v is exogenous variable such as the load level.

Considering the seasonal behavior of electricity price, one might suspect that the model of volatility performance could depend on, for instance, the time of the year. Volatility changes over the year in a wavelike fashion and, after reaching a lowest level some time during winter, it slowly increases to a maximum level during summer. A natural way to model this wavelike variation is to add a sinusoidal term to the constant term in volatility. Bystrom [55] suggests the seasonal pattern of volatility as

$$\sigma_t^2 = \alpha \sin\left(\frac{2\pi}{8760}t - \beta\right) + \gamma \tag{4-177}$$

where α, β, and γ are parameters.

The value of the above models lies in the fact that they initialize the effort to treat volatility in a deterministic manner. Due to the highly predictable components of electricity prices, a deterministic function is a good option for modeling.

Standard GARCH Model   The linear ARMA-type models assume *homoskedasticity*, that is, a constant variance and covariance function. From an empirical point of view, electricity prices present various forms of nonlinear dynamics, the crucial one being the strong dependence of variability of the series on its own past. Some nonlinearities of these series take the form of a nonconstant conditional variance. Generally, they are characterized by clustering of large *heteroskedasticity*. Up to now, GARCH models dominate this category.

The GARCH(p, q) model is given by

$$\tilde{\sigma}_t = \alpha_0 + \sum_{i=1}^{q} \alpha_i \tilde{\varepsilon}_{t-i}^2 + \sum_{i=1}^{p} \beta_i \tilde{\sigma}_{t-i}^2 \qquad (4\text{-}178)$$

where p is the order of GARCH terms $\tilde{\sigma}_t^2$, and q is the order of $\tilde{\varepsilon}_t^2$.

The model assumes that σ_t depends on its values at times $t-1, t-2, \ldots, t-q$, that is, $\sigma_{t-1}, \sigma_{t-2}, \ldots, \sigma_{t-q}$, and the outcomes at time $t-1, t-2, \ldots, t-p$, that is, $\varepsilon_{t-1}, \varepsilon_{t-2}, \ldots, \varepsilon_{t-p}$. The dependence of σ_t^2 on $\varepsilon_{t-1}, \varepsilon_{t-2}, \ldots, \varepsilon_{t-p}$ means that if a large disturbance occurs at $t-1, t-2, \ldots, t-p$, another large disturbance is more likely to occur at time t. The factor β_i governs the decay of the effect of past disturbances.

The GARCH model by itself is not very attractive for price forecasting, however, coupled with ARMA-type model presents an interesting alternative. Researchers model electricity price volatility with GARCH models. Garcia [51] applies GARCH to model prices of Spanish and California markets. They showed that an ARIMA + GARCH model outperforms a pure ARIMA model. Average forecast errors using Spanish and Californian market data are around 9%, depending on the month of the year. All forecasts have been derived from subsets of a general GARCH model applied to both markets, including all hours, days, weekends, and holidays. Guirguis et al. [102] apply GARCH model to prices of New York City and Central New York State. It is shown that an ARX + GARCH model forecasts prices with higher accuracy than an ARX model in both regions. Karakatsani and Bunn [101] test four approaches (including regression GARCH) to explain the stochastic dynamics of spot volatility and understand agent reactions to shocks. Limitations of GARCH models due to extreme values are resolved when a regression model with the assumptions of an implicit jump component for prices and a leptokurtic distribution for innovation are used.

GARCH Model with Asymmetric Effect   Threshold ARCH was originally proposed to describe the asymmetric effect of good and bad news in a stock market. In a power market, the preliminary data analysis reveals that electricity prices exhibit volatility clustering. In addition, intuition shows that it is possible that innovations to the prices series have an asymmetric impact on price volatility. In this model, expect positive price shocks to increase volatility more than negative surprises. The intuition behind this is that a positive shock to prices is really an unexpected positive demand shock. Therefore, since marginal costs are convex, positive demand shocks have a larger impact on price changes relative to negative shocks. In Knetill et al. [42], the model is given by

$$\tilde{\sigma}_t^2 = c + \alpha \tilde{\varepsilon}_{t-1}^2 + \beta \tilde{\sigma}_{t-1}^2 + \eta \tilde{\varepsilon}_{t-1}^2 \tilde{d}_{t-1} \qquad (4\text{-}179)$$

where, $d_{t-1} = \begin{cases} 1 & \text{if } \tilde{\varepsilon}_{t-1} < 0 \\ 0 & \text{if } \tilde{\varepsilon}_{t-1} > 0 \end{cases}$. The additional term $\eta \tilde{\varepsilon}_{t-1}^2 \tilde{d}_{t-1}$ is to differentiate the asymmetric effect of positive and negative shocks. If $\eta < 0$, negative shocks have smaller effects on $\tilde{\sigma}_t$ than positive shocks. The condition $\eta > 0$ indicates that negative shocks have greater effects on ... than positive shocks. Knetill [42] studied California prices and arrived at an observation that "positive shocks to prices amplify the conditional variance of the process more so than negative shocks." The parameter η is significantly negative. Knetill's explanation is that because of the convex supply stack, positive demand shocks have a larger impact on price changes than negative shocks.

Hadsell [103] examines the volatility of wholesale electricity prices for five US markets. Using data covering the period from May 1996 to September 2001, for California–Oregon Border, Palo Verde, Cinergy, Entergy, and PJM markets, they examine the volatility of electricity wholesale prices over time and across markets.

NUMERICAL EXAMPLES

The time series models, described in Time Series Methods for Electricity Prices, serve as tools to understand the inherent structure embedded in a sequence of data, such as electricity prices. Estimating parameters of stochastic processes, such as the stochastic differential models described in Stochastic Process Models for Electricity Prices, is an application of time series models. To estimate parameters precisely and effectively, numerous methods are available [2]. In this chapter, a straightforward method to estimate parameters of mean reversion process (3.36) is described. In this method, AR(1) is used. Another application of time series models, perhaps the most popular, is to fit a sequence of data. In this section, four models are compared, that is, AR, ARX, ARMA, and ARMAX, with different orders to fit a sequence of electricity prices. The methods to estimate parameters of these models are presented and the hourly electricity prices from August 1 to August 22, 1998 in California (see Figure 4-5) are utilized.

Estimate Parameters of Mean Reversion Process by AR Model

As mentioned in Stochastic Process Models for Electricity Prices, the distribution of solutions to the additive mean reversion process (4-36) is normal. The expectation and variance are described by (4-38) and (4-39), respectively. To estimate the parameters of this process by AR model, the sample path of (4-36) is necessary. According to (4-17), the sample path can be generated step-by-step. For example, from x at t, $x(t)$, to x at $t + \Delta t$, $x(t + \Delta t)$. The random variable $\tilde{x}(t + \Delta t)$ is normal. One first finds its deterministic component and then adds to it the random component. From (4-17), the deterministic component is obtained

$$x(t + \Delta t) = e^{-k\Delta t}x(t) + km \int_t^{t+\Delta t} e^{-k(t+\Delta t-\tau)}d\tau = m(1 - e^{-k\Delta t}) + e^{-k\Delta t}x(t)$$

$$(4\text{-}180)$$

After adding the random component, $\tilde{x}(t + \Delta t)$ can be written as

$$\tilde{x}(t + \Delta t) = m(1 - e^{-k\Delta t}) + e^{-k\Delta t}\tilde{x}(t) + N\left(0, \frac{\sigma^2}{2k}(1 - e^{-2k\Delta t})\right) \quad (4\text{-}181)$$

where $N(0, \frac{\sigma^2}{2k}(1 - e^{-2k\Delta t}))$ is a normal distribution.

Comparing (4-181) with AR(1) model, the relationship between mean reversion process and AR(1) can be found. It is written as

$$c = m(1 - e^{-k\Delta t}) \quad (4\text{-}182)$$
$$\alpha = e^{-k\Delta t} \quad (4\text{-}183)$$

$$\sigma_\varepsilon = \sqrt{\frac{\sigma^2}{2k}(1 - e^{-2k\Delta t})} \quad (4\text{-}184)$$

There are three equations with three unknown variables. Therefore, after the parameters of AR(1) model are obtained for a sequence of data, an equivalent mean reversion process can be built based on (4-182) to (4-184).

Note that if the step size Δt is sufficiently small, $e^{-k\Delta t} \approx 1 - k\Delta t$, (4-182) to (4-184) can be approximated by

$$c = mk\Delta t \quad (4\text{-}185)$$
$$\alpha = 1 - k\Delta t \quad (4\text{-}186)$$
$$\sigma_\varepsilon = \sigma\sqrt{\Delta t} \quad (4\text{-}187)$$

The results of this section can be generalized to the additive mean reversion process with time-varying parameters (4-42). AR with time-varying mean model (4-135), is an alternative method to estimate the parameters of (4-42).

Generally, parameters of the additive models, such as (4-36) and (4-42), can be estimated by different kinds of AR models. Furthermore, for the multiplicative model, that is, GBM and GMR process, described in Table 4-1, a similar technique can be used to estimate parameters. Due to the property of lognormal distribution, as described in Properties of a Lognormal Distribution, its logarithm is a normal distribution. Therefore, one can estimate the parameters of a corresponding additive model (normal distribution) first, and then transform it back to the parameters of the original lognormal distribution.

Numerical Examples of AR Model

The *autoregressive* (AR) model and *autoregressive with exogenous variable* (ARX) model are discussed in this section.

AR Model

Estimating Parameters of AR(p)   In order to estimate parameters of AR(p), many methods can be employed. The simplest method based on least squares regression is described here.

For an AR(p) given by (4-121), a compact form can be written as

$$Y = X \cdot A + \Theta \tag{4-188}$$

where

$$Y = (\tilde{x}_{p+1}, \tilde{x}_{p+2}, \ldots, \tilde{x}_N)^T \tag{4-189}$$

$$A = (\alpha_1, \alpha_2, \cdots, \alpha_p)^T \tag{4-190}$$

$$\Theta = (\tilde{\varepsilon}_{p+1}, \tilde{\varepsilon}_{p+2}, \cdots, \tilde{\varepsilon}_N)^T \tag{4-191}$$

$$X = \begin{pmatrix} \tilde{x}_p & \tilde{x}_{p+1} & \cdots & \tilde{x}_{N-1} \\ \tilde{x}_{p-1} & \tilde{x}_p & \cdots & \tilde{x}_N \\ \vdots & \vdots & \vdots & \vdots \\ \tilde{x}_1 & \tilde{x}_2 & \cdots & \tilde{x}_{N-p} \end{pmatrix}^T \tag{4-192}$$

According to (4-188), the residual error vector can be written as

$$\Theta = Y - X \cdot A \tag{4-193}$$

In order to minimize the error, an estimating vector $\hat{A} = (\hat{\alpha}_1, \hat{\alpha}_2, \ldots, \hat{\alpha}_p)^T$ can be calculated. The objective function is

$$J = \min\left(\sum_{k=p+1}^{N} \tilde{\varepsilon}_k^2\right) = \min\left(\Theta^T \cdot \Theta\right) \tag{4-194}$$

Substituting (4-193) into (4-194), it is obtained that

$$J = \min[(Y - X \cdot A)^T \cdot (Y - X \cdot A)] = \min(Y^T Y - A^T X^T Y - Y^T XA + A^T X^T XA) \tag{4-195}$$

To minimize J, let

$$\left.\frac{\partial J}{\partial A}\right|_{A=\hat{A}} = -2X^T Y + 2X^T X\hat{A} = 0 \tag{4-196}$$

Therefore, the estimating parameters are

$$\hat{A} = (X^T X)^{-1} X^T Y \tag{4-197}$$

If matrix $X^T X$ is of full rank, $\hat{A} = (\hat{\alpha}_1, \hat{\alpha}_2, \ldots, \hat{\alpha}_p)^T$ is existent and unique.

Furthermore, the variance can be estimated as

$$\hat{\sigma}^2 = \frac{1}{N-p} \sum_{k=p+1}^{N} \hat{\varepsilon}_k^2 = \frac{1}{N-p}(Y - X \cdot \hat{A})^T \cdot (Y - X \cdot \hat{A}) \qquad (4\text{-}198)$$

A Simple Example   To illustrate the use of least squares regression to fit the price data, a simple example is provided. In this case, the electricity prices of 4 p.m. of the first week's workdays (from August 3, 1998 to August 7, 1998) are selected. These prices are

$$162.81, 124.82, 152.21, 153.97, 54.99$$

AR(2) is employed to fit this price curve. According to (4-188) to (4-192),

$$p = 2$$
$$Y = (\tilde{x}_3, \tilde{x}_4, \tilde{x}_5)^T = (152.21, 153.97, 54.99)^T$$
$$X = \begin{pmatrix} \tilde{x}_2 & \tilde{x}_3 & \tilde{x}_4 \\ \tilde{x}_1 & \tilde{x}_2 & \tilde{x}_3 \end{pmatrix}^T = \begin{pmatrix} 124.82 & 152.21 & 153.97 \\ 162.81 & 124.82 & 152.21 \end{pmatrix}^T$$
$$A = (\alpha_1, \alpha_2)^T$$

From (4-197), the estimating parameters are

$$\hat{A} = (X^T X)^{-1} X^T Y$$
$$= \left[\begin{pmatrix} 124.82 & 152.21 & 153.97 \\ 162.81 & 124.82 & 152.21 \end{pmatrix} \cdot \begin{pmatrix} 124.82 & 152.21 & 153.97 \\ 162.81 & 124.82 & 152.21 \end{pmatrix}^T \right]^{-1} \cdot$$
$$\times \begin{pmatrix} 124.82 & 152.21 & 153.97 \\ 162.81 & 124.82 & 152.21 \end{pmatrix} \cdot (152.21, 153.97, 54.99)^T$$
$$= (0.26, 0.56)^T$$

Therefore, the AR(2) model is

$$\tilde{x}_k = 0.26\tilde{x}_{k-1} + 0.56\tilde{x}_{k-2} + \tilde{\varepsilon}_k, k = 3, 4, 5$$

Cases Analysis   To evaluate the ability to fit electricity prices, data from August 1 to August 22, 1998 in California (see Fig. 4-5) are employed. To obtain numerical values of errors with different p, compare the variance. Using the least squares regression method, the results are tabulated in Table 4-4. It confirms the fact that as the parameter p increases, the error decreases.

The actual price and the price estimated by AR(4) are compared in Fig. 4-16. The two curves are close.

Table 4-4   Comparison of Variances of AR Model

P	1	2	3	4
α_1	0.9694	1.3730	1.3500	1.3526
α_2	—	−0.4164	−0.3404	−0.3243
α_3	—	—	−0.0554	−0.1192
α_4	—	—	—	0.0472
Σ^2	156.2633	129.4114	129.2596	129.2125

ARX Model   To model the effect of demand on prices, as well as the autocorrelation of prices, the *autoregressive with exogenous variable* (ARX) model is a good option. An ARX model with distributed lags of electricity demand is described as (4-145). The parameters of ARX can be estimated based on the least squares regression method, which is similar to the method employed in AR model. In this case study, it is assumed that $n + 1 = m$ and $c = 0$.

To evaluate the ability of APX to fit electricity prices, same data as in Figure 4-5 (from August 1 to August 22, 1998 in California) are employed. To obtain numerical values of errors with different n and m, using the least squares regression method, the results are tabulated in Table 4-5.

The actual price and the price estimated by ARX with $n = 3$ and $m = 4$ are shown in Fig. 4-17.

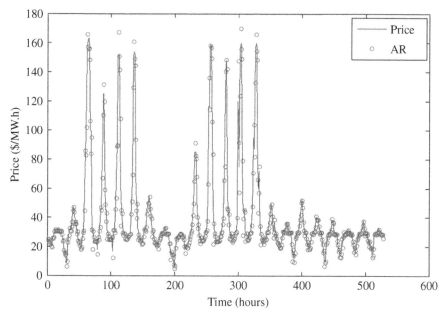

Figure 4-16   The curves of actual price versus AR(4).

Table 4-5   Comparison of Variances of ARX Model

m	1	2	3	4
α_1	0.8783	1.223	1.206	1.209
α_2	—	−0.3596	−0.2652	−0.2922
α_3	—	—	−0.0885	−0.0587
α_4	—	—	—	−0.0106
γ_0	0.2047	2.817	3.295	3.249
γ_1	—	−2.596	−3.976	−4.599
γ_2	—	—	0.9225	2.527
γ_3	—	—	—	−0.9308
σ^2	148.021	106.311	105.049	104.511

The results of Table 4-5 confirm the fact that as parameter p increases, the error decreases. Comparing errors of AR model and ARX model, ARX model gives a better fit.

Numerical Examples of ARMA Model

ARMA Model   As illustrated in Time Series Methods for Electricity Prices, combined with *moving average* (MA), an AR model forms a powerful tool—the ARMA model. In this model, MA regards time series as a moving average of a

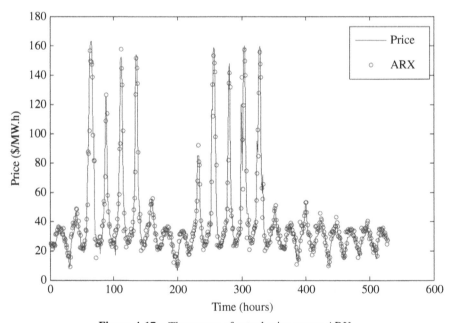

Figure 4-17   The curves of actual price versus ARX.

Table 4-6 Comparison of Variances of ARMA Model

p	1		2		8		4	
q	1	2	1	2	1	2	1	2
α_1	0.9846	0.9737	1.349	1.199	0.4506	2.523	2.249	1.535
α_2	—	—	−0.368	−0.2215	0.8322	−2.094	−1.499	0.1348
α_3	—	—	—	—	−0.3215	0.5822	0.1579	−1.015
α_4	—	—	—	—	—	—	0.09156	0.3448
β_1	0.2752	0.3458	−0.03244	0.1234	0.8628	−1.251	−0.9753	−0.2285
β_2	—	0.1229	—	0.0637	—	0.2702	—	−0.7326
σ^2	138.266	132.72	132.457	132.14	131.671	124.93	125.228	125.57

random shock series. Hence, ARMA expresses the current value of the time series in terms of its past values and in terms of previous values of the noise. The ARMA model of order (p,q), or ARMA(p,q), is written as

$$\tilde{S}_t = \sum_{i=1}^{p} \alpha_i \tilde{S}_{t-i} + \tilde{\varepsilon}_k + \sum_{i=1}^{q} \beta_i \tilde{\varepsilon}_{t-i} \tag{4-199}$$

where $\tilde{S}_t$ is the electricity price at time t, and $\tilde{\varepsilon}_t$ is error term. Many approaches [49, 50], such as least squares regression, recursive least squares regression, can be employed to estimate parameters. By these algorithms, the total error of the model can be minimized. The same set of data is used to evaluate the ability of ARMA to fit electricity prices. The results are tabulated in Table 4-6.

The data in Table 4-6 illustrate that in ARMA, comparing with the order of MA, the order of AR affect accuracy more clearly.

ARMAX Model   Similar to an ARX model, ARMA can be extended to ARMAX model to incorporate the effect of demand. The ARMAX(p, q, m) is written as

$$\tilde{S}_t = \sum_{i=1}^{p} \alpha_i \tilde{S}_{t-i} + \tilde{\varepsilon}_t + \sum_{i=1}^{q} \beta_i \tilde{\varepsilon}_{t-i} + \sum_{i=0}^{m-1} \gamma_i \tilde{d}_{t-i} \tag{4-200}$$

where $\tilde{S}_t$ is the electricity price at time t, $\tilde{d}_t$ is the demand, and $\tilde{\varepsilon}_t$ is error term. The same data set is used to evaluate the ability of ARMAX to fit electricity prices. The parameters are estimated by least squares regression. With $p = 4$ and $q = 2$, the results with different m are tabulated in Table 4-7 below.

Comparing with the results of AR, ARX, and ARMA model, ARMAX gives the best fit.

Table 4-7 Comparison of Variances of ARMAX Model

m	1	2	3	4
α_1	0.5418	1.013	1.679	2.821
α_2	0.8187	0.2257	−0.4479	−3.08
α_3	−0.626	−0.4406	−0.4655	1.442
α_4	0.00362	0.1073	0.224	−0.2094
β_1	0.6866	0.1847	−0.4719	−1.681
β_2	−0.2514	−0.2281	−0.412	0.9072
γ_0	0.4259	1.6	2.46	3.613
γ_1	—	−1.467	−4.002	−10.01
γ_2	—	—	1.557	+10.11
γ_3	—	—	—	−3.665
σ^2	112.331	110.963	106.626	101.501

CONCLUSIONS

The objective of this chapter is to summarize a set of stochastic process models that can be used for electricity prices according to the purpose of modeling. Significant features of electricity prices from various markets, such as mean reversion, seasonality, volatility, and spikes, are discussed. The nonstorable property of electricity, the need to constantly balance supply and demand, and available capabilities of the transmission network, constitute the physical constraints of electricity markets.

Continuous time stochastic models are widely used for modeling financial assets and derivatives. In this chapter, commonly used continuous time stochastic models such as Brownian motion, mean reversion process, GBM, GMR process, are described in detail. In these models, deterministic components account for regularities in the behavior of electricity prices. The stochastic components described the uncertainties involved in electricity prices. Among these models, the GMR process is well accepted for modeling of electricity prices, especially with the seasonal pattern of prices. With price-dependent volatility, GMR process is superior to other models. Moreover, the behavior of underlying electricity prices is nonlinear, implying the structural nonstationarity of fundamentals. This suggests that spot prices may be better modeled by a set of adaptive regime-switching models than from a single specification.

Time series models are used to understand the underlying context of data points, or to make predictions. Among these time series models, the stationary time series models, ARMA, is the foundation. For the system with covariance stationary property, ARMA class models can be used to model the data series well. These models can be viewed as a special class of linear stochastic difference equations. In electricity price modeling, familiar approaches include ARMA with time-varying parameters and ARMA with exogenous variables. Usually, time-varying parameters are used to describe the seasonal pattern of prices; historical and forecasted loads, fuel price, and time factors, served as exogenous variables. For the short-term price modeling, fluctuation of covariance is not prominent. Therefore, these models are good for price fitting and forecasting. However, the nonstationary property is an essential characteristic of electricity prices. ARCH and GARCH models are employed to deal with the issue. A distinctive feature of the models is that they recognize that volatilities and correlations are not constant. Similarly, these models can be extended with time-varying parameters and exogenous variables.

Modeling electricity prices has attracted the attention of a good number of researchers. Most of them address the short-term prices, that is, hourly prices in the next 24 hours or the next week. One or several methods described in this chapter can be used for short-term electricity prices models. However, due to the many sources of uncertainty that affect the future prices, and the lack of sufficient historical data, little work has been conducted to address the long-term price issues. Modeling of long-term prices remains a challenge, although it is a critical issue for investment decision and risk management.

ACKNOWLEDGMENT

This work was supported by the U.S. National Science Foundation under Grants ECS 0217701 and ECS 0424022.

REFERENCES

1. F.C. Schweppe, M.C. Caramanis, R.D. Tabors, and R.E. Bohn, *Spot Pricing of Electricity*, Kluwer Academic Publishers, Boston, MA, 1988.

2. D. Pilipovic, *Energy Risk: Valuing and Managing Energy Derivatives*, 2nd edition, The McGraw-Hill, New York, 2007.

3. M. Burger, B. Graeber, and G. Schindlmayr, *Managing Energy Risk: An Integrated View on Power and Other Energy Markets*, John Wiley & Sons, Ltd, West Sussex, England, 2007.

4. R. Weron, *Modeling and Forecasting Electricity Loads and Prices: A Statistical Approach*, John Wiley & Sons, Ltd, West Sussex, 2006.

5. M. Florio, "Electricity prices as signals for the evaluation of reforms: An empirical analysis of four European countries," *International Review of Applied Economics*, vol. 21, pp. 1–27, 2007.

6. S.M. Ross, *Stochastic Processes*, 2nd edition, John Wiley & Sons Inc., 1996.

7. G. Li, C.-C. Liu, C. Mattson, and J. Lawarrée, "Day-ahead electricity price forecasting in a grid environment," *IEEE Transactions on Power Systems*, vol. 22, pp. 266–274, 2007.

8. G. Li, C.-C. Liu, J. Lawarrée, M. Gallanti, and A. Venturini, "State-of-the-art of electricity price forecasting," presented at CIGRE/IEEE PES International Symposium, 2005.

9. R. Bjorgan, C.-C. Liu, and J. Lawarrée, "Financial risk management in a competitive electricity market," *IEEE Transactions on Power Systems*, vol. 14, pp. 1285–1291, 1999.

10. H. Song, C.-C. Liu, and J. Lawarrée, "Nash equilibrium bidding strategies in a bilateral electricity market," *IEEE Transactions on Power Systems*, vol. 17, pp. 73–79, 2002.

11. R. Dahlgren, C.-C. Liu, and J. Lawarrée, "Risk assessment in energy trading," *IEEE Transactions on Power Systems*, vol. 18, pp. 503–511, 2003.

12. I. Vehvilainen and J. Keppo, "Managing electricity market price risk," *European Journal of Operational Research*, vol. 145, pp. 136–147, 2003.

13. I. Vehvilainen, "Basics of electricity derivative pricing in competitive markets," *Applied Mathematical Finance*, vol. 9, pp. 45–60, 2002.

14. A.K. Dixit and R.S. Pindyck, *Investment Under Uncertainty*, Princeton University Press, Princeton, 1994.

15. Z. Xia, "Pricing and Risk Management in Competitive Electricity Markets," Ph.D. Thesis, School of Industrial and System Engineering, Georgia Institute of Technology, Atlanta, GA, 2005.

16. L. Chen, H. Suzuki, T. Wachi, and Y. Shimura, "Components of nodal prices for electric power systems," *IEEE Transactions on Power Systems*, vol. 17, pp. 41–49, 2002.

17. A. Boogert and D. Dupont, "When supply meets demand: The case of hourly spot electricity prices," *IEEE Transactions on Power Systems*, vol. 23, pp. 389–398, 2008.

18. A. Boogert and D. Dupont, "When supply meets demand: The case of hourly spot electricity prices," *School of Economics, Mathematics & Statistics, University of London Working paper, Birkbeck,* January 2007.

19. M.G. Lijesen, "The real-time price elasticity of electricity," *Energy Economics*, vol. 29, pp. 249–258, 2007.

20. D.S. Kirschen, G. Strbac, P. Cumperayot, and D. de Paiva Mendes, "Factoring the elasticity of demand in electricity prices," *IEEE Transactions on Power Systems*, vol. 15, pp. 612–617, 2000.

21. Y. Chang and C. Park, "Electricity market structure, electricity price, and its volatility," *Economics Letters*, vol. 95, pp. 192–197, 2007.

22. Y. Li and P.C. Flynn, "Electricity deregulation, spot price patterns and demand-side management," *Energy*, vol. 31, pp. 908–922, 2006.

23. I. Vehvilainen and T. Pyykkonen, "Stochastic factor model for electricity spot price - the case of the Nordic market," *Energy Economics*, vol. 27, pp. 351–367, 2005.

24. P.M. Schwarz, T.N. Taylor, M. Birmingham, and S.L. Dardan, "Industrial response to electricity real-time prices: Short run and long run," *Economic Inquiry*, vol. 40, pp. 597–610, 2002.

25. A. Boogert and D. Dupont, "The nature of supply side effects on electricity prices: The impact of water temperature," *Economics Letters*, vol. 88, pp. 121–125, 2005.

26. T. Ishikida and P.P. Varaiya, "Pricing of electric power under uncertainty: Information and efficiency," *IEEE Transactions on Power Systems*, vol. 10, pp. 884–890, 1995.

27. F. Black and M.S. Scholes, "The pricing of options and corporate liabilities," *Journal of Political Economy*, vol. 81, pp. 637–654, 1973.

28. J. Hull, *Options, Futures, and Other Derivatives*, 7th edition, Pearson, Upper Saddle River, NJ, 2008.

29. F.L. Alvarado and R. Rajaraman, "Understanding price volatility in electricity markets," Proceedings of the 33rd Annual Hawaii International Conference, 2000.

30. S. Deng, "Financial Methods in Competitive Electricity Markets," Ph.D. Thesis, Engineering-Industrial Engineering and Operations Research, University of California, Berkeley, CA, 1999.

31. J.J. Lucia and E. Schwartz, "Electricity prices and power derivatives: Evidence from the Nordic Power exchange," *Review of Derivatives Research*, vol. 5, pp. 5–50, 2002.

32. E.S. Schwartz, "The stochastic behavior of commodity prices: Implications for valuation and hedging," *Journal of Finance*, vol. 52, pp. 923–973, 1997.

33. B. Johnson and G. Barz, "Selecting stochastic processes for modeling electricity prices," in *Energy Modelling and the Management of Uncertainty*, E. Avril (Ed.) Risk Publications, London, 1999.

34. T. Kanamura and K. Ohashi, "A structural model for electricity prices with spikes: Measurement of spike risk and optimal policies for hydropower plant operation," *Energy Economics*, vol. 29, pp. 1010–1032, 2007.

35. C. Mari, "Regime-switching characterization of electricity prices dynamics," *Physica A: Statistical and Theoretical Physics*, vol. 371, pp. 552–564, 2006.

36. R. Weron, M. Bierbrauer, and S. Truck, "Modeling electricity prices: Jump diffusion and regime switching," *Physica A: Statistical and Theoretical Physics*, vol. 336, pp. 39–48, 2004.

37. A. Cartea and M.G. Figueroa, "Pricing in electricity markets: A mean reverting jump diffusion model with seasonality," *Applied Mathematical Finance*, vol. 12, pp. 313–335, 2005.

38. N.K. Nomikos and O. Soldatos, "Using affine jump diffusion models for modelling and pricing electricity derivatives," *Applied Mathematical Finance*, vol. 15, pp. 41–71, 2008.

39. M.T. Barlow, "A diffusion model for electricity prices," *Mathematical Finance*, vol. 12, pp. 287–298, 2002.

40. P. Villaplana, "Pricing power derivatives: A two-factor jump-diffusion approach," Departamento de Economía de la Empresa Working paper, Universidad Carlos III, January 2003.

41. H. Geman and A. Roncoroni, "Understanding the fine structure of electricity prices," *Journal of Business*, vol. 79, pp. 1225–1261, 2006.

42. C.R. Knittel and M.R. Roberts, "An empirical examination of restructured electricity prices," *Energy Economics*, vol. 27, pp. 791–817, 2005.

43. P. Skantze, M. Ilic, and J. Chapman, "Stochastic modeling of electric power prices in a multi-market environment," presented at Power Engineering Society Winter Meeting, 2000. IEEE.

44. F.E. Benth, J. Kallsen, and T. Meyer-Brandis, "A non-Gaussian Ornstein-Uhlenbeck process for electricity spot price modeling and derivatives pricing," *Applied Mathematical Finance*, vol. 14, pp. 153–169, 2007.

45. S.-J. Deng and W. Jiang, "Levy process-driven mean-reverting electricity price model: The marginal distribution analysis," *Decision Support Systems*, vol. 40, pp. 483–494, 2005.

46. R. Huisman and R. Mahieu, "Regime jumps in electricity prices," *Energy Economics*, vol. 25, pp. 425–434, 2003.

47. N. Haldrup and M.O. Nielsen, "A regime switching long memory model for electricity prices," *Journal of Econometrics*, vol. 135, pp. 349–376, 2006.

48. P. Kosater and K. Mosler, "Can Markov regime-switching models improve power-price forecasts? Evidence from German daily power prices," *Applied Energy*, vol. 83, pp. 943–958, 2006.

49. J.D. Hamilton, *Time Series Analysis*, Princeton University Press, Princeton, NJ, 1994.

50. P.J. Brockwell and R.A. Davis, *Introduction to time series and forecasting*, 2nd edition, Springer, New York, 2002.

51. R.C. Garcia, J. Contreras, M.V. Akkeren, and J.B.C. Garcia, "A GARCH forecasting model to predict day-ahead electricity prices," *IEEE Transactions on Power Systems*, vol. 20, pp. 867–874, 2005.

52. L. Hadsell and H.A. Shawky, "Electricity price volatility and the marginal cost of congestion: An empirical study of peak hours on the NYISO market, 2001-2004," *Energy Journal*, vol. 27, pp. 157–179, 2006.

53. A. Worthington, A. Kay-Spratley, and H. Higgs, "Transmission of prices and price volatility in Australian electricity spot markets: A multivariate GARCH analysis," *Energy Economics*, vol. 27, pp. 337–350, 2005.

54. K.F. Chan and P. Gray, "Using extreme value theory to measure value-at-risk for daily electricity spot prices," *International Journal of Forecasting*, vol. 22, pp. 283–300, 2006.

55. H.N.E. Bystrom, "Extreme value theory and extremely large electricity price changes," *International Review of Economics and Finance*, vol. 14, pp. 41–55, 2005.

56. H. Park, J.W. Mjelde, and D.A. Bessler, "Price dynamics among U.S. electricity spot markets," *Energy Economics*, vol. 28, pp. 81–101, 2006.

57. M. Olsson and L. Soder, "Modeling real-time balancing power market prices using combined SARIMA and Markov processes," *IEEE Transactions on Power Systems*, vol. 23, pp. 443–450, 2008.

58. D.J. Swider and C. Weber, "Extended ARMA models for estimating price developments on day-ahead electricity markets," *Electric Power Systems Research*, vol. 77, pp. 583–593, 2007.

59. A.J. Conejo, J. Contreras, R. Espinola, and M.A. Plazas, "Forecasting electricity prices for a day-ahead pool-based electric energy market," *International Journal of Forecasting*, vol. 21, pp. 435–462, 2005.

60. G. Guthrie and S. Videbeck, "Electricity spot price dynamics: Beyond financial models," *Energy Policy*, vol. 35, pp. 5614–5621, 2007.

61. A. Misiorek, S. Trueck, and R. Weron, "Point and interval forecasting of spot electricity prices: Linear vs. non-linear time series models," *Studies in Nonlinear Dynamics and Econometrics*, vol. 10, 2006.

62. B.R. Szkuta, L.A. Sanabria, and T.S. Dillon, "Electricity price short-term forecasting using artificial neural networks," *IEEE Transactions on Power Systems*, vol. 14, pp. 851–857, 1999.

63. J.-J. Guo. and P.B. Luh, "Selecting input factors for clusters of Gaussian radial basis function networks to improve market clearing price prediction," *IEEE Transactions on Power Systems*, vol. 18, pp. 665–672, 2003.

64. H.Y. Yamin, S.M. Shahidehpour, and Z. Li, "Adaptive short-term electricity price forecasting using artificial neural networks in the restructured power markets," *International Journal of Electrical Power and Energy Systems*, vol. 26, pp. 571–581, 2004.

65. A.I. Arciniegas and I.E.A. Rueda, "Forecasting short-term power prices in the Ontario electricity market (OEM) with a fuzzy logic based inference system," *Utilities Policy*, vol. 16, pp. 39–48, 2008.

66. R. Pino, J. Parreno, A. Gomez, and P. Priore, "Forecasting next-day price of electricity in the Spanish energy market using artificial neural networks," *Engineering Applications of Artificial Intelligence*, vol. 21, pp. 53–62, 2008.

67. J.P.S. Catalao, S.J.P.S. Mariano, V.M.F. Mendes, and L.A.F.M. Ferreira, "Short-term electricity prices forecasting in a competitive market: A neural network approach," *Electric Power Systems Research*, vol. 77, pp. 1297–1304, 2007.

68. P. Mandal, T. Senjyu, and T. Funabashi, "Neural networks approach to forecast several hour ahead electricity prices and loads in deregulated market," *Energy Conversion and Management*, vol. 47, pp. 2128–2142, 2006.

69. R. Gareta, L.M. Romeo, and A. Gil, "Forecasting of electricity prices with neural networks," *Energy Conversion and Management*, vol. 47, pp. 1770–1778, 2006.

70. Y.-Y. Hong and C.-F. Lee, "A neuro-fuzzy price forecasting approach in deregulated electricity markets," *Electric Power Systems Research*, vol. 73, pp. 151–157, 2005.

71. X. Lu, Z.Y. Dong, and X. Li, "Electricity market price spike forecast with data mining techniques," *Electric Power Systems Research*, vol. 73, pp. 19–29, 2005.

72. P. Mandal, T. Senjyu, N. Urasaki, T. Funabashi, and A.K. Srivastava, "A novel approach to forecast electricity price for PJM using neural network and similar days method," *IEEE Transactions on Power Systems*, vol. 22, pp. 2058–2065, 2007.

73. V. Vahidinasab, S. Jadid, and A. Kazemi, "Day-ahead price forecasting in restructured power systems using artificial neural networks," *Electric Power Systems Research*, vol. 78, pp. 1332–1342, 2008.

74. J.H. Zhao, Z.Y. Dong, Z. Xu, and K.P. Wong, "A statistical approach for interval forecasting of the electricity price," *IEEE Transactions on Power Systems*, vol. 23, pp. 267–276, 2008.

75. C. Gao, E. Bompard, R. Napoli, and H. Cheng, "Price forecast in the competitive electricity market by support vector machine," *Physica A: Statistical Mechanics and its Applications*, vol. 382, pp. 98–113, 2007.

76. C.-I. Kim, I.-K. Yu, and Y.H. Song, "Prediction of system marginal price of electricity using wavelet transform analysis," *Energy Conversion and Management*, vol. 43, pp. 1839–1851, 2002.

77. D.J. Pedregal and J.R. Trapero, "Electricity prices forecasting by automatic dynamic harmonic regression models," *Energy Conversion and Management*, vol. 48, pp. 1710–1719, 2007.

78. J. Hinz, "Modelling day-ahead electricity prices," *Applied Mathematical Finance*, vol. 10, pp. 149–161, 2003.

79. C.M. Ruibal and M. Mazumdar, "Forecasting the mean and the variance of electricity prices in deregulated markets," *IEEE Transactions on Power Systems*, vol. 23, pp. 25–32, 2008.

80. T.S. Genc and S. Sen, "An analysis of capacity and price trajectories for the Ontario electricity market using dynamic Nash equilibrium under uncertainty," *Energy Economics*, vol. 30, pp. 173–191, 2008.

81. S. Borenstein and J. Bushnell, "An empirical analysis of the potential market power in California's electricity industry," *Journal of Industrial Economics*, vol. 47, pp. 285–323, 1999.

82. H. Bessembinder and M.L. Lemmon, "Equilibrium pricing and optimal hedging in electricity forward markets," *The Journal of Finance*, vol. 57, pp. 1347–1382, 2002.

83. T. Mount, "Market power and price volatility in restructured markets for electricity," *Decision Support Systems*, vol. 30, pp. 311–325, 2001.

84. R. Weron and B. Przybylowicz, "Hurst analysis of electricity price dynamics," *Physica A: Statistical Mechanics and its Applications*, vol. 283, pp. 462–468, 2000.

85. D. Duffie, S. Gray, and P. Hoang, "Volatility in energy prices," in *Managing Energy Price Risks*, J. Robert (Ed.), 2nd edition. *Risk Books*, London, 1999.

86. B. Øksendal, *Stochastic Differential Equations*, 5th edition, Springer-Verlag, Berlin, 1998.

87. A. Etheridge, *A Course in Financial Calculus*, 1st edition, Cambridge University Press, Cambridge, 2002.

88. Y. Hou and F. Wu, "Probability Distribution Functions of Generator Profit from Spot Market," presented at 16th Power System Computation Conference, Glasgow, 2008.

89. E.L. Crow and K. Shimizu, *Lognormal Distributions: Theory and Applications*, Marcel Dekker Inc., New York, 1988.

90. K. Dowd, *Measuring Market Risk*, Wiley, Chichester, 2002.

91. R.T. Rockafellar and S. Uryasev, "Optimization of conditional value-at-risk," *Journal of Risk*, vol. 2, pp. 21–41, 2000.

92. J.C. Cuaresma, J. Hlouskova, S. Kossmeier, and M. Obersteiner, "Forecasting electricity spot-prices using linear univariate time-series models," *Applied Energy*, vol. 77, pp. 87–106, 2004.

93. S.R. Huang, "Short-term load forecasting using threshold autoregressive models," *IEE Proceedings on Generation, Transmission and Distribution*, vol. 144, pp. 477–481, 1997.

94. F.J. Nogales, J. Contreras, A.J. Conejo, and R. Espinola, "Forecasting next-day electricity prices by time series models," *IEEE Transactions on Power Systems*, vol. 17, pp. 342–348, 2002.

95. T.A. Robinson, "Electricity pool prices: A case study in nonlinear time-series modelling," *Applied Economics*, vol. 32, pp. 527–532, 2000.

96. J. Nowicka-Zagrajek and R. Weron, "Modeling electricity loads in California: ARMA models with hyperbolic noise" *Signal Processing*, vol. 82, pp. 1903–1915 2002.

97. J. Contreras, R. Espinola, F.J. Nogales, and A.J. Conejo, "ARIMA models to predict next-day electricity prices," *IEEE Transactions on Power Systems*, vol. 18, pp. 1014–1020, 2003.

98. M. Zhou, Z. Yan, Y. Ni, and G. Li, "An ARIMA approach to forecasting electricity price with accuracy improvement by predicted errors," presented at Power Engineering Society General Meeting, 2004. IEEE.

99. M. Zhou, Z. Yan, Y.X. Ni, G. Li, and Y. Nie, "Electricity price forecasting with confidence-interval estimation through an extended ARIMA approach," *IEE Proceedings on Generation, Transmission and Distribution*, vol. 153, pp. 187–195, 2006.

100. A.J. Conejo, M.A. Plazas, R. Espinola, and A.B. Molina, "Day-ahead electricity price forecasting using the wavelet transform and ARIMA models," *IEEE Transactions on Power Systems*, vol. 20, pp. 1035–1042, 2005.

101. N. Karakatsani and D.W. Bunn, "Modeling the volatility of spot electricity prices," *EMG Working Paper, London Business School*, 2004.

102. H.S. Guirguis and F.A. Felder, "Further advances in forecasting day-ahead electricity prices using time series models," *KIEE International Transactions on PE*, vol. 4-A, pp. 159–166, 2004.

103. L. Hadsell, A. Marathe, and H.A. Shawky, "Estimating the volatility of wholesale electricity spot prices in the US," *Energy Journal*, vol. 25, pp. 23–40, 2004.

Chapter 5

Short-Term Forecasting of Electricity Prices Using Mixed Models

Carolina García-Martos, Julio Rodríguez, and María Jesús Sánchez

INTRODUCTION AND PROBLEM STATEMENT

Until recently, only demand was predicted in centralized markets, but trading rules had changed and electricity is presently traded under competitive rules in the same way as any other commodity. Electricity presents some characteristics that make it different, since it cannot be stored and any unmet demand is simply lost. These special features are responsible for the extremely volatile and largely unpredictable behavior of electricity pricing. Bearing this in mind, specific tools must be developed for electricity price forecasting.

Presently, in competitive markets, there are several ways to trade electricity and different problems associated with them that must be mentioned:

1. Forward markets and options, which are well developed in some electricity markets like the EEX in Germany.

2. In the *pool*, both the generating companies and the consumers submit to the market operator, their respective generation and consumption bids for each hour of the next day. In the Spanish market, once the market operator has sorted out the bidding prices for generation or consumption bids, respectively, the marginal price is defined as the bid submitted by the last generation unit needed to satisfy the whole demand. This mechanism results in what is also known as the market clearing price and is shown in Fig. 5-1.

Advances in Electric Power and Energy Systems: Load and Price Forecasting, First Edition.
Edited by Mohamed E. El-Hawary.
© 2017 by The Institute of Electrical and Electronics Engineers, Inc. Published 2017 by John Wiley & Sons, Inc.

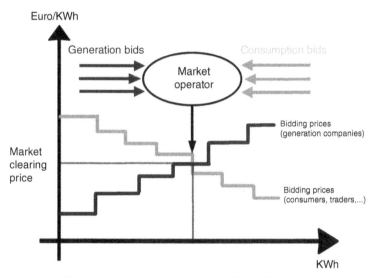

Figure 5-1 Market clearing price, Spanish market.

Accurate short-term forecasts help the producers schedule the production of their units to maximize profit. The producers assume certain risks, and the more accurate the forecasts are, the lower the risk. A generating company can better decide its bidding price when having accurate 1-day-ahead forecasts. A powerful tool for short-term forecasting is the basis on which every bidding-rule stands [1, 2]. Bearing this in mind, the availability of adequate models to predict next-day electricity prices is of great interest.

3. As far as bilateral contracts are concerned, an interesting question is how to reduce the risks that they involve. This can be done by forecasting electricity prices with a horizon that covers, at least, the duration of the contract, usually 1 year, which means long-term forecasting.

In the current environment, forecasting electricity prices (both in the short and long term) has become a necessary function for all market-participants (power generators, sellers, and consumers). This chapter is focused on short-term forecasting (forecasting horizon 24 hours). We offer a simple but accurate method and an automatic tool to compute 1-day-ahead forecasts of electricity prices.

We propose two new models, each of which deals with the 24-hourly time series of electricity prices instead of a complete one. We also determine the appropriate length of the time series used to build the forecasting autoregressive integrated moving average (ARIMA) models. We report on a computational experiment to determine the combination of the variables "Model" and "Length" with the best "global perfor-mance" for a representative span of years (1998–2003). This period has been chosen to allow comparing our results with the others presented in previous well-known works [3, 4]. This allows us to ensure that the mixed model finally selected does not only have a good performance for some selected weeks, but works properly almost always.

This chapter is organized as follows. In State of the Art, some previous work on short-term forecasting of electricity prices is presented. In Time Series Analysis and ARIMA Models, the theoretical basics of time series analysis are reviewed, since they are used throughout the chapter and the mixed models developed are based on time series models. In Development of Mixed Models and its Computational Implementation, the development of our model and its computational implementation are described. In Analysis of Forecasting Errors, a study of forecasting errors is carried out from two different points of view: descriptive and applying LOESS. In Design of Experiments, the fundamentals of the design of experiments (DOE) are presented as well as its application to develop mixed models, and selecting the preferred combination of "Model" and "Length" in terms of prediction error. Numerical Results, presents numerical results for the Spanish market and a comparison with other well-known work. Some conclusions are given at the end.

STATE OF THE ART

While electricity prices forecasting is relatively new, competitive markets have opened a new field of study and research. In the last years, a great number of publications dealing with the subject have appeared.

Various forecasting techniques such as ARIMA models, transfer function models, neural networks, spectral analysis, or multivariate dynamic models have been applied to modeling and forecasting prices. On the other hand, conditional heteroskedastic models such as auto-regressive conditional heteroscedasticity (ARCH), generalized auto-regressive conditional heteroskedasticity (GARCH), exponential general auto-regressive conditional heteroskedasticity (EGARCH), and stochastic volatility models have been applied to model volatilities. Bearing this in mind, we focus in this section, summarizing previous works on neural networks and time series models which are widely used in modeling and forecasting.

Considering the application of neural networks, Ramsay and Wang [5] treated the South England–Wales electricity market and proposed a hybrid technique that combines neural networks and fuzzy logic to forecast daily prices (instead of hourly forecasts which reduces variability), obtaining average prediction errors of about 10%. Rodríguez and Anders [6] applied a procedure similar to Ramsay and Wang [5], but it was to obtain hourly forecasts and study the influence of different inputs (such as demand, capacity shortage) on errors. They obtained errors of about 25%, but for a span of years in which the appearance of peak prices is really high. They were not able to compare their results to those in Ramsay and Wang [5], in which the number of outliers is clearly lower. Szkuta et al. [7] applied three-layered artificial neural networks with back-propagation to the case of the Victorian Market, obtaining daily prediction errors of around 15%. In addition, using neural networks for forecasting electricity prices is of additional interest, given that they are able to approximate nonlinear functions, and this can be used to process nonlinearities and obtain a more realistic output. Nicolaisen et al. [8] applied Fourier and Hartley transforms as linear filters for forecasting electricity prices.

Despite neural networks being applicable for computing electricity prices, the need for a training process presents a challenge, and the fact that they tend to over-adjust. Neural networks are extremely useful in predicting variables whose characteristics are very similar to those used during the training period, but when computing forecasts for a period during which the behavior of prices differs from that of the training period, forecasting errors increase.

On the other hand, time series, ARIMA models have been applied with great success for forecasting not only the electricity prices but also other commodities. Time series models had been successfully applied for 1-day-ahead forecasting of electricity prices. Moreover, three main categories can be considered based on the time series analysis: ARIMA models, dynamic regression, and transfer function models. All of these models are based on the Box–Jenkins methodology, but there are some differences concerning the way in which they relate prices and errors. An ARIMA model relates prices at time t with those in the past. Dynamic regression relates prices and explanatory variables at time t with both past values of prices and the explanatory variables. Transfer function models relate price at time t to past prices, explanatory variables, and innovations.

Nogales et al. [9] applied dynamic regression and transfer function models as a possible solution for the problem of correlations between errors when applying linear regression between the load and the price of electricity. In addition, they provide comparisons between numerical results obtained from both methodologies. By means of a dynamic regression model, they related electricity price at time t to past prices and demand, using the following model:

$$p_t = c + w_d(B)d_t + w_p(B)p_t + \varepsilon_t,$$

where p_t is the price at time t, $w_d(B) = 1 - w_1 B - \cdots - w_p B^p$ and $w_p(B) = 1 - w_1' B - \cdots - w_{p'}' B^{p'}$ are polynomials of degree p and p', respectively, B is the unit delay operator such that $Bd_t = d_{t-1}$, ε_t are the innovations, which are uncorrelated random shocks.

The general equation in the case of transfer functions of the model is

$$p_t = c + w_d(B)d_t + \varepsilon_t.$$

For the mainland Spain market, they obtain average prediction errors at around 5%, for some selected weeks in 2000, a period in which there is a lower proportion of outliers, which definitely influences obtaining small forecasting errors. In addition, these models include the true value of the load as an explanatory variable. Since the true value of the load is unknown for the same period of time, we must use a forecast and this may add further errors in the price forecasts.

Troncoso et al. [10] compared two single-step-ahead forecasting methods for the Spanish market. The first one is similar to k weighted nearest neighbors (kWNN), and they compute price forecasts using a linear combination of the "nearest neighbors". They also provide results obtained by using dynamic regression. For both methods, they used data corresponding to the period January–August 2001. Moreover, January and February are used to calculate the parameters of the models, which are updated daily. They considered only working day data, and noted the different behavior of prices depending on working days or weekends. Prediction errors of about 11.4% are obtained by means of kWNN for the forecasts computed for the period March–May 2001, and 9.3% for the period June–August. Using dynamic regression models, the errors in these periods are 10.1% and 7.5%, respectively.

Contreras et al. [3] produced 1-day-ahead forecasts using multiplicative seasonal ARIMA models with daily and weekly seasonality, computing forecasts for several weeks during 2000. They modeled not only Spanish prices but also Californian ones. Mean week errors (MWE) ranges from 5% to 20%.

Crespo-Cuaresma et al. [11] proposed a set of univariate models for forecasting electricity prices for the Leipzig spot market which is considered to be the most important in Germany. Historical data used to produce a forecast is about two and half months of that being forecasted. Different models are provided, including those modeling the complete time series and others that formulate different models for each hour. The higher accuracy is obtained with separate models depending on the hour being measured. Numerical results indicate that the strategy of building separate models for each hour does significantly reduce prediction errors. Those vary (depending on the model and whether the complete time series is split or not) from 11% (considering 24 dynamic processes) up to 33%. This shows the higher accuracy obtained when considering 24-hourly time series, since it reduces forecasting errors, and it seems that considering different models for the hourly time series is a viable and interesting alternative approach.

Conejo et al. [12] analyzed and compared different methodologies for forecasting prices for the subsequent 24-hourly prices for the PJM interconnection (a regional transmission organization in the United States). The time series models look like the best alternative among others such as neural networks or pre-filtering using wavelet transform. Models based on time series, dynamic regression, and transfer functions seem to produce more accurate forecasts. The true demand is used instead its forecast, which is not correct. Forecasting errors obtained are around 10%. Using ARIMA models, prediction errors are similar to those calculated from forecasts calculated using wavelet transform (ranging from 6.6% to 24%). Applying neural networks produces higher errors.

Nogales et al. [9] applied transfer function models to one-step-ahead forecasting of electricity prices in the PJM interconnection in 2003. They used a transfer function between price and demand and offered a detailed explanation of the process of transfer function building and parameter estimation. They evaluated the results when the load is included and not included. Incorporating the load in the model reduces the error but not significantly. This can be explained taking into account the instantaneous relationship between prices and load. From the numerical results obtained, prediction errors for the period July–August 2003 are around 11% and 13.3% without considering demand as an explanatory variable. Moreover, they compared the results with those based on some models used as benchmark:

1. A simple model that forecasts each hourly price for the next day to be the same as that of the corresponding hour of the previous day. Forecasting error is 16%.

2. Exponential smoothing considering daily and weekly seasonality. The errors obtained with this model are around 17%.

Researches reviewed so far deal with short-term forecasting of electricity prices, which is useful for scheduling power generation units. On the other hand, medium- and long-term forecasting of electricity prices as well as load forecasting are also of interest. Long-term forecasting is more difficult and not commonly addressed, but accurate one-year-ahead forecasts of electricity prices are necessary for reducing the risk implied in bilateral contracts trading. A recent work on long-term forecasting of electricity prices is given by Alonso et al. [13].

For load forecasting, Cottet and Smith [14] proposed a vector auto-regressive (VAR) model. Load forecasting is easier and in general, prediction errors are lower as the load time series are less volatile. In this work, the correlation structure is carefully analyzed and an adequate set of regressors is selected using Bayesian methodology. They proposed a detailed model that considers past 3 years' prices, so the trend, seasonality, and influence of the temperature and cloud cover, as well as the correlation between the components varies depending on the hour of the day. This justifies disaggregating in 48-hourly time series (since this model was applied to the New South Wales market, in which there is a market clearing price and load data every half an hour). In addition, considering working days and weekends separately reduces the prediction errors. When considering a 6-month forecasting horizon, prediction errors are even lower than 5%. Including the temperature as an explanatory variable does incorporate a lot of information. However, including demand as an explanatory variable for prices does not reduce the error significantly (see [9]).

Vehviläinen and Pyykkönen [15] presented a model based on stochastic factors for calculating medium-term forecasts of electricity prices. Stochastic factors affecting spot prices are modeled separately and then a model for the equilibrium of the market is considered for combining them adequately and then obtaining a forecast of the price. The main advantage is that all the factors influencing the price can be modeled and studied in detail. This work provides numerical results for Nordpool. When forecasting for the year 2001, using data from years 1996–2000, the average prediction error obtained is about 30% when no updating explanatory variables. When doing this, the results improved but the forecasting horizon is reduced to 1 month.

Models Presented in this Chapter

Once we covered some important references on short-term forecasting of electricity prices, both in the field of neural networks and time series, we proceed to justify the application of time series methodology.

Neural networks tend to over-adjust, and some training is needed. The models perform extremely well when the forecast variable has a very similar behavior both in the forecasting period and that used in training. Given the features of electricity prices (high proportion of outliers) and the special ones in the Spanish market (lower level of competition making prices even less predictable), neural networks do not seem to be the preferred forecasting technique.

Moreover, considering the time series analysis literature dealing with electricity markets, it appears that including demand as an explanatory variable in transfer function models or dynamic regression does not reduce prediction errors significantly. This is because price and load are instantaneously related. On the other hand, considering the parallel approach, in which different models are fitted for each hourly time series, or different kind of days (working days or weekends), (see [11, 14, 16]) produces smaller prediction errors.

Moreover, for short-term forecasting, several lengths have been considered by many authors for the time series used to build the models used to forecast. It would be of interest to determine whether there is a significant influence of this factor on the accuracy of the forecasts. It would be of interest to determine the best length in terms of prediction errors. In this work different lengths are considered, ranging from 8 to 80 weeks.

Moreover, not only do we offer some numerical results for selected weeks but we present an extensive analysis of forecasting errors computed for all the hours in the period 1998–2003, considering all the possible combinations of "Model" and "Length". The final mixed model selected among all combinations has been obtained by means of a computational experiment considering the two factors studied.

TIME SERIES ANALYSIS AND ARIMA MODELS

A time series is a realization of a stochastic process and is the result of observing the values of a variable over time during regular intervals (every day, every month, every year, etc.).

ARIMA processes are a class of stochastic processes used to model and forecast time series. The application of the ARIMA methodology for the study of time series analysis is due to Box and Jenkins [17].

Introduction to Time Series Analysis

The dynamic phenomena observed in a time series can be grouped into two classes:

- Those that take stable values in time about a constant level, without showing a long-term increasing or decreasing trend. For example, the yearly rainfall in a region, the average yearly temperatures or the proportion of births corresponding to males. These processes are called stationary.

- Nonstationary processes are those that can show trend, seasonality and other evolutionary effects over time. For example, energy demand series.

A stochastic process is a set of random variables Z_t, where the index t takes values in a certain set C. In this case, this set is ordered and corresponds to instants of time (days, months, years, etc.).

For each value of t in the set C (for each point in time), a random variable, Z_t, is defined and the observed values of the random variables at different times form a time series. That is, a series of T data, $(Z_1, \ldots, Z_T)$, is a sample of size one of the vector of T random variables ordered sequentially in time corresponding to the time instants $t = 1, \ldots, T$, and the observed series is considered a result or trajectory of the stochastic process.

The mean function of the process refers to a function of time representing the expected value of the marginal distributions Z_t for each time instant t: $E[z_t] = \mu_t$. An important particular case arises when the mean function is a constant. The realizations of the process show no trend and we say that the process is stable in the mean $E[z_t] = \mu, \forall t$.

On many occasions, only one realization of the stochastic process is available and the decision about the mean function of the process is (constant or not over time) must be obtained based on this information. Figure 5-2 shows a nonstationary process in mean as well as a stationary one.

The variance function of the process gives the variance at each point in time: $\text{Var}(z_t) = \sigma_t^2$. The process is stable in variance if the variability is constant over time.

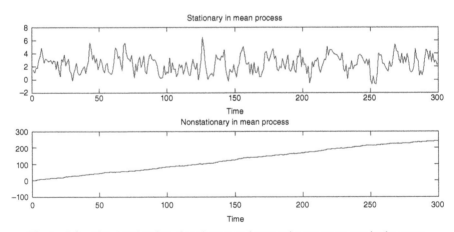

Figure 5-2   Simulated series of stationary and nonstationary processes in the mean.

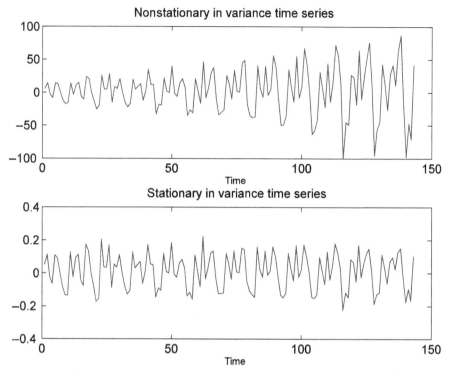

Figure 5-3   Stationary and nonstationary in variance processes.

Figure 5-3 shows stationary and nonstationary in the variance processes. In the first, the variance increases with time.

Strict stationarity is a very strong condition, since to prove it, we need the joint distributions for any set of variables in the process. A weaker property, but one which is easier to prove, is weak sense stationarity. A process is stationary in the weak sense if, for all t:

1. $\mu_t = \mu = cte$,
2. $\sigma_t^2 = \sigma^2 = cte$
3. $\gamma(t, t - k) = E[(z_t - \mu)(z_{t-k} - \mu)] = \gamma_k, k = 0, \pm 1, \pm 2, \ldots$

The first two conditions indicate that the mean and variance are constant. The third indicates that the covariance between two variables depends only on their separation, k.

As a result, in a stationary process: $\text{Cov}(z_t, z_{t+k}) = \text{Cov}(z_{t+j}, z_{t+k+j})$, $j = 0, \pm 1$, $\pm 2 \ldots$ and for autocorrelations as well, $\rho_k = \dfrac{\text{Cov}(z_t, z_{t-k})}{\sqrt{\text{Var}(z_t)\text{Var}(z_{t-k})}}$.

To summarize, in stationary processes we have that $\gamma_0 = \sigma^2, \gamma_k = \gamma_{-k}$ and for the autocorrelations, $\rho_k = \rho_{-k}$.

We use the term covariance matrix of the stationary process of order k, Γ_k, for the square and symmetric matrix of order k that has the variances in its principal diagonal and the autocovariances in the off-diagonals:

$$
\Gamma_k = \begin{pmatrix}
\gamma_0 & \gamma_1 & \cdots & \gamma_{k-1} \\
\gamma_1 & \gamma_0 & \cdots & \gamma_{k-2} \\
\vdots & \vdots & \ddots & \vdots \\
\gamma_{k-1} & \gamma_{k-2} & \cdots & \gamma_0
\end{pmatrix}.
$$

We use the term autocorrelation function (ACF) to refer to the representation of the autocorrelation coefficients of the process as a function of the lag and the term autocorrelation matrix for the square and symmetric Toeplitz matrix with ones in the diagonal and the autocorrelation coefficients in the off-diagonal:

$$
R_k = \begin{pmatrix}
1 & \rho_1 & \cdots & \rho_{k-1} \\
\rho_1 & 1 & \cdots & \rho_{k-2} \\
\vdots & \vdots & \ddots & \vdots \\
\rho_{k-1} & \rho_{k-2} & \cdots & 1
\end{pmatrix}.
$$

From this point on, for simplicity's sake, we will use the term stationary process to refer to a stationary process in the weak sense.

An estimator of the population autocovariance is $\hat{\gamma}_k = \frac{1}{T} \sum_{t=k+1}^{T} (z_t - \bar{z})(z_{t-k} - \bar{z})$.

Using the estimator $\hat{\gamma}_k$, the sample autocovariance matrix $\hat{\Gamma}_k$ is always non-negative definite.

$$
\hat{\Gamma}_k = \begin{pmatrix}
\hat{\gamma}_0 & \hat{\gamma}_1 & \cdots & \hat{\gamma}_{k-1} \\
\hat{\gamma}_1 & \hat{\gamma}_0 & \cdots & \hat{\gamma}_{k-2} \\
\vdots & \vdots & \ddots & \vdots \\
\hat{\gamma}_{k-1} & \hat{\gamma}_{k-2} & \cdots & \hat{\gamma}_0
\end{pmatrix}.
$$

Then, the autocorrelations are estimated as $r_k = \hat{\gamma}_k / \hat{\gamma}_0$.

Figure 5-4 shows Series A in Box and Jenkins [17] and its sample ACF. The data in series A are chemical process concentration readings, and the data are collected every 2 hours.

Autoregressive (AR) Processes

The study of models for stationary processes starts with those which are useful in representing the dependency of the values of a time series on its past. The simplest family of these models is the autoregressive (AR), which generalize the idea of

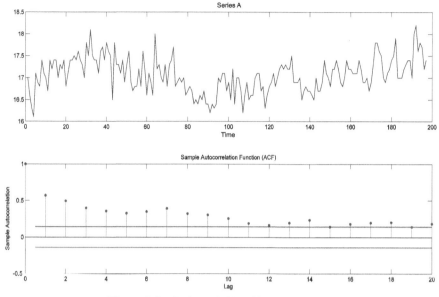

Figure 5-4   Series A [17] and its sample ACF.

regression to represent the linear dependence between a dependent variable y (z_t) and an explanatory variable x (z_{t-1}), using the relationship:

$$z_t = c + b z_{t-1} + a_t, \tag{5-1}$$

where c and b are constants to be determined and a_t are i.i.d. $N(0, \sigma^2)$. Equation (5-1) defines the first order autoregressive process (AR(1)). This linear dependence can be generalized so that the present value of the series, z_t, depends not only on z_{t-1}, but also on the previous p lags, $z_{t-2}, \dots, z_{t-p}$. Thus, an autoregressive process of order p is obtained.

A series z_t follows an AR(1), if it has been generated by

$$z_t = c + \phi \cdot z_{t-1} + a_t, \tag{5-2}$$

where c and $-1 < \phi < 1$ (for the stationarity of the process), are constants and a_t is a white noise process with variance σ^2. The variables a_t, which represent the new information that is added to the process at each instant, are known as innovations. If $|\phi| > 1$, then (5-2) is an explosive process and the values of the variable increases with no limit. If $\phi = 1$, (5-2) is a random walk.

An AR(1) process can be written using the lag operator B, such that $Bz_t = z_{t-1}$. Letting $\tilde{z}_t = z_t - \mu$ and since $B\tilde{z}_t = \tilde{z}_{t-1}$, we have $(1 - \phi B)\tilde{z}_t = a_t$. The operator $(1 - \phi B)$ can be considered as a filter that transforms a series into a white noise process.

The covariance between observations separated by k periods, or the autocovariance of order k for these processes are obtained: $\gamma_k = E[(z_{t-k} - \mu)(z_t - \mu)] =$

$E[\tilde{z}_{t-k}(\tilde{z}_{t-1} + a_t)]$, and since $E[\tilde{z}_{t-k}a_t] = 0$, since the innovations are uncorrelated with the past values of the series, we have the following recursion: $\gamma_k = \phi\gamma_{k-1}, k = 1, 2, \ldots,$ where $\gamma_0 = \sigma^2$. This equation shows that since $|\phi| < 1$ the dependence between observations decreases when the lag increases.

The autocorrelations contain the same information as the autocovariances, with the advantage of not depending on the units of measurement. From here on, we will use the term simple ACF to denote the ACF of the process in order to distinguish it from other functions linked to the autocorrelation that are defined at the end of this section.

Let k be the autocorrelation of order k, defined by $\rho_k = \phi\gamma_{k-1}/\gamma_0 = \phi\rho_{k-1}$. Since $\rho_1 = \phi$, we conclude that $\rho_k = \phi^k$, and when k is large, ρ_k goes to zero at a rate that depends on the value of ϕ. Figure 5-5 shows an AR(1) process whose parameter is $\phi = 0.5$ and -0.5, respectively, and their ACF.

In general, we say that a stationary time series z_t follows an autoregressive process of order p, AR(p), if $\tilde{z}_t = \phi_1\tilde{z}_{t-1} + \cdots + \phi_p\tilde{z}_{t-p} + a_t$. And using the lag operator, it can be written like this: $\phi(B)\tilde{z}_t = (1 - \phi_1 - \cdots - \phi_p)\tilde{z}_t = a_t$. The characteristic equation of the process, $\phi(B) = \sum_{i=1}^{P}(1 - G_iB) = 0$, and if $|G_i| < 1, \forall i = 1, \ldots, p$, then the process is stationary. The ACF of an AR(p) process is a mixture of exponentials, due to the terms with real roots, and sinusoids, due to the complex conjugates. As a result, their structure can be very complex. Because of this reason, determining the order of an autoregressive process from its ACF is difficult. To solve this problem, the partial ACF is introduced. In general, an AR(p) has direct effects on observations

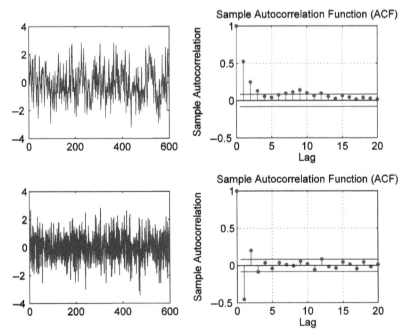

Figure 5-5   Simulated AR(1) processes and their ACF.

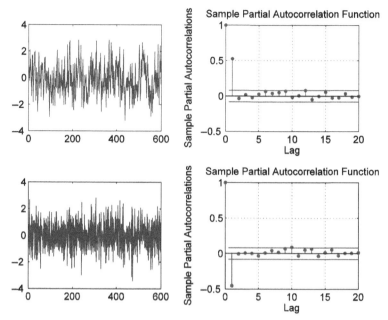

Figure 5-6 AR(1) processes and their PACF.

separated by $1, 2, \dots, p$ lags and the direct effects of the observations separated by more than p lags are zero. The partial autocorrelation coefficient of order k, denoted by ρ_k^p is defined as the correlation coefficient between observations separated by k periods, when the linear dependence due to intermediate values is removed. From this definition, it is clear that an AR(p) process will have the first p nonzero partial autocorrelation coefficients and, therefore, in the partial autocorrelation function (PACF), the number of nonzero coefficients indicates the order of the AR process.

This property will be a key element in identifying the order of an autoregressive process. Furthermore, the partial correlation coefficient of order p always coincides with the parameter ϕ_p. Figure 5-6 shows PACF for AR(1) processes.

Moving Average (MA) Processes

The autoregressive processes have, in general, infinite nonzero autocorrelation coefficients that decay with the lag. The AR processes have a relatively "long" memory, since the current value of a series is correlated with all previous ones, although with decreasing coefficients.

This property means that we can write an AR process as a linear function of all its innovations, with weights that tend to zero with the lag. The AR processes cannot represent short memory series, where the current value of the series is only correlated with a small number of previous values.

A family of processes that have this "very short memory" property are the moving average (MA) processes. The MA processes are a function of a finite, and generally small, number of its past innovations.

A first order moving average, MA(1), is defined by a linear combination of the last two innovations, according to the equation $\tilde{z}_t = a_t - \theta a_{t-1}$, where $\tilde{z}_t = z_t - \mu$, with μ being the mean of the process and a_t a white noise process with variance σ^2. The MA(1) process can be written with the operator notation: $\tilde{z}_t = (1 - \theta B)a_t$. This process is the sum of the two stationary processes, a_t and a_{t-1} and, therefore, will always be stationary for any value of the parameter, unlike the AR processes.

In these processes, we will assume that $|\theta| < 1$, so that the past innovation has less weight than the present. Then, we say that the process is invertible and has the property whereby the effect of past values of the series decreases with time.

Thus, since $|\theta| < 1$, there exists an inverse operator $(1 - \theta B)^{-1}$ and we can write the equation of the MA(1) process as

$$(1 + \theta B + \theta^2 B^2 + \cdots)\tilde{z}_t = a_t. \tag{5-3}$$

Equation (5-3) represents a moving average process of order 1, like an autoregressive process of infinite order, AR (∞), with coefficients that decay in geometric progression.

It can be shown [18] that the ACF of an MA(1) process has the same properties as the PACF of an AR(1) process, that is, there is a first coefficient different from zero and the rest are zero.

This duality between the AR(1) and the MA(1) is also seen in PACF. Therefore, the PACF has all nonzero coefficients and they decay geometrically with k.

Generalizing on the idea of an MA(1), we can write processes whose current value depends not only on the last innovation but also on the last q innovations. Thus, the MA(q) process is obtained, with general representation:

$$\tilde{z}_t = \left(1 - \theta_1 B - \theta_2 B^2 - \cdots - \theta_q B^q\right) a_t = \theta(B)a_t. \tag{5-4}$$

The process is invertible if the roots of the operator $\theta(B) = 0$ are, in modulus, greater than the unit. To compute the PACF of an MA(q), Eq. (5-4) is expressed as an autoregressive process of infinite order, $\theta^{-1}(B)\tilde{z}_t = \pi(B)\tilde{z}_t = a_t$. When $\theta(B)$ is invertible, then $\pi(B)$ converges.

Then, the PACF of an MA is nonzero for all lags, since a direct effect of $\tilde{z}_{t-i}$ on $\tilde{z}_t$ exists for all i. The PACF of an MA process thus has the same structure as the ACF of an AR process of the same order. Thus there is a duality between the AR and MA processes such that the PACF of an MA(q) has the structure of the ACF of an AR(q) and the ACF of an MA(q) has the structure of the PACF of an AR(q).

Figure 5-7 shows a MA(1) process and its ACF and PACF.

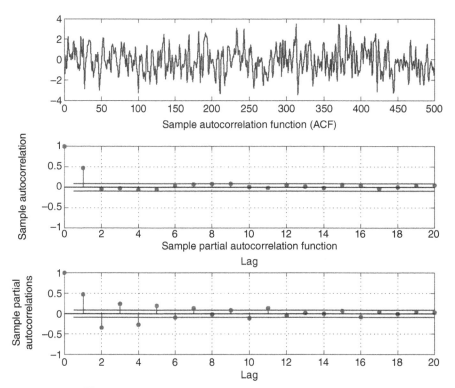

Figure 5-7   ACF and PACF of an MA(1) simulated process.

ARMA Processes

The autoregressive moving average (ARMA) processes combine the properties of AR and MA processes and allow us to represent using few parameters, those processes whose first q coefficients can be any, whereas subsequent ones decay according to simple rules. ARMA processes give us a very broad and flexible family of stationary stochastic processes useful in representing many practical time series.

The simplest process of these, is the ARMA(1,1), with the following expression:

$$(1 - \phi_1 B)\tilde{z}_t = (1 - \theta_1 B)a_t,$$

where $|\phi_1| < 1$ and $|\theta_1| < 1$ for the process being stationary and invertible, respectively. In addition, it is assumed that $\phi_1 \neq \theta_1$, if not, the process would be white noise, so it is always assumed that there are no common roots between the autoregressive and moving average components.

When both coefficients ϕ_1 and θ_1 are positive and $\phi_1 > \theta_1$, it can be proved that the correlation increases with $\phi_1 - \theta_1$ (see [18]).

Table 5-1   ACF and PACF, AR(p), MA(q), and ARMA(p, q)

	ACF	PACF
AR(p)	Many nonzero coefficients	0 except for the first p ones
MA(q)	0 except for the first q ones	Many nonzero coefficients
ARMA(p, q)	Many nonzero coefficients	Many nonzero coefficients

On the other hand, to calculate the PACF, the model is written in its AR(∞) form, $(1 - \phi_1 B)(1 - \theta_1 B)^{-1}\tilde{z}_t = a_t$.

The direct effect of $\tilde{z}_{t-k}$ on $\tilde{z}_t$ decays geometrically with θ_1^k and, therefore, the PACF will have a geometric decay starting from an initial value.

In an ARMA(1,1) process, the ACF and the PACF have a similar structure: an initial value, whose magnitude depends on $\phi_1 - \theta_1$, followed by geometric decay. The rate of decay in the ACF depends on ϕ_1, whereas in the PACF, it depends on θ_1.

Generalizing the idea of the ARMA(1,1), the ARMA(p, q) process is defined by the equation

$$\left(1 - \phi_1 B - \phi_2 B^2 - \cdots - \phi_p B^p\right) \tilde{z}_t = \left(1 - \theta_1 B - \theta_2 B^2 - \cdots - \theta_q B^q\right) a_t, \quad (5\text{-}5)$$

Eq. (5-5) can be expressed compactly as $\phi(B)\tilde{z}_t = \theta(B)a_t$. The process is stationary and invertible respectively, if the roots of $\phi(B) = 0$ and $\theta(B) = 0$ are outside the unit circle.

The ACF and PACF of the ARMA processes are the result of combining AR and MA properties. In the ACF, there are certain initial coefficients that depend on the order of the MA part and later a decay dictated by the AR part. In the PACF, initial values dependent on the AR order followed by the decay due to the MA part.

In practice, it is difficult to select the order of the AR and MA part of the process. But, the main important features of ACF and PACF depending on the order of AR and MA part are summarized in Table 5-1.

In Fig. 5-8, an ARMA (1,1) process and its ACF and PACF are shown.

Integrated Processes and ARIMA Models

A process can be nonstationary in the mean, in the variance, or in other characteristics of the distribution of variables. When the level of the series is not stable in time, in particular, showing increasing or decreasing trends, we say that the series is not stable in the mean. When the variability or autocorrelations change with time, we say that the series is not stationary in the variance or autocovariance.

The most important nonstationary processes are the integrated ones, which have the basic property that stationary processes are obtained when they are differentiated. In the integrated processes, the autocorrelations diminish linearly over time and it is possible to find significant autocorrelation coefficients for very high lags. Nonstationary processes could be really useful when describing the behavior of many

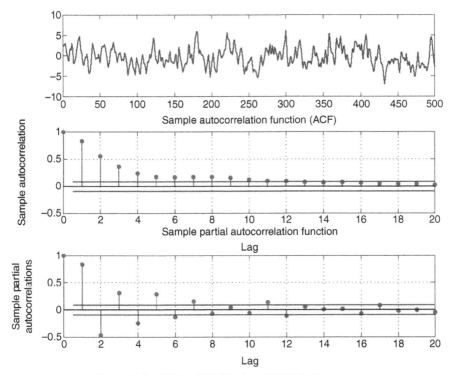

Figure 5-8   ACF and PACF of an ARMA(1,1) process.

climatological or financial time series. When a nonstationary process in the mean is differentiated and a stationary process is obtained, the original process is known as an integrated one.

In Fig. 5-9, a nonstationary process in the mean and its first difference are represented.

If in Eq. (5-1), $\phi = 1$, then the process is known as random walk with drift and $\omega_t = z_t - z_{t-1} = c + a_t$ is a stationary process.

The constant, c, is important in nonstationary processes, and it has a great influence in the long-term forecasting. Nevertheless, concerning stationary processes, the constant is not important, and we are able to subtract its mean from the observations and work with zero mean processes. Figure 5-10 shows a random walk with drift process as well as its ACF, which decays slowly.

The idea of unit roots in the autoregressive part, can be generalized to any ARMA process, allowing one or several, d, unit roots of the AR operator to be the unit. The general equation of an ARIMA(p, d, q) model is

$$\left(1 - \phi_1 B - \phi_2 B^2 - \cdots - \phi_p B^p\right)(1 - B)^d \tilde{z}_t = \left(1 - \theta_1 B - \theta_2 B^2 - \cdots - \theta_q B^q\right) a_t,$$

(5-6)

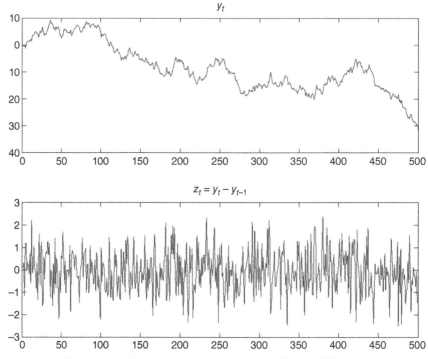

Figure 5-9 Integrated process of order 1 and its first difference.

where p is the order of the autoregressive part, q is the order of the moving average part and d is the number of unit roots. Equation (5-6) can be written compactly using the difference operator $\nabla = 1 - B$, $\phi(B)$, and $\theta(B)$, which gives

$$\phi(B)\nabla^d \tilde{z}_t = c + \theta(B)a_t.$$

All nonstationary processes present a common feature: the slow decay of the coefficients of the ACF.

Seasonality and Seasonal ARIMA Models

A particular case of nonstationarity is seasonality. A time series presents a seasonal pattern when the mean is not constant and evolves according to a cyclical pattern. For example, data coming from electricity markets present a variable mean, and it varies depending on the hour of the day and also depending on the day of the week. It is an example of data with two seasonal patterns: daily and weekly. The consumption of electricity (and also the price) is higher in some determined hours and lower in others, especially during the night. Moreover, the demand is different depending on the day

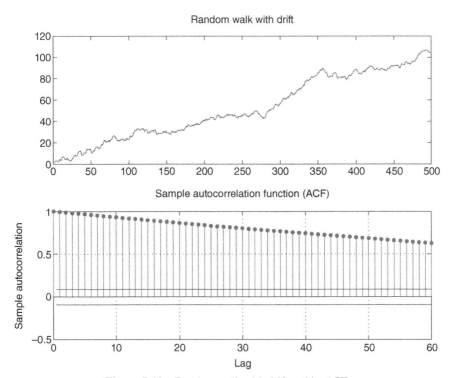

Figure 5-10 Random walk with drift and its ACF.

of the week, similar patterns are expected in Mondays or Sundays, respectively, since domestic, commercial, and industrial consumption depends on the day of the week.

Figure 5-11 shows the hourly load in the third and fourth week in May 2004 in the Spanish Market, and the ACF. A strong daily and weekly cyclical behavior can be observed. The order of daily seasonality is 24, and $24 \times 7 = 168$ is the order of the weekly seasonality in hourly data. The order or period, s, defines the number of observations that make up a seasonal cycle, in this case 24 and 168, respectively, for days and weeks.

Figure 5-12 shows the hourly prices for the period May 18–31, 2004, depending on the day of the week. A higher level in prices is shown for working days.

The seasonal pattern would be similar when dealing with time series of prices in the Spanish market, due to the simultaneous causality between demand and price in electricity markets.

Figure 5-13 shows the hourly time series of electricity prices in the Spanish market, in November 2004. Note that the price during a midnight hour, such as the third hour is lower than that in a peak-hour, like the 13th hour.

To model this type of data, while additive models to deal with seasonality are available, it is usual to incorporate seasonality into the ARIMA model via a multiplicative seasonal ARIMA model.

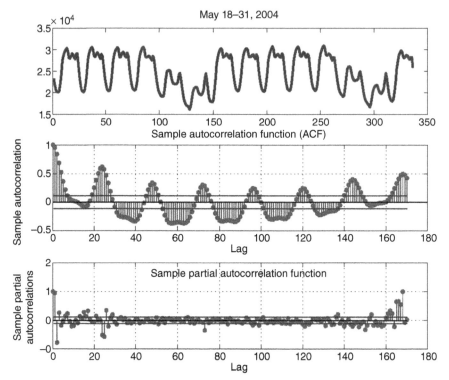

Figure 5-11 Hourly load and ACF and PACF, May 18–31, 2004, Spanish market.

The general expression of a seasonal multiplicative ARIMA model is given by

$$\phi(B)\Phi(B^s)(1-B)^d(1-B^s)^D \tilde{z}_t = \theta(B)\Theta(B^s)a_t, \tag{5-7}$$

where $\phi(B) = (1 - \phi_1 B - \cdots - \phi_p B^p)$, $\Phi(B^s) = (1 - \Phi_1 B^s - \cdots - \Phi_P B^{Ps})$, $\theta(B) = (1 - \theta_1 B - \cdots - \theta_q B^q)$, and $\Theta(B^s) = (1 - \Theta_1 B^s - \cdots - \Theta_Q B^{Qs})$. P and p are the orders of the seasonal and regular autoregressive parts, respectively, and q and Q are the orders of the regular and seasonal moving average components, respectively. The order of integration for the stationary process, d and D, the number of regular and seasonal differences.

This class of models, introduced by Box and Jenkins [17], offers a good representation of many seasonal time series that are found in practice, and Eq. (5-7) can be written in simplified form as the ARIMA model $(p, d, q) \times (P, D, Q)_s$.

When describing the ACF and PACF of seasonal ARIMA processes, some considerations apply:

- For small lags ($j = 1, \dots, 6$), only the regular part is observed.
- For seasonal lags basically the seasonal part is observed.

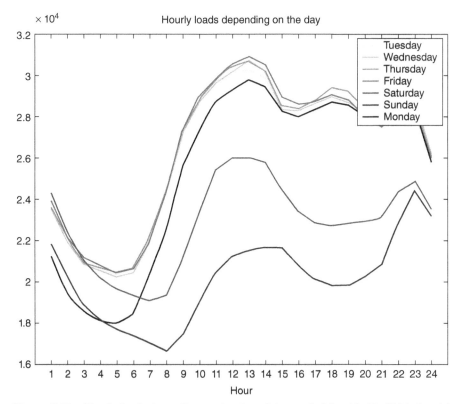

Figure 5-12   Hourly loads depending on the day of the week, May 18–31, 2004, Spanish market.

- Around the seasonal lags the interaction between the regular and seasonal parts is observed, which shows the regular part of the ACFon both sides of each seasonal lag.

The orders of the autoregressive and moving average parts of regular and seasonal dynamics (p, d, P, Q) can be selected in a way similar to that explained in Table 5-1, once it is confirmed that the process is stationary both in the variance and mean. The usual transformation to stabilize the variance is taking the logarithms and then seasonal or regular differences for stabilizing the mean.

In Fig. 5-14, the ACF and PACF of electricity prices of the 13th hour in the period 1998–2003 are shown.

Estimation of ARMA Models

The estimation procedure is presented assuming stationarity. The notation ω_t is used because in many cases, this series is a transformation of the original one, z_t.

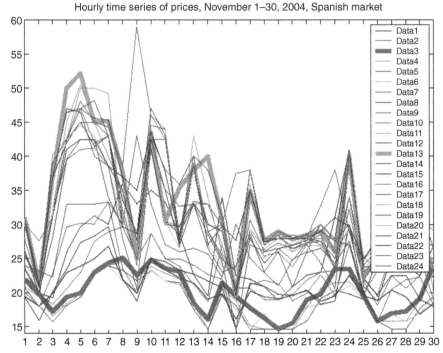

Figure 5-13 Hourly time series of electricity prices, November 2004, Spanish market.

For example, and considering an hourly time series of prices in which a weekly seasonal pattern is present, ω_t is a transformation of the original series, such as $\omega_t = (1 - B)(1 - B^7) \log(z_t)$.

Here, the maximum likelihood estimation procedure is presented. For this purpose, the joint density function must be written and then maximized with respect to the parameters, assuming that the data is fixed.

To write the joint density of the time series considered, the following relation is used:

$$f(\mathbf{x}, \mathbf{y}) = f(\mathbf{x}) \cdot f(\mathbf{y}|\mathbf{x}),$$

which is also valid if all the density functions are conditional on another variable $\mathbf{z}$.

Applying this property iteratively, when considering the joint density function of the T observations ω_T, gives

$$f(\omega_T) = f(\omega_1) \cdot f(\omega_2|\omega_1) \cdot f(\omega_3|\omega_2, \omega_1) \cdot \cdots \cdot f(\omega_T|\omega_{T-1}, \ldots, \omega_1). \quad (5\text{-}8)$$

Equation (5-8), allows us to write the joint density function of the T variables as a product of T univariate distributions, and the likelihood of an ARMA model can be derived since a Gaussian distribution is assumed, so all the conditional distributions

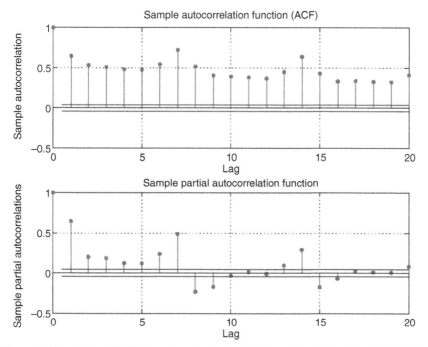

Figure 5-14   ACF and PACF, hourly prices in the 13th hour. Spanish market, 1998–2003.

are also Gaussian. Its expectation is the one-step-ahead prediction which minimizes the quadratic prediction error.

Thus, the joint density function of the sample for a general ARMA process can be written as

$$f(\omega_T) = \prod_{t=1}^{T} \sigma^{-1} v_{t|t-1}^{-1/2}(2\pi)^{-1/2} \exp\left\{-\frac{1}{2\sigma^2}\sum_{t=1}^{T}\frac{(\omega_t - \omega_{t|t-1})^2}{v_{t|t-1}}\right\}. \qquad (5\text{-}9)$$

Taking logarithms of expression (5-9), and including all the parameters that must be estimated in $\beta = \{\mu, \phi_1, \ldots, \phi_p, \theta_1, \ldots, \theta_q, \sigma^2\}$, the expression for the support function gives

$$L(\beta) = -\frac{T}{2}\ln\sigma^2 - \frac{1}{2}\sum_{t=1}^{T}\ln v_{t|t-1} - \frac{1}{2\sigma^2}\sum_{t=1}^{T}\frac{e_t^2}{v_{t|t-1}},$$

where both the conditional variances $v_{t|t-1}$ as well as the one-step-ahead prediction errors, $e_t = \omega_t - \omega_{t|t-1}$, depend on the parameters.

As a result, the likelihood can be evaluated once the one-step-ahead errors, e_t, and their variances have been calculated. Then, the maximization of the exact likelihood function is carried out using a nonlinear optimization algorithm, and the parameters of the model are estimated, $\hat{\beta} = \{\hat{\mu}, \hat{\phi}_1, \ldots, \hat{\phi}_p, \hat{\theta}_1, \ldots, \hat{\theta}_q, \hat{\sigma}^2\}$.

In addition, ARMA models can be estimated using the state-space formulation of ARMA models and then using the Kalman filter and smoother, see Durbin and Koopman [19].

Diagnostic Checking for Time Series Models

The hypotheses of the model must be validated. These assumed error hypotheses are that $a_t \sim iid \, N(0, \sigma^2)$. So, we must check that the residuals are

- uncorrelated for all lags,
- having zero mean,
- having constant variance,
- normally distributed.

Additionally, diagnostics include the detection of any deterministic terms, whenever present. If the hypotheses are validated, the model could be used to forecast, otherwise, the model must be refined. We present next some of the most referenced tests for independence, zero mean, homoscedasticity, and meeting the Gaussian requirement the residuals.

Autocorrelation Tests   Let z_t be a zero mean process generated by the ARMA(p, q) model $\phi(B)\tilde{z}_t = \theta(B)a_t$, and let $\hat{a}_1, \dots, \hat{a}_T$ be the residuals obtained after estimating the model in a sample of size T, and let

$$\hat{r}_j = \frac{\sum_{t=j+1}^{T} \hat{a}_t \, \hat{a}_{t-j}}{\sum_{t=1}^{T} \hat{a}_t^2} \quad \text{for} \quad j = 1, 2, \dots, T \tag{5-10}$$

be the estimated residual autocorrelation coefficients. Here some well-known references concerning goodness of fit test are reported.

A family of Portmanteau goodness of fit tests for the analysis of the independence of the residuals can be obtained based on the expression (see [20]):

$$Q = T \left\{ \delta \sum_{i=1}^{m} w_i g\left(\hat{r}_i^2\right) + (1 - \delta) \sum_{i=1}^{m} \omega_i g\left(\hat{\pi}_i^2\right) \right\}, \tag{5-11}$$

where $\hat{\pi}_i$ are the estimated residual partial autocorrelation coefficients (see [17]), $0 \le \delta \le 1$, $m < T$, $w_i 0$, $\omega_i 0$, and g is a nondecreasing smooth function with $g(0) = 0$. Some well-known members of this class when $g(x) = x$ are the tests proposed by Box and Pierce [21], where $\delta = 1$ and $w_i = 1$, Ljung and Box [22], where $\delta = 1$ and $w_i = (T + 2)/(T - i)$, and Monti [23], where $\delta = 0$ and $\omega_i = (T + 2)/(T - i)$. This family includes also many tests obtained in the frequency domain by measuring the distance between the spectral density estimator and the one corresponding to a white noise process. The test proposed by Anderson [24], is a member of this class, where $\delta = 1$, $w_i = 1/(\pi i^2)$, and $m = T - 1$, and it was improved by Velilla [25], who replaced the vector of autocorrelations with a vector of modified autocorrelation which is free of the unknown parameters. Finally, Hong [26] proposed a general class of these statistics, where $\delta = 1$ and $w_j = k^2(j/m)$, where k is a symmetric function, $k : R \rightarrow [-1, 1]$, that is continuous at zero and at all but a finite number of points, with $k(0) = 1$ and $\int_{-\infty}^{\infty} k^2(z)dz < \infty$. Hong [26] shows that within a suitable class of kernel functions, the Daniell kernel, $(k(z) = sin(\pi z)/\pi z, z \in (-\infty, \infty))$, maximizes the power of the test under both local and global alternatives.

Peña and Rodríguez [27], based on a general measure of multivariate dependence, the effective dependence, see Peña and Rodríguez [28], proposed a Portmanteau test by applying this measure to the autocorrelation matrix, leading to the statistic

$$\hat{D}_m = T[1 - |\hat{R}_m|^{1/m}], \tag{5-12}$$

where $\hat{R}_m$ is

$$\hat{R}_m = \begin{bmatrix} 1 & \hat{r}_1 & \cdots & \hat{r}_m \\ r_1 & 1 & \cdots & \hat{r}_{m-1} \\ \vdots & \vdots & \ddots & \vdots \\ \hat{r}_m & \hat{r}_{m-1} & \cdots & 1 \end{bmatrix}. \tag{5-13}$$

and $\hat{r}_j$ is estimated by Eq. (5-10). They show that this test is more powerful than the ones proposed by Ljung and Box [22] and Monti [23]. This test is highly asymmetric with respect to the autocorrelation coefficients and large weights are assigned to lower order lags and smaller weights to higher lags.

Finally, the test statistic proposed by Peña and Rodríguez [20]:

$$D_m^* = -\frac{T}{m + 1} \log |\hat{R}_m|, \tag{5-14}$$

where the standardized estimated correlation matrix (5-13) is standardized by its dimension. This statistics can be considered as a modification of D_m as given by Eq. (5-12). We prefer to standardize by the dimension of the matrix instead of using the number of autocorrelation coefficients in order to obtain the following interpretations of this statistic: (i) $D_m^* = -T \log \bar{\delta}$, where $\bar{\delta} = \prod_{i=1}^{m+1} (\delta_i)^{1/(m+1)}$ is the geometric mean of the eigenvalues of $\hat{R}_m$. (ii) As the eigenvalues of the covariance matrix are approximately equal to the power spectrum ordinates at the frequencies $\delta_i = 2\pi i/m$, (see [29]), the statistic is also approximately $D_m^* \approx -T(m + 1)^{-1} \sum \log f(\delta_i)$, where $f(\delta_i)$ is the spectral density. (iii) $D_m^* = -T \log(1 - \bar{R}^2)$, where $(1 - \bar{R}^2) = \prod_{i=1}^{m} (1 - R_i^2)^{1/(m+1)}$. (iv) This statistic is a member of the class (5-11), because as shown in Ramsey [30], $|\hat{R}_m| = \prod_{i=1}^{m} (1 - \hat{\pi}_i^2)^{((m+1-i))}$, and thus we have

$$D_m^* = -T \sum_{i=1}^{m} \frac{(m + 1 - i)}{(m + 1)} \log \left(1 - \hat{\pi}_i^2\right), \tag{5-15}$$

which is in the form (5-11) with $g(x) = -\log(1 - x)$, $\delta = 0$, and $\omega_i = (m + 1 - i)/(m + 1)$. Thus this statistic is proportional to a weighted average of the squared partial autocorrelation coefficients with larger weights given to low order coefficients and smaller weights assigned to high order coefficients.

Zero Mean Test   In principle, the estimated residuals for an ARIMA model are not subject to the constraint $\sum_{t=1}^{T} \hat{a}_t = 0$. To test for the hypothesis of zero mean in the general case of an ARMA (p, q) model, in which there are $p + q$ parameters and T residuals, their mean and variance are calculated by the expressions

$$\bar{a} = \frac{\sum_{t=1}^{T} \hat{a}_t}{T}, \quad \hat{\sigma}^2 = \frac{\sum_{t=1}^{T} (\hat{a}_t - \bar{a})^2}{T - p - q}.$$

We conclude that the expectation of the residuals is not zero, if $\bar{a}/(\hat{\sigma}/\sqrt{T})$ is significantly large, comparing it with the standard Normal distribution. This test is applicable once we have checked that the residuals are uncorrelated.

Constant Variance Test   This can be checked by analyzing the plot of the residuals versus time. Moreover, there are specific tests for certain types of heteroskedasticity. Some well-known tests are those related to conditional heteroskedasticity, testing for ARCH effects. McLeod and Li [31] and Rodríguez and Ruiz [32] provided tests for conditional heteroskedasticity in time series.

Moreover, there are tests for detecting other forms of nonlinearity, different from conditional heteroskedasticity, such as threshold autoregressive (TAR) effects, among others. Many references can be cited in this regard. The Tsay test [33] checks for the inclusion of added variables to represent the nonlinear behavior, whereas the BDS test by Brock et al. [34] tests were based on smoothness properties. This test has become quite popular. On the other hand, the McLeod and Li [31] test uses the asymptotic sample distribution of the estimated autocorrelations, whereas the Peña and Rodríguez [20] test uses the determinant of their correlation matrix.

Normal Distribution of the Residuals   The Gaussian assumption is tested using any of the available tests for this purpose. The Jarque–Bera test (1980) is a goodness-of-fit measure of departure from normality, based on the coefficient of asymmetry, α_3, and kurtosis, α_4 of the residuals.

$$\alpha_3 = \frac{\sum_{t=1}^{T} (\hat{a}_t - \bar{a})^3}{\hat{\sigma}^3}, \quad \alpha_4 = \frac{\sum_{t=1}^{T} (\hat{a}_t - \bar{a})^4}{\hat{\sigma}^4}.$$

Then, under the hypothesis of normality, the condition that, the variable $X = \frac{T \cdot \alpha_3^2}{6} + \frac{T \cdot (\alpha_4 - 3)^2}{24}$ is a χ^2 with two degrees of freedom. It is always advisable to study the graph of the estimated residuals $\hat{a}_t$ over time.

Model Selection Criteria

Suppose that we want to select the autoregressive order for a given time series. We cannot select the order by using the residual variance because this measure cannot increase if we increase the order of the autoregression. Similar problems arise with other measures of fit, as the deviance. Model selection criteria were introduced to solve this problem. The most often used criterion can be written as

$$ k \min \left\{ \log \hat{\sigma}_k^2 + k \times C(T, k) \right\}, \tag{5-16} $$

where $\hat{\sigma}_k^2$ is the maximum likelihood estimate of the residual variance, k is the number of estimated parameters for the mean function of the process, T is the sample size, and the function $C(T, k)$ converges to 0 when $T \to \infty$. These criteria have been derived from different points of view. Akaike [35], a pioneer in this field, proposed selecting the model with the smallest expected out-of-sample forecast error, and derived an asymptotic estimate of this quantity. This led to the final prediction error criterion, FPE, where $C(T, k) = k^{-1} \log(\frac{T+k}{T-k})$. This criterion was further generalized, using information theory and Kullback–Leibler distances, by Akaike [36], in the well-known AIC criterion, where $C(T, k) = 2/T$. Shibata [37] proved that this criterion is efficient, which means that if we consider models of increasing order with the sample size, the model selected by this criterion is the one which produces the least mean square prediction error (MSPE). The AIC criterion performs poorly for small samples because it tends to over-parametrize too much. To avoid this problem, Hurvich and Tsay [38] introduced the corrected Akaike's information criterion (AICC), where $C(T, k) = \frac{1}{k} \frac{2(k+1)}{T-(k+2)}$.

From the Bayesian point of view, it is natural to choose between models by selecting the one with the largest posterior probability. Schwarz [39] derived a large sample approximation to the posterior probability of the models assuming the same prior probabilities for all of them. The resulting model selection criterion is called the Bayesian information criterion (BIC) and in Eq. (5-16), it corresponds to $C(T, k) = \log(T)/T$. As the posterior probability of the true model will go to 1 when the sample size increase, it can be proved that BIC is a consistent criterion, that is, under the assumption that the data come from a finite order ARMA process, we have a probability of obtaining the true order that goes to 1 when $T \to \infty$. Another often used consistency criterion is the one due to Hannan and Quinn [40], called HQC, where $C(T, k) = 2m \log(T)/T$ with $m > 1$.

Galeano and Peña [41] proposed to look at model selection in time series as a discriminant analysis problem. We have a set of possible models, $M_1, \ldots, M_\alpha$, with prior probabilities $P(M_i)$, $\sum P(M_i) = 1$, and we want to classify a given time series,

$\mathbf{y} = (y_1, \dots, y_n)$ as generated from one of these models. The standard discriminant analysis solution to this problem is to classify the data in the model with highest posterior probability and, if the prior probabilities are equal, this leads to the BIC criterion. From a frequency domain point of view, the standard discriminant analysis solution when the parameters of the model are known is to assign the data to the model with the highest likelihood. If the parameters of the models are unknown, we can estimate them using maximum likelihood, substitute them in the likelihood function, and again select the model with the highest estimated likelihood. However, although this procedure works well when we are comparing models with the same number of unknown parameters, it cannot be used when the number of parameters is different. As the estimated likelihood cannot decrease by using a more general model, the maximum estimated likelihood criterion will always select the model with more parameters. To avoid this problem, Galeano and Peña [41] proposed to select the model which has the largest *expected* likelihood, as follows. Compute the expected value of the likelihood over all possible sequences generated by the model and choose the model with largest expected likelihood. These authors proved that the resulting procedure is equivalent to the AIC criterion.

Forecasting with ARIMA Models

Predictions of the estimated model can be calculated using the estimated parameters as if they were the true ones. The optimal predictors are those that minimize the MSPEs.

When considering the prediction function of an ARIMA model, the nonstationary operators (the differences) determine the long-term prediction. The short-term forecast is determined by the stationary operators AR and MA.

Let $z_t = (z_1, z_2, \dots, z_T)$ be a realization of the time series to be forecast k steps ahead. Let $\hat{z}_{T+k}$ be a predictor of z_{T+k} obtained as a function of the T values observed, so the forecast origin is at T, and the forecast horizon at k. The prediction error of this predictor is given by $e_{T+k} = z_{T+k} - \hat{z}_{T+k}$, and we want it to be as small as possible. To be able to compare several predictors, a criterion must be defined and the predictors compared using it. For example, if the main objective is to minimize the prediction error e_{T+k}, it does not matter whether it is positive or negative, the squared or the absolute value of the prediction error can be considered. If there is a preference for the error in one direction (positive or negative), an asymmetric loss function can be considered.

The most frequently utilized loss function is the quadratic, which leads to the criterion of minimizing the MSPE of z_{T+k}, given the information $\mathbf{z}_T$. The expression $E[z_{T+k} - \hat{z}_{T+k}|\mathbf{z}_T] = E[e_{T+k}|\mathbf{z}_T] = \text{MSPE}(z_{T+k}|\mathbf{z}_T)$. The predictor that minimizes this mean square error is the expectation of the variable z_{T+k} conditional on the available information in T, $\mathbf{z}_T$ [18].

For example, if the process is ARMA(1,1), the innovations are obtained by means of the equation

$$a_t = \tilde{z}_t - c - \phi \tilde{z}_{t-1} + \theta a_{t-1}, \ t = 2, \dots, T.$$

The innovation for $t = 1$ is given by $a_1 = \tilde{z}_1 - c - \phi\tilde{z}_0 + \theta a_0$. Since neither z_0 nor a_0 are known, they are replaced by their expectations and the rest of the innovations are calculated considering this initial condition.

In addition, when considering seasonal ARIMA processes, see Eq. (5-7), it must be taken into account that the seasonal operator $(1 - B^s)$ incorporates two operators:

- The difference operator, $1 - B$, and
- The pure seasonal operator $S_s(B) = 1 + B + \cdots + B^{s-1}$,

since $(1 - B^s) = (1 - B)(1 + B + \cdots + B^{s-1})$, and Eq. (5-7) can be written when $D = 1$ in the form

$$\phi(B)\Phi(B^s)(1 - B)^{d+1}S_s(B)z_t = c + \theta(B)\Theta(B^s)a_t,$$

where now the four operators $\phi, \Phi, (1 - B)$ and S_s do not have any roots in common. For an ARIMA $(p, d, 0) \times (P, D, 0)_s$, the prediction equation is given by

$$\phi(B)\Phi(B^s)(1 - B)^{d+1}S_s(B)\hat{z}_{T+k} = c.$$

Finally, the most often used seasonal ARIMA model (the airline passenger model, whose name is inspired by the data which was firstly adjusted to this model, the G series in Box and Jenkins [17] that contains monthly data of the number of airline passengers). The model, a multiplicative seasonal ARIMA $(0, 1, 1) \times (0, 1, 1)$ is given by the equation

$$(1 - B)(1 - B^{12})z_t = (1 - \theta B)(1 - \Theta B^{12})a_t. \tag{5-17}$$

The equation for calculating the forecasts for model (5-17) is as follows:

$$\hat{z}_{T+k} = \hat{z}_{T+k-1} + \hat{z}_{T+k-12} - \hat{z}_{T+k-13} - \theta\,\hat{a}_{T+k-1} - \Theta\,\hat{a}_{T+k-12} + \theta\Theta\,\hat{a}_{T+k-13}.$$

Figure 5-15 shows the airline passenger series as well as the predictions.

DEVELOPMENT OF MIXED MODELS AND ITS COMPUTATIONAL IMPLEMENTATION

In this section, two new models are proposed to deal with this problem. Both models deal with the 24-hourly time series of electricity prices, and consider multiplicative ARIMA models for modeling and forecasting the hourly time series. For this reason, a brief overview on time series analysis and general methodology for building ARIMA models for price forecasting were given in the previous section.

The methodology will be described taking the Spanish market as an example for illustration. The main idea is to compute accurate forecasts, building a mixed model that combines the advantages of the ones used to build it.

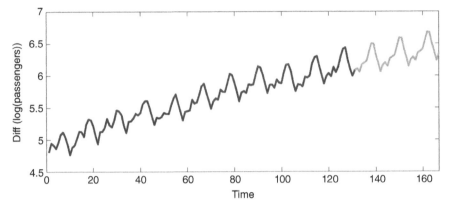

Figure 5-15 Data and forecasts. Airline passenger data after taking logs and both regular and seasonal differences.

Models Proposed for Short-Term Forecasting of Electricity Prices

A brief descriptive analysis of the prices corresponding to the period 1998–2003 has been done. A boxplot (also known as Box and Whisker plot) is a graphical representation of a distribution built to show its most important features and the outliers. The limits of the rectangle are quantiles 25 and 75 (Q_{25} and Q_{75}). The position of the median, Q_{50}, is indicated by drawing a line. By construction, 50% of the data in the sample is inside the rectangle, and 75% of the data is smaller than Q_{75}. In addition, Q_{50} is the median, so it is the center of the sample. Admissible values for the upper (UL) and lower limit (LL) are calculated, where $UL = Q_{75} + 1.5(Q_{75} - Q_{25})$ and $LL = Q_{75} - 1.5(Q_{75} - Q_{25})$. These limits can be used to identify outliers. Box plots are especially useful for providing a general idea about the distribution of a variable. Figure 5-16 presents a boxplot of the hourly prices for the period under study.

In this work, we have used boxplots to study the distribution of hourly prices (level and variability), and also to better visualize forecasting errors. Not only does the level of the prices but also their variability depend on the corresponding hour of the day. This conclusion can be extended to other markets since it is a consequence of the instantaneous effect of the demand on the price.

The prediction errors obtained by separately studying and modeling each of these 24-hourly time series, far from being affected negatively by the loss of information, are reduced. The larger homogeneity of the 24-hourly time series in comparison with the complete one, as well as the fact that 24 one-step-ahead forecasts are calculated everyday instead of 24 individual forecasts with prediction horizons varying from 1 to 24, will allow an improvement in the accuracy of the forecasts.

Two new different models are proposed. The first one, which we will refer to from now on as Model 24, forecasts electricity prices for each of the 24 hours of

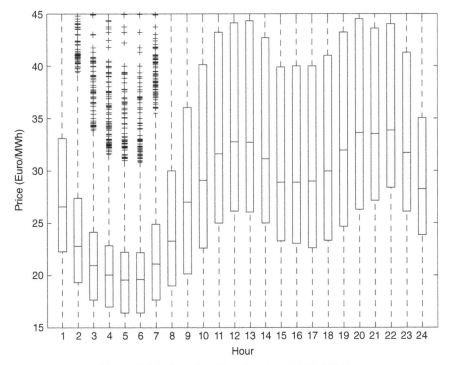

Figure 5-16   Boxplot of hourly prices (1998–2003).

the next day using the ARIMA models built for each of the 24-hourly time series. So, to produce a forecast with Model 24 for tomorrow's Hth hour, we use the model estimated for this hour with the previous L complete weeks (7-day week, considering both workday and weekend data).

The second model computes the forecasts for the working days using the 24-workday time series and the forecasts of the prices in the weekends with the weekend data. This second model is hereafter referred to as Model 48, because of the number of series with which it works (24 series for working days and 24 for weekends). To produce a forecast for tomorrow's Hth hour (if it is a weekday), we use the estimation of the model for this hour built with the previous L weeks (considering 5-day weeks, only weekday data). On the other hand, to produce a forecast for tomorrow's Hth hour (if it is Saturday or Sunday), we use the estimation of the model for this hour built with the previous L weeks (considering a 2-day week, only the weekend data). In addition, the disaggregation of the price time series into 24-hourly time series allows the use of ARIMA methodology in better conditions, because the frequency of the data is reduced.

A computational experiment has been carried out to determine which model leads to more accurate forecasts. Some relevant factors in short-term prediction have been included. We thus have obtained the appropriate length of the time series used to

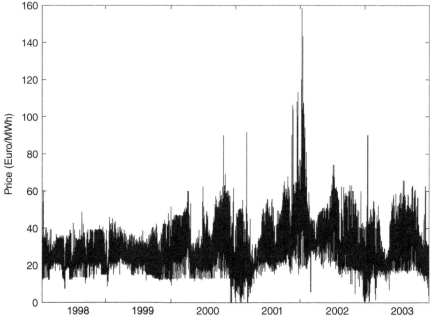

Figure 5-17   Hourly prices (January 1, 1998–December 31, 2003).

build the forecasting models. Until now, a default length of 2 or 3 months had been used to build the models, but no published study analyzes this issue.

It is of interest to analyze the sources of variability affecting the prediction error. We thus chose the following factors and levels:

- Model: Model 48 or Model 24.
- Length of the time series used to build the forecasting model: 8, 12, 16, 20, 24, 28, 32, 44, 52, and 80 weeks.

Forecasts have been computed for all the prices in the period from January 1, 1998 to December 31, 2003, both for Model 24 and Model 48, for the 10 possible levels of the factor "length of the time series used to build the model." The prices for this period are shown in Fig. 5-17. The size of the sample as well as its representativeness, as it includes the years 1998–1999 in which the variability of the prices is smaller because of the initialization of the competitive markets, years 2000–2003 when larger variability in the prices can be appreciated and the great increase in the prices that occurred at the end of 2001 and in the beginning of 2002, will allow inferences and extend the conclusions reached in forecasting prices in the future. We want the combination of "Model" and "Length" that best fits in "global terms" for a very long period of time in which different patterns in level and variability of the prices were registered. This mixed model would be able to produce accurate forecasts for future periods, depending neither on the level of the prices nor on their variability.

We would like to highlight the great number of models to be identified and estimated (bear in mind that the models are refitted every day). Forecasting prices for 1 day using a different multiplicative ARIMA model for each hour (as we propose) implies the identification and estimation of 24 models. If the objective is extended to computing forecasts for the prices in 1 week, $24 \times 7 = 168$ models must be identified and estimated. Given our ambitious objectives, trying to compute forecasts for the 6 years considered ($6 \times 365 \times 24 = 52,854$ hours) with the $20 = 2 \times 10$ possible combinations of the levels of the two factors considered (two levels for the factor "Model" and 10 levels for the factor "Length" of the time series) required more than $1,051,200 (= 6 \times 365 \times 24 \times 20)$ models to be identified and estimated.

With the very large amount of models to identify, more than $1,000,000$, it is impossible to manually fit all of them using the ACF and PACF, as well as using diagnostic checking, and requires the automation of the procedure.

Scientific computing associates (SCA) software or EViews can deal with the automatic identification of ARIMA models and also includes the option of outlier identification and intervention, but none of them are free, and expensive licenses are required to use them. In this work, we have used time series regression with ARIMA noise, missing observations and outliers (TRAMO), a software developed by Caporello and Maravall [42] for the estimation and subsequent forecasting with ARIMA models. Identification and intervention of outliers, as well as the estimation of models can be done with an automatic procedure. Models are selected using the BIC, whose expression is provided in (5-18). This criterion takes into account both the likelihood of the model (by means of its residual variance) and the parsimony of the model (including a term that penalizes models with a large number of parameters).

The expression of the BIC is

$$\mathrm{BIC} = n \log \left(\hat{s}_R^2 \right) + k \log(n), \tag{5-18}$$

where n is the length of the time series used to estimate the model, $\hat{s}_R^2$ is the residual variance, and k is the number of parameters estimated. Among those presented in Model Selection Criteria, BIC is one of the model selection criteria whose application is more extended.

The capabilities of TRAMO are similar to those in SCA or EViews, but the main advantage of TRAMO is that it is a free software, it can be downloaded from the webpage: (http://www.bde.es/bde/en/secciones/servicios/Profesionales/Programas_estadi/Programas_estad_d9fa7f3710fd821.html), and it is easy to use for applied statisticians and engineers.

Finally, it is important to bear in mind that the results of this chapter are not restricted to obtaining forecasts for several days or weeks with small prediction errors. We have computed forecasts for all the hours in a long period of time (6 years) and it is of interest to find out the appropriate levels of the two factors considered ("Model" and "Length" of the series) in order to minimize prediction errors in "global terms."

ANALYSIS OF FORECASTING ERRORS

Once the forecasts had been computed for all possible combinations of factor levels, it is of great interest to have a prior idea of the convenience of using Model 24 or Model 48 to forecast prices and whether the decision depends on the duration of the time series used to build the model or on whether the forecasts are for a working day or a weekend

Descriptive Analysis of Forecasting Errors

Some conclusions can be drawn from a brief descriptive analysis of the prediction errors. The prediction errors decrease as the length of the time series increases, and this occurs up to a length of around 44 weeks. Building models with a longer series produces less accurate forecasts for weekends and not significantly smaller prediction errors for weekdays. It can also be observed that Model 24 provides better fitting for weekends while Model 48 does likewise for working days. This can be observed in Fig. 5-18, which shows the boxplot of the prediction errors for Models 24 and 48 and lengths 44 weeks and 80 weeks for all the days of the week.

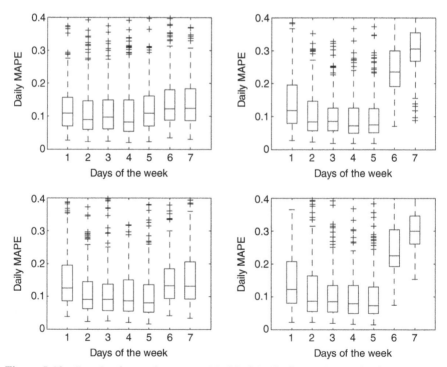

Figure 5-18 Boxplot forecasting errors. Models 24, 48 (first and second column, respectively). Lengths 44, 80 weeks (first and second row, respectively).

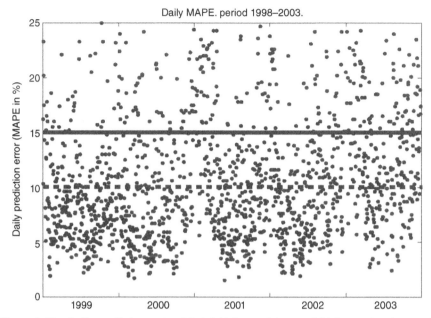

Figure 5-19   Daily prediction errors. Model 48 for workdays and 24 for weekends. Length 44 weeks.

This brief descriptive analysis of the prediction errors indicates that it would be reasonable to carry out the DOE separately for working days and weekends. In addition, a rough a priori estimate of the "Model" and the "Length" of the time series for both weekends and working days can be obtained from Fig. 5-18.

In addition, taking into account the considerations mentioned earlier, Fig. 5-19 shows the daily prediction errors obtained for the Spanish market in the period 1998–2003, when using Model 24 for weekends and Model 48 for weekdays, in both cases when using the previous 44 weeks to the day for which we are computing the forecast.

Analysis of Forecasting Errors by Means of LOESS

The visual i*nformation provided in* Fig. 5-19 could be improved *by using* smoothed values *obtained using locally* weighted robust regression *which* is a technique that *is applied to* a scatter plot ( ). The main idea is to provide an estimate for the conditional expectation of the error. *The* locally weighted regression (LOESS) was developed by Cleveland [43], and subsequently Cleveland and Devlin [44] incorporated some additional improvements.

Theoretical Basis of Locally Weighted Regression (LOESS)   Poly-
nomials local adjustment has been used to smooth time series plots in which the data are equally spaced. *LOESS results in* robust adjustment that avoids outliers that distort the time series.

A scatterplot is smoothed by means of a procedure with some subsequent steps. Let W be a weight function with the following properties:

- $W(x) > 0$, if $|x| < 1$,
- $W(-x) = W(x)$,
- $W(x)$ is a non-increasing function for $x \geq 0$,
- $W(x) = 0$, if $|x| < 1$.

Let $0 < f < 1$ and let r be the f_n rounded to the nearest integer. Roughly, the procedure is as follows:

For each x_i, weights $w_k(x_i)$ are defined using the weight function $W(x)$. For each x_i, its r nearest neighbors are considered and we fit the local polynomials regression model. The weight $w_k(x_i)$ is given to each data point. This weight will be bigger for nearer points. This is done by centering $W(x)$ at each x_i and scaling it in such a way that the first point for which $W(x) = 0$ is the rth nearest neighbor of x_i. The value initially fitted for y_i (hereafter referred as $\hat{y}_i$), for each x_i, is the adjusted value for a dth degree polynomial, adjusted using weighted least squares with weights $w_k(x_i)$. This procedure for calculating the initially adjusted values is known as LOESS. Then, some additional steps are followed in order to obtain robust estimates.

Another set of weights, δ_i, is defined, based on the size of the residuals, $y_i - \hat{y}_i$. Large residuals result in small weights and small residuals result in big weights and by doing this, the weights assigned to outliers are small.

The smoothing procedure has been designed to express y_i as a function of x_i, when $y_i = g(x_i) + \varepsilon_i$, where g is a smoothing function and ε_i are random variables with zero mean and constant scale. Then, $\hat{y}_i$ is an estimate of $g(x_i)$. The idea of smoothing consists of using the nearest neighbors of (x_i, y_i) for computing $\hat{y}_i$. When considering a weight function $W(x)$, non-increasing for $x \geq 0$, the weights decrease when the distance increases. Thus, those points whose abscissas are next to x_i have an important weight when computing $\hat{y}_i$, otherwise, when their abscissas are far from the point considered, their weights when computing $\hat{y}_i$, are smaller.

We shall now give the details about the procedure. For each i, let h_i be the distance from x_i to the rth nearest neighbor, that is, h_i is the rth smallest value for the expression $|x_i - x_j|$, with $j = 1, \ldots, n$. For $k = 1, \ldots, n$, let $w_k(x_i) = W(h_i^{-1}(x_i - x_j))$.

LOESS and robust locally weighted regression are defined by the following sequence of operations:

1. For each i, compute the estimates $\hat{\beta}_j(x_i)$, with $j = 0, \ldots, d$, parameters of the polynomial regression of d degree of y_k on x_k, which is fitted by means of weighted least squares, using the weights $w_k(x_i)$ for (x_k, y_k). Thus, the values $\hat{\beta}_j(x_i)$ are those that minimize the expression $\sum_{k=1}^{n} w_k(x_i)(y_k - \beta_0 - \beta_1 x_1 - \cdots - \beta_d x_k^d)^2$. The smoothed point at x_i is obtained using the LOESS of d degree, $(x_i, \hat{y}_i)$, where $\hat{y}_i$ is the fitted value by the regression at x_i. Thus, $\hat{y}_i = \sum_{j=0}^{d} \hat{\beta}_j(x_i)x_i^j$.

2. Let B be the bi-square weight function defined as $B(x) = (1 - x^2)^2$ if $|x| < 1$, and $B(x) = 0$, if $|x| \geq 1$, and let $e_i = y_i - \hat{y}_i$ be the residuals from the current fitted values. Let s be the median of $|e_i|$. Robust weights are defined as $\delta_k = B(e_k/s)$.

3. Compute new $\hat{y}_i$ for each i, by fitting a d th degree polynomial using weighted least squares with weight $\delta_k w_k(x_i)$ at (x_k, y_k).

4. Subsequently, carry out steps 2 and 3 t times. The final $\hat{y}_i$ are robust locally weighted regression fitted values.

The final goal of the iterative adjustment carried out in steps 2–4 is the robust estimation of the smoothed points, in which a very small proportion of outliers do not distort the final estimates. The weights corresponding to these outliers will be small and because of this, they do not affect the estimation of the smoothed values. Several authors have proved that the function $B(x)$ provides a good robust estimation for weights.

Once the adjusted values have been obtained, they can be plotted for equally spaced x_i. Furthermore, successive smoothed values $\hat{y}_i$ for the selected x_i, can be joined using straight lines.

Finally, it must be pointed out that for applying LOESS, four parameters must be fixed: $(d, W, t, \text{and} f)$. Now, some guidelines concerning their selection are as follows:

- d is the degree of the adjusted polynomial. Selecting $d = 1$ is a trade-off between the flexibility needed to reproduce the behavior of data and the desired ease of computations. The case for which $d = 0$ is the simplest one, but it is less flexible. However, when $d = 2$, computational considerations begin to override the need for having flexibility. In most cases, taking $d = 1$ results in adequate smoothed points.

- When choosing the kernel $W(x)$, the four required characteristics given at the beginning of this section, must be taken into account. Bear in mind that negative weights do not make sense, and there is no reason to treat the points on the right or on the left of x_i, in a different manner and the fact that the points which are situated farther from x_i should weigh less. The tri-cube function $W(x) = (1 - |x|^3)^3$ for $|x| < 1$ and $W(x) = 0$ elsewhere, gives an adequate smoothing in most situations.

- Moreover, a convergence criterion must be defined, and iterations continue until this criterion has been met. Experimentation with real or simulated data indicates that iterating twice ($t = 2$) is enough almost always.

- Finally, when selecting an appropriate value for f, one must consider that when increasing f, the smoothness of the smoothed values is also increased. The goal of choosing f is to pick a value as large as possible to minimize the variability in the smoothed points without distorting the pattern in the data. Choosing f in the interval $[0.2, 0.8]$ is adequate for most cases, and 0.5 can be used as an initial value for f when there is no previous idea about its value.

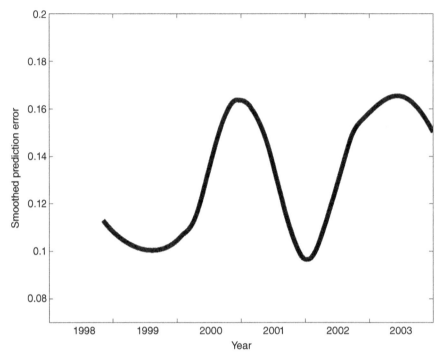

Figure 5-20   Smoothed prediction error for the whole period considered. Model 24 for weekends and Model 48 for weekdays. Length equal to 44 weeks in both cases. First forecast computed for the 45th week in 1998 since the first 44 weeks are used to identify and estimate the model.

Application to Forecasting Errors in the Period 1998–2003   Here, the application of LOESS to better summarize the information given in Fig. 5-19 is presented. In Fig. 5-20, a nonparametric estimation of the conditional mean of the errors that takes into account the time evolution is provided. This estimation has been computed using Model 24 for weekends and Model 48 for weekdays, and the "Length" is equal to 44 weeks in both cases, considering information obtained from Fig. 5-18.

Preliminary Conclusions

Some preliminary conclusions can be drawn from this descriptive analysis. Forecasts computed using Model 48 are more accurate for working days. Nevertheless, for 1-day-ahead forecasts of electricity prices on weekends, Model 24 leads to lower prediction errors. The optimal length to build the models seems to be around 44 weeks.

Based on these preliminary considerations, we can make some conclusions. Since the two new models, 24 and 48, fit better for weekends and working days, respectively, in Design of Experiments, the DOE will be carried out for working days and weekends separately.

The average daily prediction error for the full period considered (November 1998–December 2003) is 12.6%. Bearing in mind that forecasts have been calculated for a representative and long period of time, the results, in terms of prediction error, reflect the accuracy of the proposed new mixed model.

DESIGN OF EXPERIMENTS

The DOE methodology studies the way of making comparisons as homogenous as possible, so as to increase the probability of detecting changes or identifying influential variables. The main ideas presented here were developed by R.A. Fisher in [45], as a consequence of his work in the agricultural field. The early applications included biology, medicine, and natural sciences in general. Subsequently, following World War II and due to George Box, these ideas started being applied to industrial processes and engineering [46].

The main idea of the DOE is to investigate the effect of several factors on a dependent variable. In empirical research, we need to consider several factors simultaneously. The main objective of DOE is to compare several treatments. For this purpose, we must determine

1. the treatments to be compared (defined by one or more factors),
2. the response variable to be measured,
3. the type and number of experimental units to be used, as well as
4. the allocation of these experimental units to each treatment.

For example, in this work, a DOE is carried out to determine the influence of the factors "Model" and "Length" on the response or dependent variable "prediction error," in order to select the appropriate treatment, the level for each factor, and design an optimal model in terms of forecasting error.

Principles of the Design of Experiments

The objective is to study the effect on a variable of interest, hereafter referred as dependent or response variable of a set of other ones which are the factors. Some examples can be the efficiency of an industrial process depending on the temperature and pressure, or the time of connection to the Internet depending on the type of computer and the hour of the day.

The basic idea of factorial design is to obtain values of the response variable for all possible combinations of the levels of the factors under study. Each combination is named "treatment," and each one may be assigned to one experimental unit or several ones. When there is more than one experimental unit for each treatment, the factorial design has replications. For example, if there is a factor with I levels

Factor 1

		1	2	$\cdots$	I
		y_{111}	y_{211}		y_{I11}
	1	y_{112}	y_{212}	$\cdots$	y_{I12}
		$\vdots$	$\vdots$		$\vdots$
		y_{11m}	y_{21m}		y_{I1m}
		y_{121}	y_{221}		y_{I21}
Factor 2	2	y_{122}	y_{222}	$\cdots$	y_{I22}
		$\vdots$	$\vdots$		$\vdots$
		y_{12m}	y_{22m}		y_{I2m}
	$\vdots$	$\vdots$	$\vdots$	$\ddots$	$\vdots$
		y_{1J1}	y_{2J1}		y_{IJ1}
	J	y_{1J2}	y_{2J2}	$\cdots$	y_{IJ2}
		$\vdots$	$\vdots$		$\vdots$
		y_{1Jm}	y_{2Jm}		y_{IJm}

Figure 5-21 Structure of the data corresponding to an experiment with two factors and replications.

(e.g., I temperatures), and another one, for example, the pressure, with J levels, then, there are $I \times J$ possible treatments. The data could be represented in a double entry table as shown in Fig. 5-21, where y_{ijk} corresponds to the value of the response variable for the k th replication of the combination corresponding to the first factor fixed at level i, and the second one fixed to its j th level. It is important to distinguish between repeating each observation and repeating the experiment. If two measures are obtained consecutively for each treatment or experimental condition (each combination of factors), then two repetitions are obtained, but not necessarily two replications, since the variability of two consecutive measures is smaller than that corresponding to separate observations. For this reason, it is convenient to carry out the experiment once, and then, repeat it from the beginning, to avoid underestimating the variance.

The structure of the data corresponding to a factorial design with three factors is shown in Fig. 5-22. The factors considered have 6, 5, and 3 levels, respectively.

In any experiment in which the effect of a factor is investigated, there are a large number of variables that may influence the final results. There are several ways to eliminate the effect of a variable: fixing it during the realization of the experiment, reorganizing the structure of the experiment for interesting comparisons being carried out for a constant value of this variable or trying to eliminate its effect by randomizing its appearance in the experiment (which is applied in the case we do not control the effect of the variable and its effect is included in the error term).

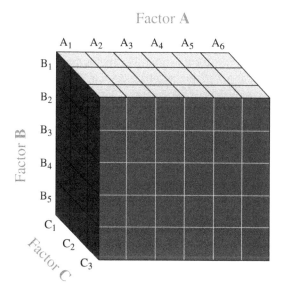

Figure 5-22   Structure of the data corresponding to an experiment with three factors.

The concept of Block   The ability to identify differences between treatments depends on several causes, one of which is the variability of the observations. The higher the variability, the more difficult it is to discover significant differences between them. An alternative for solving this problem is to use blocks. The idea is to group together the experimental units, so as to make these groups as homogeneous as possible (blocks). Then, the comparisons are carried out for each block. Since the experimental units are more homogeneous in those blocks, a great reduction in the experimental error is achieved.

A variable or factor whose effect on the dependent variable is not of interest *per se*, but is included to make homogeneous comparisons, and because it may explain a portion of the variability of the dependent variable whose inclusion in the residual variance would not be correct. For example, when trying to decide which printer is faster among five with similar price and features, four different photos are used. There is no doubt that one photo with higher resolution takes a longer period of time when being printed, whatever printer is used. In this case, the photograph is considered as a block, since it is not of our interest, but it may explain a part of the variability of the dependent variable. Moreover, considering the photograph as a block allows the comparisons between printers, since the effect of the photo is eliminated.

In this work, the objective is to develop a mixed model, obtaining the preferred combination (treatment) of "Model" (24 or 48) and "Length" (ranging from 8 to 80 weeks) in terms of prediction error. "Model" and "Length" are the factors under study. Moreover, the "Day" may affect the prediction error, and plays the same role of the "photograph" in the previous example, since it is not of interest. For each hour of the days in the period 1998–2003, a prediction of the price has been obtained with the ARIMA model fitted with each combination of "Model" ($I = 2$) and "Length"($J = 10$). Moreover, if there is an unexpected price in one of these days, none of the models built with any of the combinations of the levels of the factors will provide an accurate forecast for this hour. For the purpose of eliminating the effect of the day, we include this variable as a block. Thus, the factorial design considered in these case studies, the effect of two factors ("Model" and "Length") and one block (day) on the response variable: daily prediction error.

Factorial Design with Two Factors and One Block

In this case, we consider the logarithm of the average daily prediction error as the variable under study. The factors considered are the model (24 or 48) and the length of the time series (8, 12, 16, 20, 24, 28, 32, 44, 52, and 80 weeks). As mentioned earlier, the day is included as a block, which means not taking the interactions between the block and the other two factors ("Model" and "Length") into account. By doing so, apart from removing the effect of the day, the possible correlation between forecasting errors in consecutive days is also eliminated when including the day as a block. The nonexistence of correlations in the variable under study is one of the hypotheses assumed for DOE. The other hypotheses assumed are Gaussianity and homoscedasticity. The equation for the linear model of our computational experiment is given by

$$y_{ijt} = \mu + \alpha_i + \beta_j + \gamma_t + (\alpha\beta)_{ij} + u_{ijt},$$
$$u_{ijt} \sim \text{NID}(0, \sigma), \ i = 1, 2, \ j = 1, \ldots, 10 \text{ and } t = 1, \ldots, T, \qquad (5\text{-}19)$$

where y_{ijt} is the response, which is the logarithm of the daily mean absolute percentage error (MAPE) and μ is a global effect, that is, the average level of the response (logarithm of the prediction error). α_i are the main effects of the model. It measures the increase/decrease of the average response for model i with respect to the average level, thus $\sum_{i=1}^{I} \alpha_i = 0$.

β_j are the main effects of the length of the time series. It measures the increase/decrease of the average response for length j with respect to the average level, so $\sum_{j=1}^{J} \beta_j = 0$.

γ_t are likewise the main effects of the block (the day in this particular case), so $\sum_{t=1}^{T} \gamma_k = 0$, where T is the number of days.

$(\alpha\beta)_{ij}$ measures the difference between the expected value of the response and the one computed using a model that does not include the interactions, with $\sum_{i=1}^{I} \sum_{j=1}^{J} (\alpha\beta)_{ij} = 0$. The inclusion of this term in the model allows for the possibility of the effect of factor "Model" on the response variable depending on the level of the other factor, in this case the "Length."

The random effect, u_{ijt}, includes the effect of all other causes.

The parameters that must be estimated are the grand mean (μ) (the mean of the means of several subsamples), the main effects of factor "Model" and "Length" and block day, (α_i, β_j, and γ_t, respectively), the interactions between "Model" and

"Length" $(\alpha\beta)_{ij}$, and the variance σ^2. Thus, the number of parameters to be estimated is $1 + (I - 1) + (J - 1) + (T - 1) + (I - 1)(J - 1) + 1$, corresponding respectively to the grand mean, main effects, and interactions as well as the variance of the error term u_{ijt}.

The estimation is carried out by maximum likelihood. The estimates are obtained by minimizing the expression $\sum_i \sum_j \sum_t (y_{ijt} - \mu - \alpha_i - \beta_j - \gamma_t - (\alpha\beta)_{ij})^2$, which gives $\hat{\mu} = \bar{y}_{...}, \hat{\alpha}_i = \bar{y}_{i..} - \bar{y}_{...}, \hat{\beta}_j = \bar{y}_{.j.} - \bar{y}_{...}, \hat{\gamma}_t = \bar{y}_{..t} - \bar{y}_{...}$, and $\widehat{(\alpha\beta)}_{ij} = \bar{y}_{ij.} - \bar{y}_{i..} - \bar{y}_{.j.} + \bar{y}_{...}$.

Then, $\hat{y}_{ijt}$ is the forecast for the variable y_{ijt} and it is calculated using the expression $\hat{y}_{ijt} = \bar{y}_{ij.} + \bar{y}_{..t} - \bar{y}_{...}$.

e_{ijt} are residuals of the model, and they are the estimation of u_{ijt}, obtained as the difference between the observed and the expected value, then, $e_{ijt} = y_{ijt} - \bar{y}_{ij.} - \bar{y}_{..t} + \bar{y}_{...}$.

The estimation of the variance σ^2 is given by $\hat{s}_R^2$, which is calculated as follows:

$$\hat{s}_R^2 = \sum\sum\sum e_{ijt}^2, \quad \text{D.f.}\,(e_{ijt}) = \sum\sum\sum e_{ijt}^2 (T-1) \times (IJ-1).$$

The main objective is to test the significance of main effects and second order interactions, so the following tests are performed:

$H_0 : \alpha_1 = \alpha_2 = 0$ versus H_1 : any of the α_i different from zero.

$H_0 : \beta_1 = \beta_2 = \cdots = \beta_{10} = 0$ versus H_1 : any of the β_j different from zero.

$H_0 : \gamma_1 = \gamma_2 = \cdots = \gamma_t = 0$ versus H_1 : any of the γ_t different from zero.

$H_0 : (\alpha\beta)_{11} = (\alpha\beta)_{12} = \cdots = (\alpha\beta)_{ij} = 0$ versus H_1 : any of the $(\alpha\beta)_{ij}$ different from zero.

Finally, using the expression $\hat{y}_{ijt} = \hat{\mu} + \hat{\alpha}_i + \hat{\beta}_j + \hat{\gamma}_t + \widehat{(\alpha\beta)}_{ij} + e_{ijt}$, the decomposition of the variability in terms of its independent components is obtained. $SS_{\text{TOTAL}} = SS_A + SS_B + SS_C + SS_{AB} + SS_{\text{ERROR}}$. The terms SS_A, SS_B, SS_C, and SS_{AB} are respectively, the variability explained by the factors A and B, the block C, and the interaction between A and B. The structure of the table of the analysis of variance (ANOVA) is shown in Table 5-2.

The resulting tests are

For $H_0 : \alpha_i = 0 \quad \forall i;\ F_{(I-1),(T-1)\times(IJ-1)} = \hat{s}_\alpha^2 / \hat{s}_R^2$

$H_0 : \beta_j = 0 \quad \forall j;\ F_{(J-1),(T-1)\times(IJ-1)} = \hat{s}_\beta^2 / \hat{s}_R^2$

$H_0 : \gamma_t = 0 \quad \forall t;\ F_{(T-1),(T-1)\times(IJ-1)} = \hat{s}_\gamma^2 / \hat{s}_R^2$

$H_0 : (\alpha\beta)_{ij} = 0 \quad \forall i,j;\ F_{(I-1)\times(J-1),(T-1)\times(IJ-1)} = \hat{s}_{\alpha\beta}^2 / \hat{s}_R^2$

If the F-statistics obtained in the table with the corresponding degrees of freedom are smaller than those obtained in the corresponding tests, then the zero hypothesis is rejected and we conclude that the corresponding main effect or interaction is significant. If not, the null hypothesis is not rejected. Only in case the test for the interaction is clearly not significant (F-statistic smaller or even equal to one), it is

Table 5-2 ANOVA Table Structure

Source of Variability	Sum of Squares	Degrees of Freedom	Mean Square	F-Stat
Main effect A	$SS_A = JT \sum_i \hat{\alpha}_i^2$	$I - 1$	$\hat{s}_\alpha^2$	$\hat{s}_\alpha^2 / \hat{s}_R^2$
Main effect B	$SS_B = IT \sum_j \hat{\beta}_j^2$	$J - 1$	$\hat{s}_\beta^2$	$\hat{s}_\beta^2 / \hat{s}_R^2$
Main effect C	$SS_C = IJ \sum_t \hat{\gamma}_t^2$	$T - 1$	$\hat{s}_\gamma^2$	$\hat{s}_\gamma^2 / \hat{s}_R^2$
Interaction AB	$SS_{AB} = T \sum_i \sum_j \widehat{\alpha\beta}_{ij}^2$	$(I - 1) \times (J - 1)$	$\hat{s}_{\alpha\beta}^2$	$\hat{s}_{\alpha\beta}^2 / \hat{s}_R^2$
Residual	$SS_{ERROR} = \sum_i \sum_j \sum_t e_{ijt}^2$	$(T - 1) \times (IJ - 1)$	$\hat{s}_R^2$	
Total	$SS_{TOTAL} = \sum_i \sum_j \sum_t (y_{ijt} - y_{...})^2$	$IJT - 1$		

convenient to sum up the sum of squares corresponding to both the interaction and the residuals.

In our case, the response, y_{ijt}, which is the logarithm of the daily MAPE (logs are usually taken in the dependent variable in order to avoid heteroskedasticity), has been calculated as follows:

$$y_{ijt} = \log(\text{daily MAPE}) = \log \left(\frac{1}{24} \sum_{h=1}^{24} \frac{|\hat{p}_t^h - p_t^h|}{p_t^h} \right), \tag{5-20}$$

where the forecast of the price $\hat{p}_t^h$ for the day t in the hour h has been calculated using model i and by estimating the ARIMA model using the previous j observations. In the well-known formulation of the ARIMA $(p, d, q) \times (P, D, Q)$ models

$$\phi_p(B)\,\Phi_P(B^s)\,\nabla^d\,\nabla_S^D\,p_t^h = \theta_q(B)\,\Theta_Q(B^s)\,a_t, \tag{5-21}$$

where $\phi_p(B) = (1 - \phi_1 B - \phi_2 B^2 - \cdots - \phi_p B^p)$, $\Phi_P(B^s) = (1 - \Phi_1 B^s - \Phi_2 B^{2s} - \cdots - \Phi_P B^{Ps})$, $\nabla^d = (1 - B)^d$, $\nabla_S^D = (1 - B^s)^D$, $\theta_q(B) = (1 - \theta_1 B - \theta_2 B^2 - \cdots - \theta_p B^p)$, and $\Theta_Q(B^s) = (1 - \Theta_1 B^s - \Theta_2 B^{2s} - \cdots - \Theta_P B^{Ps})$, where $Bp_t^h = p_{t-1}^h$.

As explained in Time Series Analysis and ARIMA Models, the roots of polynomials $\phi_p(B)$, $\Phi_P(B^s)$, $\theta_q(B)$, and $\Theta_Q(B^s)$ must be outside the unit circle for stationarity.

The forecasts, $\hat{p}_h^t$, have been calculated minimizing the expression $E\left[(p_t^h - \hat{p}_h^t)^2\right]$, which means that the best predictor is the expectation of the conditional distribution:

$$\hat{p}_h^t = E\left[p_t^h \mid p_{t-1}^h, p_{t-2}^h, \ldots, p_1^h\right] = E\left[p_t^h \mid \mathbf{p}_{t-1}^h\right],$$

where $\mathbf{p}_{t-1}^h = (p_{t-1}^h, p_{t-2}^h, \ldots, p_1^h)$.

The expression for the response is

$$
y_{ijt} = \log \left(\frac{1}{24} \sum_{h=1}^{24} \frac{\left| E\left[p_t^h \mid \mathbf{p}_{t-1}^h \right] - p_t^h \right|}{p_t^h} \right).
$$

Designing a Mixed Model

We will determine whether it is useful or not to identify, estimate, and forecast using different models for working days and weekends, as well as the optimal number of observations in the time series used to build the models. To select the appropriate "Model" and "Length", we carried out a computational experiment.

Bearing in mind the results of Factorial Design with Two Factors and One Block, working days and weekends are studied separately. Bonferroni adjustments have been applied to solve multiple comparisons problem [47].

This adjustment relies on Bonferroni's inequality. Let c be the total number of comparisons that must be carried out (when having G groups, $c = \begin{pmatrix} G \\ 2 \end{pmatrix}$). Let $\overline{A}_i$ be the event of rejecting that the values of the two means μ_i, μ_j are equal when they are. It can be assumed that the comparisons between these means is carried out for a significance level α, that is, $P(\overline{A}_i) = \alpha$. Let $\mathbf{B}$ be the event "rejecting one or more equality tests when all the means are equal." Thus, $\mathbf{B}$ will be the union of the events $\overline{A}_1, \mathbf{B} = \overline{A}_1 + \cdots + \overline{A}_c$. The events $\overline{A}_i$ are not mutually exclusive, so, prob($\mathbf{B}$) = prob $(\overline{A}_1 + \cdots + \overline{A}_c) \le \sum \text{prob} (\overline{A}_i) = c\alpha$.

If we want to guarantee a total type I error α_T, the probability of the event $\mathbf{B}$ cannot be larger than α_T. This can be done taking for each individual test, a significance level α, such that $\alpha = \alpha_T/c$. When the total number of comparisons, c, is very large, then we may use very small α that the corresponding values t_v^α needed for the calculations were not tabulated. In that case, an approximation is used:

$$
t_v^\alpha = z_\alpha \left(1 - \frac{z_\alpha + 1}{4v} \right)^{-1}.
$$

Table 5-3 shows the ANOVA table for weekends. Interactions are not significant, but the main effects are significant. Moreover, since the F-statistics corresponding to the interaction is $0.50 < 1$, the model can be re-estimated without the interaction term as explained before. Table 5-4 provides the ANOVA with the interactions removed. Note that $SS_{BC} + SS_{ERROR} = 0.34 + 121.22 = SS_{ERROR} = 121.56$.

Figures 5-23 and 5-24 show the main effects of "Model" and "Length" for weekends, respectively. The prediction error is significantly smaller using Model 24, that is, building the models using the electricity prices of the complete week. There is no significant difference between the prediction errors obtained building the models with prices of the previous 32, 44, or 80 weeks, as the intervals overlap; nevertheless,

Table 5-3   ANOVA for Weekends Including Interactions

Source	Sum of Squares	Degrees of Freedom	Mean Square	F-Stat	p-Value
A: day	190.89	82	2.32	29.24	0.0
B: length	0.79	9	0.08	0.98	0.45
C: model	142.17	1	142.17	1785.91	0.0
BC: length-model	0.34	9	0.04	0.50	0.89
Residual	121.22	1518	0.08		
Total	452.88	1619			

Table 5-4   ANOVA for Weekends Without Interactions

Source	Sum of Squares	Degrees of Freedom	Mean Square	F-Stat	p-Value
A: day	190.89	82	2.32	29.24	0.0
B: length	0.79	9	0.08	0.98	0.45
C: model	142.17	1	142.17	1785.91	0.0
Residual	121.56	1527	0.08		
Total	452.88	1619			

a lower level can be observed for 44 weeks. From this "Length" on, daily prediction errors increase.

For weekdays, the results of the ANOVA are shown in Table 5-5. For weekends, the interaction is not significant and its F-statistic is $0.64 < 1$, so the model can be re-estimated without it. By doing so, the ANOVA table obtained is shown in Table 5-6. Note that $SS_{BC} + SS_{ERROR} = 0.25 + 414.63 = SS_{ERROR} = 414.88$.

Figure 5-25 shows the convenience of Model 48 for working days. It can be observed from Fig. 5-26 that the daily prediction error decreases when we increase the

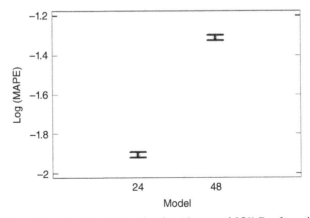

Figure 5-23   Main effect Model (weekends). Means and 95% Bonferroni intervals.

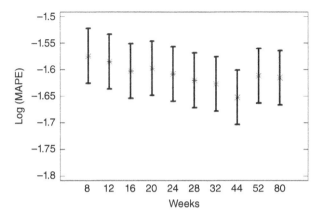

Figure 5-24   Main effect Length (weekends). Means and 95% Bonferroni intervals.

Table 5-5   ANOVA for Working Days, Including Interactions

Source	Sum of Squares	Degrees of Freedom	Mean Square	F-Stat	p-Value
A: day	4310.46	506	8.52	194.3	0.0
B: length	12.21	9	1.35	30.95	0.0
C: model	16.69	1	16.69	380.80	0.0
BC: length-model	0.25	9	0.03	0.64	0.76
Residual	414.63	9454	0.04		
Total	4753.01	9979			

number of weeks used to build the ARIMA models for the hourly time series. Using more than 44 weeks reduces the prediction error but not significantly (the intervals corresponding to 52 and 80 weeks almost completely overlap and the increase in the number of weeks is very large). Bearing this in mind and for simplicity, it would also be adequate to use the prior 44 weeks (as recommended for weekends) to build the models for weekdays. Although a decrease is shown in terms of prediction error when using the previous 80 weeks, the improvement is not significant.

Table 5-6   ANOVA for Working Days, Without Interactions

Source	Sum of Squares	Degrees of Freedom	Mean Square	F-Stat	p-Value
A: day	4310.46	506	8.52	194.3	0.0
B: length	12.21	9	1.35	30.95	0.0
C: model	16.69	1	16.69	380.80	0.0
Residual	414.88	9463	0.03		
Total	4753.01	9979			

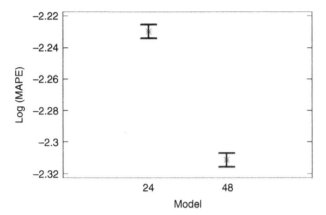

Figure 5-25   Main effect Model (weekdays). Means and 95% Bonferroni intervals.

Till now and by default, the ARIMA models proposed to forecast short-term electricity prices used the previous 10–12 weeks, but no study about the effect of the length has been published.

The main difference between our proposed methodology and the previous ones is that we propose fitting different models for each hour. The fact that prediction errors became always smaller when adding more information indicates that the 24 processes that generate hourly prices are much more homogeneous than the process that generates the complete time series. This is relevant in workdays when there is always a decrease in terms of prediction error while computing forecasts with models that have been estimated with longer time series.

The main conclusions of DOE are

- The convenience of using about 44 weeks prior to the day for which we forecast for weekends.

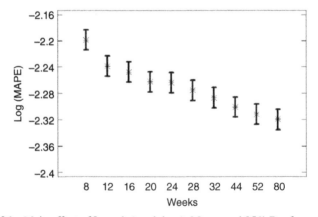

Figure 5-26   Main effect of Length (weekdays). Means and 95% Bonferroni intervals.

- The convenience of using 44 or more weeks for working days (although the mean is lower for 80 weeks, there are no significant differences between them).
- For weekdays, better estimations of the parameters are obtained when including more data, this means that the generating processes of the 24-hourly time series are very homogeneous, much more than the process generating the complete time series.
- The development of a mixed model that forecasts for working days with Model 48 and weekends using Model 24.

NUMERICAL RESULTS

The main goal of this work is to develop an alternative simple model for obtaining accurate 1-day-ahead electricity forecasts for electricity prices.

Using the new mixed model developed taking into account the conclusions of the Design of Experiments, some results are shown now to evaluate the accuracy of the forecasts.

We remark that the major difference between our approach and others is that we have computed forecasts for every hour in the period of 1998–2003, and we have done this for the 20 possible combinations of the two factors under study and then selected the appropriate levels for these factors. That is why we could say that the mixed model that was developed is the one with the best "general performance" for the whole period considered.

In this section, we first show numerical results for 8 specific weeks, 6 of them have been chosen to be comparable with the results from the previous work [3, 12].

Our mixed model computes accurate forecasts in these specific weeks, but we want to highlight the global results we obtain for such a very long period of time (1998–2003). The first week selected is the last one in May 2000 (25–31). Figure 5-27 shows the real values and the forecasts.

The daily mean errors for this week appear in Table 5-7, as well as the prediction errors obtained with the model proposed by Contreras et al. [3]. They dealt with the complete time series of the electricity prices instead of the hourly ones, so not only weekly seasonality but also daily one must be modeled. For this purpose, they fitted a multiplicative ARIMA model, as in Table 5-7, but with double seasonality ($s_1 = 24$ and $s_2 = 7 \times 24 = 168$), which gives

$$\phi(B)\Phi_1(B^{s_1})\Phi_1(B^{s_2})\Phi_2(B^{s_2})(1 - B)^d(1 - B^{s_1})^{D_1}(1 - B^{s_2})^{D_2} \tilde{z}_t$$
$$= \theta(B)\Theta_1(B^{s_1})\Theta_2(B^{s_2})a_t.$$

Contreras et al. [3] used the SCA system to compute 24-hour ahead forecasts, using exact maximum likelihood as well as the automatic detection and intervention of outliers. The order of the parameters was selected by the authors using the ACF

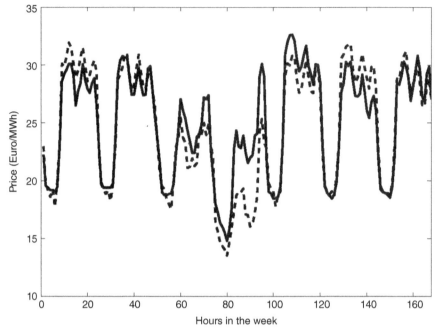

Figure 5-27 Forecasts and real prices (May 25–31, 2000).

and PACF, as explained in Time Series Analysis and ARIMA Models. The model finally estimated by these authors was

$$
\begin{aligned}
&\left(1 - \phi_1 B - \phi_2 B^2 - \phi_3 B^3 - \phi_4 B^4 - \phi_5 B^5\right) \times \left(1 - \phi_{23} B^{23} - \phi_{24} B^{24} - \phi_{47} B^{47} -\right.\\
&- \phi_{48} B^{48} - \phi_{72} B^{72} - \phi_{96} B^{96} - \phi_{47} B^{47} - \phi_{120} B^{120} - \phi_{144} B^{144}\left.\right) \times \left(1 - \phi_{168} B^{168} -\right.\\
&- \phi_{336} B^{336} - \phi_{504} B^{504}\left.\right) \log p_t = c + \left(1 - \theta_1 B - \theta_2 B^2\right) \times \left(1 - \theta_{24} B^{24}\right) \times\\
&\times \left(1 - \theta_{168} B^{168} - \theta_{336} B^{336} - \theta_{504} B^{504}\right) a_t.
\end{aligned}
$$

Since using our mixed model for computing forecasts for one complete week, $24 \times 7 = 168$, different models are estimated. Here, only the results corresponding

Table 5-7 Prediction Errors May 25–31, 2000

Prediction Error (Day)	Mixed Model Proposed (%)	Contreras et al. [3] (%)
1	4.5	4.73
2	1.9	4.13
3	4.2	3.71
4	16.5%	6.84
5	4.0	6.09
6	5.6	6.09
7	1.7	3.41

Table 5-8   Models and Diagnosis Checking

Hour	Model	Q-Val	$h-n$	χ^2_{h-n}
5	ARIMA $(0,1,1) \times (0,1,1)_5$	44.31	34	48.6
10	ARIMA $(0,1,2) \times (0,0,0)_5$	25.06	35	49.8
14	ARIMA $(1,0,1) \times (0,0,0)_5$	43.24	34	48.6
18	ARIMA $(0,1,1) \times (0,0,0)_5$	37.50	35	49.8
22	ARIMA $(1,1,1) \times (0,0,0)_5$	38.52	34	48.6

to some selected hours on May 26, 2000 are provided in Tables 5-8 and 5-9. Results for Hour 5, Hour 10, Hour 14, Hour 18, and Hour 22 are provided. They have been chosen as representative hours, as it could be observed in Fig. 5-16.

As it is a Friday (working day), Model 48 has been used. So, the models were built using the previous 5-day weeks (considering only weekdays). The notation used was introduced in (5-21).

Table 5-8 also includes the value of the Q-statistic, Ljung–Box (1978), used to check for the remaining correlations between the residuals after fitting the model. The Ljung–Box test is a global one for the h first autocorrelation coefficients.

The Q-statistic for the h first autocorrelation coefficients, $Q(h)$, is asymptotically distributed as a chi-squared with $h - n$ degrees of freedom, where n is the number of the estimated parameters.

In general, we will reject the null hypothesis of uncorrelated residuals when the probability $\Pr((\chi^2_{h-n}) > Q(h))$ is small, usually smaller than a significant level of 0.05 or 0.01.

The alternative mixed model proposed in this chapter increases the accuracy of the forecasts, since for 5 days (out of 7), our daily mean error is smaller.

The second week for which results are shown is the last one in August 2000 (25–31). It is usually a low demand week. Figure 5-28 shows the real prices and the forecasts. Table 5-10 compares the prediction errors obtained by applying our alternative methodology and the ones that appear in Contreras et al. [3].

The other four weeks selected are February 18–24, 2002, May 20–26, 2002, August 19–25, 2002, and November 18–24, 2002. We provide MWE instead of the MAPE calculated for the other weeks shown in this section, since MWE is the accuracy

Table 5-9   Estimated Parameters

Hour	Estimation of the Parameters	
5	$\theta_1 = -0.47$	$\Theta_1 = -0.99$
10	$\theta_1 = -0.37$	$\theta_2 = -0.32$
14	$\phi_1 = -0.81$	$\theta_1 = -0.31$
18	$\theta_1 = -0.57$	
22	$\phi_1 = -0.35$	$\theta_1 = -0.80$

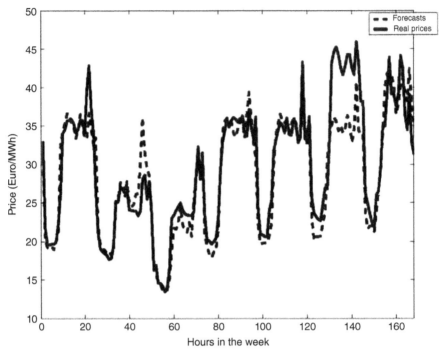

Figure 5-28   Forecasts and real prices (August 25–31, 2000).

metric provided for these weeks by Conejo et al. [12]. The MWE been calculated as proposed in

$$\text{MWE} = \frac{1}{168} \sum_{h=1}^{168} \frac{|p^h - \hat{p}^h|}{\bar{p}^h}$$

The results obtained with our proposed mixed model and those obtained by Conejo et al. [12] with ARIMA methodology are given in Table 5-11.

Table 5-10   Prediction Errors, August 25–31, 2000

Prediction Error (Day)	Mixed Model Proposed (%)	Contreras et al. [3] (%)
1	4.8	4.3
2	7.3	7.99
3	5.4	4.57
4	4.6	10.81
5	5.1	6.12
6	14.9	17.34
7	7.2	6.05

Table 5-11   MWE for Some Weeks in 2002

Week	Mixed Model (%)	ARIMA Model [12] (%)
February 18–24, 2002	6.15	6.32
May 20–26, 2002	4.46	6.36
August 19–25, 2002	14.90	13.39
November 18–24, 2002	11.68	13.78

Figure 5-29 provides the results for the week February 18–24, 2002.

The results for the week May 20–26, 2002 are shown in Fig. 5-30.

The results obtained for the week August 19–25, 2002 are given in Fig. 5-31. MWE is larger than in other weeks selected since some outliers can be observed, especially during 11th and 22nd hour of Wednesday and Thursday.

Figure 5-32 shows the forecasts obtained for the third week in November 2002 (18–24)

The seventh week selected is the one before the last in December 2001 (15–21). It has been chosen because of the existence of a peak in demand, which influences the prices, as can be observed in Fig. 5-33, where we show the hourly prices in the

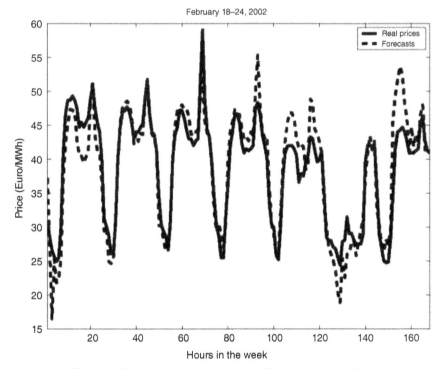

Figure 5-29   Forecasts and real prices, February 18–24, 2002.

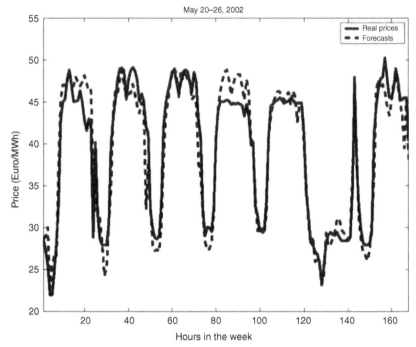

Figure 5-30 Forecasts and real prices, May 20–26, 2002.

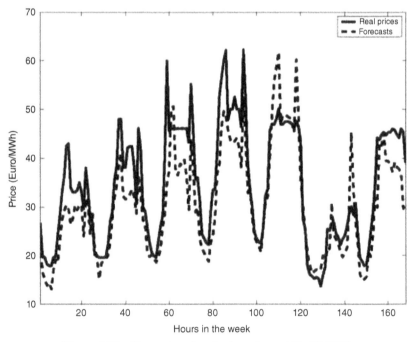

Figure 5-31 Forecasts and real prices, August 19–25, 2002.

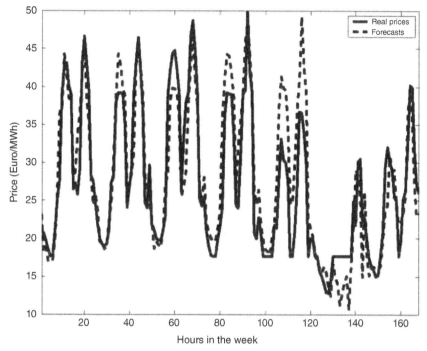

Figure 5-32   Forecasts and real prices, November 18–24, 2002.

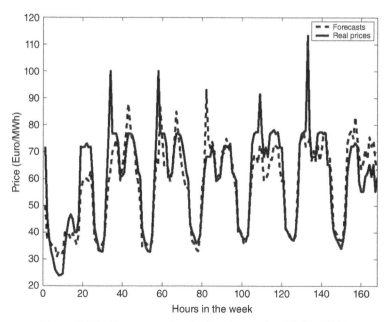

Figure 5-33   Forecasts and real prices (December 15–21, 2001).

Table 5-12   Prediction Errors
December 15–21, 2001

Day	Prediction Error (%)
1	17.9
2	8.7
3	9.4
4	5.7
5	9.5
6	5.3
7	11.5

whole period under study. The behavior of the time series at the end of 2001 and in the beginning of 2002 is clearly affected by the great increase in demand, caused by extremely low temperatures.

We emphasize the accuracy of the forecasts of the mixed model. Even when the behavior of the time series of the prices is rather unexpected, the alternative model proposed in this paper works properly, that is, the prediction errors are small. The average prediction error for the whole week is 9.7%. The prediction errors appear in Table 5-12. Figure 5-33 shows the real prices and the forecasts for this week (December 15–21, 2001).

Finally, the last week selected is the third one in April 2002. Since the ARIMA forecasts for 1 day are built using the 20 previous weeks. It is clear that the

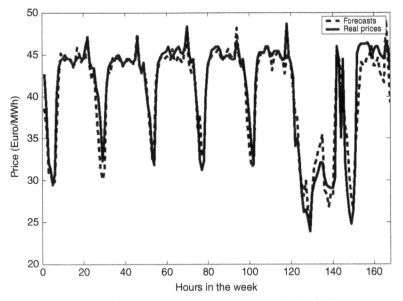

Figure 5-34   Real prices and forecasts April 15–21, 2002.

Table 5-13 Prediction Errors,
April 15–21, 2002

Day	Prediction Error (%)
1	2.2
2	2.8
3	2.5
4	2.1
5	2.5
6	7.5
7	7.6

observations corresponding to the end of 2001 and the beginning of 2002, when the prices were extremely high and volatile, are being taken into account to build the model. Nevertheless, this does not affect negatively the accuracy of the forecasts. Real prices and forecasts are shown in Fig. 5-34. Prediction errors appear in Table 5-13.

As we have computed forecasts for every hour in the period under study, Table 5-14 (content is illustrated in Fig. 5-35), includes the quantiles Q_{25}, Q_{50}, and Q_{75} of the daily MAPE, depending on the day of the week. Using 80 weeks to build weekday models does not allow the computation of forecasts until week 81 in the period under study, which means that the first day for which we can compute a forecast is July 21, 1999. However, although the prediction error corresponding to "Length" equal to 80 weeks is lower, there are no statistically significant differences between using 44 or 80 weeks. Here, we provide global results for "Length" 44 weeks in order to be able to show results for a longer period since if we use 44 weeks, the first week for which we can compute forecasts is week 45 of the period under study (so, November 8, 1998).

The results indicate that the mixed model designed provides accurate forecasts, not only for specific weeks but also in general conditions, as this period (1998–2003) includes stages of different levels and variability in prices.

Table 5-14 Global Results, Mixed Model. Quantiles of Daily MAPE (1998–2003)

	Q_{25} (%)	Q_{50} (%)	Q_{75} (%)
Monday	7.9	11.8	19.4
Tuesday	5.7	8.5	14.4
Wednesday	5.7	8.6	12.5
Thursday	4.9	7.3	12.4
Friday	5.1	7.5	12.1
Saturday	8.7	11.8	17.7
Sunday	8.7	12.3	18.0

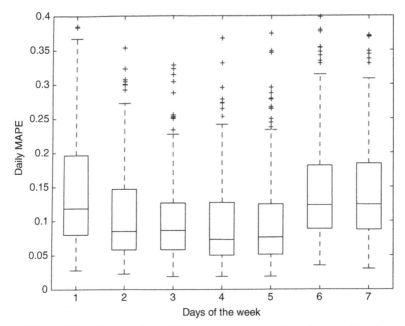

Figure 5-35 Boxplot of prediction errors for the whole period considered.

We have developed a "global model" that almost always computes accurate forecasts. Indeed, as the errors obtained for the 6 years under study are not higher or even lower than the ones obtained by other authors for specific weeks, it seems that we reached our goal.

CONCLUSIONS

This chapter develops a straightforward methodology for building mixed models for forecasting 1-day-ahead electricity prices. Several methods corresponding to different combinations of factor levels (convenience or not of analyzing separately the prices in working days and weekends and the length of the time series used to forecast) are compared. The mixed model is built combining the advantages of several of these methods. Moreover, a tutorial review of the theoretical basis of the topics used to develop the models presented in the chapter (time series analysis, LOESS, and DOE) has been included.

A complete study was carried out to choose between Model 24 and 48 and to determine the optimal length of the time series. The analysis, given the size and representativeness of the sample (1998–2003), is exhaustive and allows one to draw valid conclusions for price forecasting in the future.

A mixed model to forecast next-day electricity prices is developed. We recommend computing forecasts for working days with Model 48 (using only weekdays) and with Model 24 for weekends (using complete weeks). We have also determined the optimal length of the time series used to estimate the models for weekends (about 44 weeks) and an appropriate length starting from which, prediction errors do not decrease significantly for weekdays. Splitting the complete time series into 24-hourly time series leads to a much more homogeneous generating process, which means that adding more information (longer series) allows the computation of better estimates of the parameters of the ARIMA models and this is reflected in terms of prediction errors.

These conclusions have required a great computational effort, as the price for every hour in the period under study has been computed using the 20 models adjusted and estimated for the 20 possible combination of the factors "Model" (2 levels) and "Length" (10 levels).

Once this computational effort has been done to design the proposed mixed model, computing a forecast for the 24-hourly prices of tomorrow having the prices of the prior weeks, takes less than 20 seconds. In addition, these forecasts are computed using TRAMO, a free and easy-to-use software. The model could be useful for different agents interested on having accurate next-day forecasts for electricity prices. Once the mixed model has been proposed, no prior knowledge about time series analysis is required for producing a new forecast.

Forecasts have been computed for every hour in the years 1998–2003, and the average error for the whole period is 12.61%. Weightings were calculated adequately (5 : 2), 11.9% obtained for weekdays and 14.4% obtained for weekends. These results reflect the excellent performance and accuracy of the new mixed model, not only for a few weeks or days but for a very long and significant sample size.

The idea of building mixed models to forecast 1-day-ahead electricity prices can also be applied to other electricity markets, like the PJM interconnection, for which Nogales and Conejo [48] used ARIMA models and transfer function models to obtain forecasting errors of similar magnitude as the ones we have for the Spanish market.

For future research it would be of interest building mixed transfer function models by taking into account not only the past values of the price for forecasting the hour, but also past values of the price in the adjacent hours, or including explanatory variables such as demand or temperature using the methodology developed in Cottet and Smith [14] or in Nogales and Conejo [48].

This is an extended version of the paper García-Martos, C., Rodríguez, J. and Sánchez, M. J. (2007). "Mixed models for short-run forecasting of electricity prices: application for the Spanish market" IEEE Transactions on Power Systems, 2, 544–552

ACKNOWLEDGMENTS

This work was supported in part by Projects MTM2005-08897, MTM2009-12419, and ECO2009-10287, Ministry of Science, Spain. The authors would like to thank Andrés M. Alonso, José Mira, Javier Nogales, and Daniel Peña for their help and comments, as well as Juan Bógalo for his modifications on the interface TS of TRAMO for MATLAB. Carolina García-Martos and María Jesús Sánchez acknowledge financial support by Project DPI2011-23500, Ministry of Economy and Competitiveness,

Spain. Julio Rodríguez acknowledges financial support by Projects CTM2016-79741-R (MINECO) and MAdECo-CM(S2015/HUM-3444).

REFERENCES

1. A.J. Conejo, F.J. Nogales and J.M. Arroyo, "Price-taker bidding strategy under price uncertainty," *IEEE Transactions on Power Systems*, vol. 17, no. 4, pp. 1081–1088, 2002.

2. A.J. Conejo, F.J. Nogales, J.M. Arroyo and R. García-Bertrand, "Risk constrained self scheduling of a thermal power producer," *IEEE Transactions on Power Systems*, vol. 19, no. 3, pp. 1569–1574, 2004.

3. J. Contreras, R. Espínola, F.J. Nogales, and A.J. Conejo "ARIMA models to predict next-day electricity prices," *IEEE Transactions on Power Systems*, vol. 18, no. 3, pp. 1014–1020, 2003.

4. A.J. Conejo, J. Contreras, R. Espínola, and M.A. Plazas, "Forecasting electricity prices for a day-ahead pool-based electric energy market," *International Journal of Forecasting*, vol. 21, no. 3, pp. 435–462, 2005.

5. B. Ramsay and A.J. Wang, "A neural network based estimator for electricity spot-pricing with particular reference to weekend and public holidays," *Neurocomputing*, vol. 23, no. 1–3, pp. 47–57, 1998.

6. C.P. Rodríguez and G.J. Anders, "Energy price forecasting in the Ontario competitive power system market," *IEEE Transactions on Power Systems*, vol. 19, no. 1, pp. 366–374, 2004.

7. B.R. Szkuta, L. A. Sanabria and T.S. Dillon, "Electricity price short-term forecasting using artificial neural networks," *IEEE Transactions on Power Systems*, vol. 14, no. 3, pp. 851–857, 1999.

8. J.D. Nicolaisen, C.W. Richter and G.B. Sheblé, "Price Signal Analysis for Competitive Electric Generation Companies Using Artificial Neural Networks," Conference on Electricity Utility Deregulation and Restructuring and Power Technologies, 2000.

9. F.J. Nogales, J. Contreras, A.J. Conejo and R. Espínola, "Forecasting next-day electricity prices by time series models," *IEEE Transactions on Power Systems*, vol. 17, no. 2, pp. 342–348, 2002.

10. A. Troncoso, J. Riquelme, J. Riquelme, A. Gómez and J.L. Martínez "A comparison of two techniques for next-day electricity price forecasting," *Lecture Notes in Computer Science*, vol. 3177, pp. 384–390, 2002.

11. J. Crespo-Cuaresma, Kossmeier S. Hlouskova and M. Obersteiner, "Forecasting electricity spot-prices using linear univariate time-series models," *Applied Energy*, vol. 77, no. 1, pp. 87–106, 2004.

12. A.J. Conejo, M.A. Plazas, R. Espínola and A.B. Molina, "Day-ahead electricity price forecasting using wavelet transform and ARIMA models," *IEEE Transactions on Power Systems*, vol. 20, no. 2, pp. 1035–1042, 2005.

13. A.M. Alonso, C. García-Martos, J. Rodríguez and M.J. Sánchez, "Seasonal Dynamic Factor Analysis and Bootstrap Inference: Application to Electricity Market Forecasting," Working Paper 08–14, Statistic and Econometric Series, Universidad Carlos III de Madrid, 2008.

14. R. Cottet and M. Smith, "Bayesian modeling and forecasting of intraday electricity load," *Journal of the American Statistical Association*, vol. 98, no. 464, pp. 839–849, 2003.

15. I. Vehviläinen and T. Pyykkönen, "Stochastic factor model for electricity spot price: the case of the Nordic market," *Energy Economics*, vol. 27, no. 2, pp. 351–367, 2005.

16. A.C. Harvey and S.J. Koopman, "Forecasting hourly electricity demand using time varying splines," *Journal of the American Statistical Association*, vol. 88, pp. 1228–1236, 1993.

17. G.E.P. Box and G.M. Jenkins, *Time Series Analysis: Forecasting and Control*, revised edition, Holden Day, San Francisco, 1976.

18. D. Peña, G.C. Tiao, and R.S. Tsay, *A Course in Time Series Analysis*, Wiley, 2000.

19. J. Durbin and S.J. Koopman, *Time Series Analysis by State Space Methods*, Oxford University Press, 2001.

20. D. Peña and J. Rodríguez, "The log of the determinant of the autocorrelation matrix for testing goodness of fit in time series," *Journal of Statistical Planning and Inference*, vol. 136, no. 8, pp. 2706–2718, 2006.

21. G.E.P. Box and D.A. Pierce, "Distribution of residual autocorrelations in autoregressive-integrated moving average time series models," *Journal of the American Statistical Association*, vol. 65, no. 332, pp. 1509–1526, 1970.

22. G.M. Ljung and G.E.P. Box, "On a measure of lack of fit in time series," *Biometrika*, vol. 65, no. 2, pp. 297–303, 1978.

23. A.C. Monti, "A proposal for residual autocorrelation test in linear models," *Biometrika*, vol. 81, no. 4, pp. 776–780, 1994.

24. T.W. Anderson, "Goodness of fit tests for spectral distributions," *Annals of Statistics*, vol. 21, no. 2, pp. 830–847, 1993.

25. S. Velilla, "A goodness-of-fit test for autoregressive moving-average models based on the standardized sample spectral distribution of the residuals," *Journal of Time Series Analysis*, vol. 15, no. 6, pp. 637–647, 1994.

26. Y. Hong, "Consistent testing for serial correlation of unknown form," *Econometrica*, vol. 64, no. 4, pp. 837–864, 1996.

27. D. Peña and J. Rodríguez"A powerful portmanteau test of lack of fit for time series," *Journal of the American Statistical Association*, vol. 97, no. 458, pp. 601–610, 2002.

28. D. Peña and J. Rodríguez, "Descriptive measures of multivariate scatter and linear dependence," *Journal of Multivariate Analysis*, vol. 85, no. 2, pp. 361–374, 2003.

29. E. Hannan, *Time Series Analysis*, Wiley, New York, 1960.

30. F.L. Ramsey, "Characterization of the partial autocorrelation function," *Annals of Statistics*, vol. 2, no. 6, pp. 1296–1301, 1974.

31. A.I. McLeod and W.K. Li, "Diagnostic checking ARMA time series models using squared-residual autocorrelations," *Journal of Time Series Analysis*, vol. 4, no. 4, pp. 269–273, 1983.

32. J. Rodríguez and E. Ruiz, "A powerful test for conditional heteroscedasticity for financial time series with highly persistent volatilities," *Statistica Sinica*, vol. 15, no. 2, pp. 505–526, 2005.

33. R.S. Tsay, "Nonlinear time series analysis, in *A Course in Time Series Analysis*," Peña, D., Tiao, G.C., and Tsay, R.S. (eds), John Wiley & Sons, Inc., 2001.

34. W. Brock, D. Dechert, J. Scheinkman, and B. LeBaron, "A test for independence based on the correlation dimension," *Econometric Reviews*, vol. 15, pp. 197–235, 1996.

35. H. Akaike, "Fitting autoregressive models for prediction," *Annals of the Institute of Statistical Mathematics*, vol. 21, no. 1, pp. 243–247, 1969.

36. H. Akaike, "Information Theory and an Extension of the Maximum Likelihood Principle," Proceedings of the 2nd International Symposium on Information Theory, pp. 267–281, Budapest, 1973.

37. R. Shibata, "Asymptotically efficient selection of the order of the model for estimating parameters of a linear process," *Annals of Statistics*, vol. 8, no. 1, pp. 147–164, 1980.

38. C.M. Hurvich and C. Tsai, "Regression and time series model selection in small samples," *Biometrika*, vol. 76, no. 2, pp. 297–307, 1989.

39. G. Schwarz, "Estimating the dimension of a model," *Annals of Statistics*, vol. 6, no. 2, 1978.

40. E.J. Hannan and B.G. Quinn, "The determination of the order of an autoregression," *Journal of the Royal Statistical Society B*, vol. 41, pp. 190–195, 1979.

41. P. Galeano and D. Peña (2004), "Model selection criteria and quadratic discrimination in time series," Working Paper 04–14, Universidad Carlos III de Madrid.

42. G. Caporello and A. Maravall, *TSW Revised Reference Manual*, Banco de España, 2004.

43. W.S. Cleveland, "Robust locally weighted regression and smoothing scatterplots," *Journal of American Statistical Association*, vol. 74, pp. 829–836, 1979.

44. W.S. Cleveland and S.J. Devlin, "Locally weighted regression: an approach to regression analysis by local fitting," *Journal of the American Statistical Association*, vol. 83, no. 403, pp. 596–610, 1988.

45. R.A. Fisher, *The Design of Experiments*, Hafner Publishing, 1935. Available at http://www.phil.vt.edu/dmayo/PhilStatistics/b%20Fisher%20design%20of%20experiments.pdf

46. D. Pena, "An interview with George Box," *International Journal of Forecasting*, vol. 17, pp. 1–9, 2001.

47. D.C. Montgomery, *Design and Analysis of Experiments*, Wiley, 1984.

48. F.J. Nogales and A.J. Conejo, "Electricity price forecasting through transfer function models," *Journal of the Operational Research Society*, vol. 57, no. 4, pp. 350–356, 2006.

Chapter 6

Electricity Price Forecasting Using Neural Networks and Similar Days

Paras Mandal, Anurag K. Srivastava, Tomonobu Senjyu, and Michael Negnevitsky

INTRODUCTION

Background

Forecasting electricity prices is one of the most important issues in the competitive environment of the power industry. The price of electricity influences important decisions by market participants and plays a crucial role in establishing a proper economical operation. The main objective of an electricity market is to reduce the cost of electricity through competition [1–5]. Nowadays, electricity has been turned into a traded commodity, to be sold and bought at market prices. In general, loads and prices in wholesale markets are mutually intertwined activities [3]. However, electricity has its own distinct characteristics since it cannot be queued and stored economically with the exception of pumped-storage hydro plants when appropriate conditions are met, and the power system stability requires a constant balance between generation and load [1, 6, 7]. Furthermore, electricity is affected by power network limitations. In many cases, electric power cannot be transported from one region to another because of existing bottlenecks or limited transmission capacity of the grid. Hence, prices are local and differ among regions [2, 6, 8]. Electricity price can rise to tens or even hundreds of times of its normal value showing one of the greatest volatilities among all commodities. Application of forecasting methods commonly used in other commodity markets can have a large error in forecasting the price of electricity.

Advances in Electric Power and Energy Systems: Load and Price Forecasting, First Edition.
Edited by Mohamed E. El-Hawary.

Electricity price forecasting is required for different time horizons: short-term, medium-term and long-term forecasting. Market participants need to forecast short-term (mainly one day-ahead) prices to maximize their profits in spot markets. Accurate medium term price forecasts are necessary for successful negotiations of bilateral contracts between suppliers and consumers. Long-term price forecasts influence decisions on transmission expansion and enhancement, generation augmentation, distribution planning, and regional energy exchange decisions [2, 9–12].

This chapter focuses on short-term price forecasting in the PJM electricity market. The PJM competitive market is a regional transmission organization (RTO) that plays a vital role in the US electric system. PJM ensures reliability of the largest centrally dispatched control areas in North America by coordinating the movement of electricity in all or parts of Delaware, Illinois, Indiana, Kentucky, Maryland, Michigan, New Jersey, North Carolina, Ohio, Pennsylvania, Tennessee, Virginia, West Virginia, and the District of Columbia [13]. The PJM market that operates the largest competitive wholesale electricity market in the world is co-ordinated by an independent system operator (ISO) who ensures a secured, economical, and efficient operation as well as determining all locational marginal prices (LMP) according to voluntary bids and bilateral transactions [14]. Forecasting these LMPs is becoming more and more important as it does not only help market participants who bid into the spot price market to determine the bidding strategy of their generators, but also, if more accuracy in forecasting is obtained, it provides better risk management [15]. In the PJM market, it is observed that daily power demand curves have similar patterns, whereas the daily LMP curves are volatile. The LMP and load values are low at night, with LMP usually reaching their peak around 9 a.m. and 8 p.m.

Electricity price forecasting models include statistical and nonstatistical models. Time-series models, econometric models, and intelligent system methods are the three main categories of statistical methods. Nonstatistical methods include equilibrium analysis and simulation methods. Methods based on time-series or artificial neural network (ANN) are more common for electricity price forecasting due to their flexibility and ease of implementation. Time-series models such as autoregressive integrated moving average (ARIMA) have been used to predict price in the Spanish and Californian electricity markets [16]. Nogales et al. [10] have proposed time-series models, including dynamic regression and linear transfer function models for short-term price forecasting. The main drawback of time-series model is that they are usually based on the hypothesis of stationarity; however, the price series violates this assumption. Another kind of time-series models like generalized autoregressive conditional heteroskedastic (GARCH) [5] and input–output hidden Markov models (IOHMM) [17] have been developed but their application to electricity price prediction encounters difficulties. A rapid variation in load can have a sudden impact on the hourly price. The time-series techniques are successful in areas where the frequency of the data is low, such as weekly patterns, but they can be problematic when there are rapid variations and high-frequency changes of the target signal. Hence, there is a need of more efficient forecast tool capable of learning complex and nonlinear relationships that are difficult to model with conventional techniques.

Several publications reported forecasting methods based on artificial intelligence techniques, and especially, ANNs have received much attention as they provide good solutions to model complex nonlinear relationships much better than traditional linear models. ANNs were used for solving the short-term load forecasting problems [18–22]. ANNs were also used to forecast system marginal prices (SMP) [9], and market-clearing price (MCP) [4, 23]. A typical neural network for electricity price forecasting is either a multi-layer perceptron (MLP) neural network or a recurrent neural network. Szkuta et al. [9] proposed a three layer feed-forward NN for short-term price forecasting in the Victorian Electricity market. Hong and Hsiao [14] have proposed recurrent neural network structure to forecast LMP in the PJM electricity market. Rodriguez and Anders [11] proposed back propagation (BP) NN with one hidden layer and one output layer and a Neuro-Fuzzy system for predicting hourly MCP in the Ontario energy market. Doulai and Cahill [24] proposed a fully connected multi-layer perception (MLP) neural network for forecasting electricity price in an Australian electricity market (New South Wales). It has also been reported that multiple neural networks can be cascaded as streamline [25] or grouped into a committee machine [26].

The work described in this chapter focuses on day-ahead forecasts of electricity price in the PJM market using ANN model based on the similar days (SD) method. In the proposed price prediction method, day-ahead electricity price is obtained from the neural network that modifies the price curves obtained by averaging a selected number of similar price days corresponding to the forecast day, that is, two procedures are analyzed: (a) prediction based on averaging prices of SD and (b) prediction based on averaging prices of SD plus neural network refinement.

ANN's technique has been applied to forecasting prices in the England–Wales pool [27], the Australian market [9], the PJM Interconnection [14], and the New England ISO [25]. The advantages of the proposed price prediction technique with respect to other techniques reported in the literature are (a) better accuracy, (b) simplified ANN with lesser number of inputs, (c) performs well for low and high demand, and (d) works well for multiple seasons.

This study contributes to forecast electricity prices in the day-ahead market. The fundamental and novel contribution is the choice of input factors based on SD technique and the development of a novel forecasting technique based on neural networks in order to translate the available market information into price forecasts.

In addition to the integration of SD and ANN method mentioned earlier, which is termed as Case 1 throughout the chapter, this study also proposes a new technique to forecast hourly electricity prices in the PJM market using a recursive neural network (RNN), which is based on the SD method. The work described using integration of RNN and SD method, which is termed as Case 2 throughout this chapter, is an extension of Case 1 with better error analysis for a different data set of the PJM market to forecast hourly prices. RNN is a multi-step approach based on one output node, forecasting price a single step ahead ($t + 1$), and the network is applied recursively using the previous prediction as input for the subsequent forecasts [28]. In this way, it is carried out recursively for 24 steps to predict the next 24-hour electricity prices. The proposed recursive neural network model is also applied to generate the next three-day price forecasts. The advantage of RNN-based price prediction techniques is that the applied recursive algorithm offers adaptive predictions that compare favorably with other techniques.

To evaluate the performance of the proposed neural networks, the mean absolute percentage error (MAPE), mean absolute error (MAE), and forecast mean square error (FMSE) are calculated. Robustness of the proposed models is also measured through the estimation of the variance of the error. These values show that the proposed neural network models are capable of forecasting day-ahead or short-term electricity prices efficiently and accurately at different time-horizon level (daily, weekly, and annual seasonality).

Importance of Price Forecasting

The results of price forecasting are very much useful for the transmission company for a variety of purposes to schedule short-term generator outages, design load response programs, etc. They can also be used by generation companies to bid into the market strategically as well as to optimally manage its assets. Price forecasting plays an important role in power system planning and operation, risk assessment, and other decision-making [1–3, 12]. Furthermore, the importance of forecasting electricity price is that a producer needs day-ahead price forecasts to optimally self-schedule and to derive its bidding strategy in the pool. Similarly, once a good next day price forecast is available, consumers can derive a plan to maximize its own profit using the electricity purchased from the pool [10]. Hence, a feasible and practical method for price forecasting will certainly bring out safe and reliable supply of electricity at competitive prices.

Problem Statement and Objectives

A good price forecasting tool in deregulated markets should be able to capture the uncertainty associated with those prices. Some of the key parameters governing uncertainties are fuel prices, future additions to generation and transmission capacity, regulatory structure and rules, future demand growth, plant operations, and climate change [7]. Many factors influence the electricity price, such as time (the hour of the day, the day of the week, month, year, and special days), operating reserves, historical prices and demand, bidding strategies, temperature effect, predicted power shortfall, and generation outages. Electricity load is mainly affected by weather parameters. The load curve is relatively homogeneous and its variations are cyclic. However, the price curve is nonhomogeneous and its variations show a little cyclic property [2, 3, 7, 9].

This chapter deals mainly with short-term price forecasting in the environment of competitive electricity market where price forecasting aims at providing estimates of electricity prices for the future several days. This study has the following objectives:

- Investigating existing and developing new techniques for accurate short-term price forecasting.

- Presenting a simple and easy-to-use technique for forecasting the hourly electricity prices in an energy market.
- Developing price forecasting techniques based on SD, artificial neural network, and recursive neural network.
- Demonstrating the superiority of the developed techniques based on publicly available data acquired from the PJM electricity market, USA. Sufficient simulation experiments are conducted to validate the data to be used in training a short-term price forecasting system.
- Carrying out comparative studies with other efficient price forecasting techniques in order to show effectiveness and assess the prediction capacity of the developed techniques.

SOLUTION METHODOLOGIES

Price Forecasting Based on Similarity Technique

The present study of price forecasting is based on similarity technique in which we select similar price days corresponding to the forecast day based on the Euclidean norm, that is, using the minimum-distance Euclidean criterion. In general, the Euclidean distance between a pair of vectors X and W_j is defined by [29, 30]

$$\|D\| = \|X - W_j\| = \sqrt{\sum_{i=1}^{n} (x_i - w_{ij})^2} \tag{6-1}$$

where x_i and w_{ij} are the ith elements of the vectors X and W_j, respectively. In other words, Euclidean distance or simply distance examines the root of square differences between coordinates of a pair of objects. The similarity between the vectors X and W_j is determined as the reciprocal of the Euclidean distance $\|D\|$.

In this work, the Euclidean norm with weighting factors is used in order to evaluate the similarity between forecast day and searched previous days. The evaluation determined by the Euclidean norm makes us understand the similarity by using an expression based on the concept of norm [18, 30]. The smaller the Euclidean distance, the better the evaluation of SD and details of which are described in Selection of Similar Price Days section.

Selection of Similar Price Days

Before the application of Euclidean norm criteria, the correlation between price and load was evaluated. The relation between load and price can be seen in Fig. 6-1, which is simply an example to show the correlation between demand and price for the period January 1, 2006 to May 31, 2006. Correlation coefficient of determination (R^2) value is obtained as 0.6744, and a correlation coefficient (R) of 0.822 is reasonable to predict

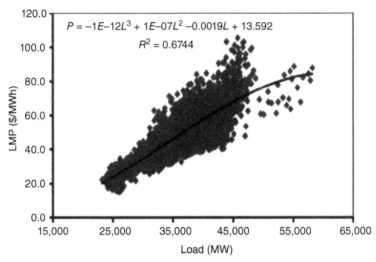

Figure 6-1 Relationship between LMP and load in the PJM (January–May, 2006).

electricity price. Note that the correlation coefficient is obtained from approximate curve fitting of a third order equation. Hence, load is a natural choice as a parameter to use in predicting day-ahead electricity price and in order to make our in-house-built C-programming simple, other influential aspects such as transmission congestion, generator availability, generator bidding strategy on the generation side, outages, and spinning reserves. have not been taken as inputs in this study.

In general, the following equation is used as the Euclidean norm with weighted factors for the selection of similar price days.

$$\|D\| = \sqrt{(\hat{w}_1 \Delta L_t)^2 + (\hat{w}_2 \Delta L_{t-1})^2 + (\hat{w}_3 \Delta P_t)^2 + (\hat{w}_4 \Delta P_{t-1})^2} \qquad (6\text{-}2)$$

$$\Delta L_t = L_t - L_t^p \qquad (6\text{-}3)$$

$$\Delta L_{t-1} = L_{t-1} - L_{t-1}^p \qquad (6\text{-}4)$$

$$\Delta P_t = P_t - P_t^p \qquad (6\text{-}5)$$

$$\Delta P_{t-1} = P_{t-1} - P_{t-1}^p \qquad (6\text{-}6)$$

Where L_t and P_t are the load and price on forecast day, L_t^p and P_t^p are load and price on SD in past, respectively. ΔL_t is the load deviation between forecast day and SD. ΔP_t is the price deviation between forecast day and SD. ΔP_{t-1} is the deviation of price at $(t-1)$ on forecast day and SD. The weighting factor, $\hat{w}_i (i = 1 - 4)$ is determined by using least square method based on regression model that is constructed by using historical load and price data [18]. Accordingly, regression model has been created and then simplified. Error was calculated based on the difference of regression analysis-based predicted value and actual value. Least cost square method was used to minimize this error. Details of the calculation of the weighting factors are mentioned in the Appendix. Therefore, a selection of similar price days that considers a trend

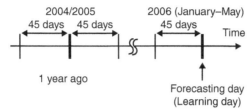

Figure 6-2   Time framework for the selection of similar price days corresponding to forecast day (ANN approach).

of price and load is performed. The above equations give the hourly calculation of Euclidean norm.

Equation (6-2) is a very important equation for the selection of similar price days based on historical data. To minimize the number of input data while maximizing the accuracy of the proposed algorithm, a comprehensive statistical analysis of the historical data was conducted. Finally, it was observed that load and price at the same hour and previous hour are appropriate input to find SD based on correlation analysis. Combining weighting factor obtained with regression model in the Euclidean norm improved the performance of the SD method and Euclidean norm equation finally evolved as given in (6-2).

SD are selected within the same season. The time framework for the selection of similar price days corresponding to forecast day is shown in Fig. 6-2. The past 45 days from the day before a forecast day and past 45 days before and after the forecast day in the previous year are considered for the selection of SD. Each hour has a separate set of SD based on the algorithm. If the forecast day is changed, similar days are selected in the same manner.

Procedure for Selecting Similar Days

> **Step 1.** Select similar price days by calculating $\|D\|$ (Euclidean norm) from (2).
>
> **Step 2.** Price data of SD at time $t + 1$ are selected in Step 1 and assumed to be forecast price $\bar{P}_{t+1}$ at time $t + 1$.
>
> **Step 3.** Step 2 is repeated. Similarly, the price forecast is obtained at time $t + h - 1$. Then, it is assumed to be actual price, and using this data, select similar price day at time $t + h$, where h represents hour.

NEURAL NETWORK APPROACH FOR PRICE PREDICTION

Artificial Neural Network

An artificial neural network consists of a number of very simple and highly interconnected processors, called neurons, which are analogous to the biological neurons in the brain. The neurons are connected by weighted links that pass signals from one neuron to the other. ANNs are capable of inferring hidden relationship (mapping) in data [29, 30], and are a class of flexible nonlinear models that discover patterns adaptively from the data. Theoretically, it has been shown that given an appropriate number of nonlinear processing units, An ANN can learn from experience and estimates any complex functional relationship with high accuracy [30].

In power systems, the ANN have been used to solve problems such as load forecasting, component and system fault diagnosis, security assessment, unit commitment. In price forecasting applications, the main function of the ANN is to predict price for the next half-hour, day(s) or week(s). In general, demand is the main variable that drives price. The correlation of temperature and price was found to be very much similar to the demand-price correlation, which revealed that the temperature effect is included in the power demand. Another variable that drives price is the hour during the day, however, its impact is also reflected in the demand [11].

In this study, we propose a multi-layer feed-forward neural network for forecasting next-day 24-hour electricity prices. A feed-forward neural network consists of an input layer of source neurons, at least one middle or hidden layer of computational neurons, and an output layer of computational neurons. The input layer accepts input signals from the outside world and redistributes these signals to all neurons in the hidden layer. It is the neurons in the hidden layer that allow neural networks to detect the feature, to capture the pattern in the data, and to perform complicated nonlinear mapping between input and output variables. The output layer establishes the output pattern of the entire network.

The proposed ANN is shown in Fig. 6-3. The network model is composed of one input layer, one hidden layer, and one output layer. The sigmoidal function has been used as an activation function since it helps to improve computational efficiency and ANN can learn much faster. The hidden layer and neurons play very important roles for many successful applications of neural networks. It has been proven that only one layer of hidden units is sufficient for NNs to approximate any complex nonlinear function with any desired accuracy. In this study, $2n + 1$ hidden neurons are chosen for better forecasting accuracy, where n is the number of input nodes.

In general, the price based on several selected SD is averaged to improve the accuracy of price forecasting. Since the ANN yields corrections which are simple data, it is not necessary for the ANN to learn all the SD data. Therefore, the ANN can forecast prices by a simple learning process. The correction from the ANN in the context of this study is referred to as the modification of the price curves obtained by averaging similar price data corresponding to the forecast day. The input variables of the proposed ANN structure for price forecasting as shown in Fig. 6-3 are mean SD data $\bar{P}$, which is an average of 5 similar price days; actual hourly price P_t; actual hourly load L_t; and h represents hour (for day-ahead price forecasting, $h = 24$).

In the proposed price prediction method, we forecast the price ($\hat{P}_{t+24}$) by using the neural network (Fig. 6-3) to modify the price curve obtained by averaging 5 similar price days corresponding to the forecast day, that is, the ANN corrects the output obtained from the SD approach. The neural network uses the nominal values of load and price as learning data.

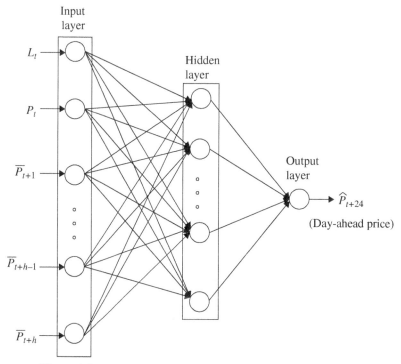

Figure 6-3 Proposed ANN model for day-ahead price forecasting.

Recursive Neural Network

In addition to the ANN approach to forecast prices as described in the previous sub-section entitled Artificial Neural Network, this study also proposes a new approach to forecast hourly electricity prices in the PJM market using RNN, which is based on SD method. The RNN, a three-layer feed-forward model, is adopted for forecasting the next-24 h and -72 h electricity prices. RNN is a multi-step approach based on one output node, forecasting price a single step ahead ($t + 1$), and the network is applied recursively using the previous prediction as input for subsequent forecasts [28].

The proposed recursive NN price forecasting technique performs well in volatile price cases; performs well for multiple seasons; and the recursive algorithm applied offers adaptive predictions that compare favorably with respect to other techniques.

The architecture of the proposed RNN model is shown in Fig. 6-4. The proposed RNN is capable of modeling nonlinear and fast variations, as well as complicated input/output relationships through training processes with historical data. In Artificial Neural Network section, the multi-layer perceptron (MLP) neural network was adopted, whereas in this section, a RNN, that is, a series of 24 cascaded feed-forward model is proposed. The input variables of the proposed RNN structure for price

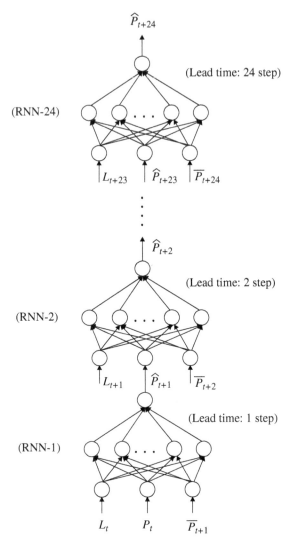

Figure 6-4   Recursive neural network model for price forecasting.

forecasting as shown in Fig. 6-4 are mean SD price data $\bar{P}$, which is an average of 5 similar price days; actual hourly price P_t data; and actual hourly load L_t data. Since the RNN is a multi-step approach based on one output node, price is forecast a single step ahead $(t + 1)$ to obtain $\hat{P}_{t+1}$, and the network is applied recursively using the previous prediction as input for the subsequent forecasts. For the price forecast of the lead time of second step $(t + 2)$ to obtain $\hat{P}_{t+2}$ as shown in Fig. 6-4, the input to RNN-2 are L_{t+1}, $\hat{P}_{t+1}$ and $\bar{P}_{t+2}$, where L_{t+1} is the forecasted load for next hour and this study assumes that forecasted values are available; $\bar{P}_{t+2}$ are similar price day data. In this way, it is carried out recursively for 24 steps to predict next 24-hour prices.

In the proposed price prediction method, we forecast hourly prices by using RNN (Fig. 6-4) to modify the price curve obtained by averaging 5 similar price days corresponding to the forecast day. The proposed RNN uses the nominal values of load and price as learning data.

The time framework for the selection of SD in this case is the past 30 days from the day before a forecast day, and past 30 days before and after the forecast day, which will be discussed in detail later in this chapter. The price forecasting results obtained from ANN and RNN, both based on SD method, are discussed in detail in Result Analysis and Discussion section.

Neural Network Training   We have adopted the same training procedure for both ANN and RNN. The neural network is trained by using the data of the past 45 days from the day before the forecast day, and the past 45 days before and after the forecast day in the previous year. The training algorithm for the proposed neural network used in this study is the well-known error back-propagation (BP) training algorithm [10, 18, 30]. In the learning process of BP, the interconnection weights are adjusted using error convergence technique to obtain a desired output for a given input. In general, the error at the output layer in the BP model propagates backward to the input layer through the hidden layer in the network to obtain the final desired output. The steepest gradient descent method is utilized to calculate the weight of the network and adjusts the weight of interconnections to minimize the output error. The error minimization process is repeated until the error converges to a predefined small value. The error function at the output neuron is defined as [30],

$$E = \frac{1}{2} \sum_k (T_k - A_k)^2, \tag{6-7}$$

where T_k and A_k represent the target and actual values of output neuron, k, respectively.

In order to accelerate the learning process, two parameters of back-propagation algorithm, the *learning rate* (η) and another, *momentum* (α) can be adjusted. The learning rate is the proportion of error gradient by which the weights are adjusted. Large value of η can give a faster convergence to the gradient minimum but may produce oscillations around the minimum. The momentum determines the proportion of the change of past weights that is used in the calculation of the new weights [30]. Both η and α affect the error gradient minimization process. For the proposed neural networks, the values chosen for η and α are 0.8 and 0.1, respectively.

Through learning algorithms, the network tries to minimize the error between the output produced and the desired output, adjusting the weights and biases. The error minimization process is repeated until the error converges to a predefined small value. After the error becomes constant, the learning procedure terminates. The initial conditions for the neural network parameters are the same when the forecast time is changed. If the forecast time is changed, neural network is retrained to obtain the relationship between 'load and price' around the forecast day.

Learning and Forecasting Procedure for the Neural Network

The flowchart for the developed price forecasting methodology is shown in Fig. 6-5. The major steps for the developed algorithm to forecast electricity price including learning and forecasting procedure of the proposed neural networks are summarized as follows:

Step 1. Get data.

Step 2. *Determine the learning range of the proposed neural network:*
In this study, the proposed neural network (ANN or RNN) is trained by using the data of the past 45 days from the day before the forecast day, and the past 45 days before and after the forecast day in the previous year, that is, total available learning days (d_{NN}) are 135 days. Note that the ANN and RNN approaches are two different cases in this study.

Step 3. *Determine the time framework for the selection of similar days for one learning day:*
In the *ANN approach*, the limits on the selection of SD for one learning day are the past 45 days from the day before a learning day, and past 45 days before and after the learning day in the previous year.

In the *RNN approach*, the chosen value in the time framework is different, that is, the limits on the selection of SD for one learning day are the past 30 days from the day before a learning day, and the past 30 days before and after the learning day in the previous year.

Step 4. *Select similar days for first learning day:*
From the time framework, **N** number of SD is selected for the first learning day. In this study, five number of SD (**N** = 5) are selected and we take the average of these five similar days.

Step 5. *BP learning for **N** similar days:*
The neural network is trained by using **N** similar days for one learning day.

Step 6. *BP learning for all the days of learning range:*
The neural network is trained for all the days of learning range in the same way as in steps 4 and 5.

Step 7. *Number of iterations of BP learning within the specific range:*
BP learning within the specific range consists of a BP learning set. The neural network is trained by repeating the BP learning set for 1500 times.

Step 8. *Select the similar days for the forecast day:*
Before forecasting the hourly prices, we select **M** similar days corresponding to the forecast day. Five number of similar days are selected in this study. The price data obtained by averaging **M** price corresponding to **M** similar days are then considered as the input variables of the neural network.

Step 9. *Forecast using similar days and neural network:*
In the *ANN approach*, after getting all required input data using SD method and the trained ANN, 24-hour-ahead price ($\hat{P}_{t+24}$) forecasting was performed.

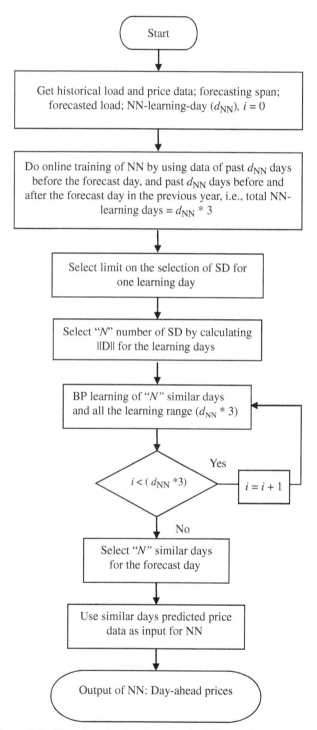

Figure 6-5   Flowchart for developed methodology to forecast prices.

This process was repeated for all 24 forecasting hours using SD technique and ANN (Fig. 6-3) to produce day-ahead price forecasts.

In the *RNN approach*, after getting all required input data, that is, load data L_t, price data P_t, mean SD data $\bar{P}$, they are used in the trained RNN and we obtain the next-hour price $\hat{P}_{t+1}$. This process was repeated for all 24 forecasting hours using SD technique and RNN (Fig. 6-4) to produce next-24 h price forecasts.

RESULT ANALYSIS AND DISCUSSION

This chapter describes a price forecasting technique based on averaging the prices of SD and refining the results through a standard neural network procedure. This section is divided into two major cases:

Case 1: Forecasting results obtained from ANN, that is, integration of SD and ANN.

Case 2: Forecasting results obtained from RNN, that is, integration of SD and RNN.

Cases 1 and 2 are described in detail in Artificial Neural Network and Recursive Neural Network, respectively.

Price Forecasts Using Artificial Neural Network

Accuracy Measures   The proposed ANN model as shown in Fig. 6-3 was trained and tested using the data (from January 1, 2004 to May 31, 2006) derived from the PJM electricity market [13]. Sets of data include hourly price and load data. For all the days and weeks under study, two types of average percentage errors are computed: the one corresponding to the 24 h of each day and the one corresponding to the 168 h of each week. The performance was measured mainly using a mean absolute percentage error (MAPE). For daily error, MAPE is defined as,

$$\text{MAPE}_{\text{day}}(\%) = \frac{1}{24} \sum_{i=1}^{24} \frac{|P_i^{\text{true}} - P_i^{\text{est}}|}{\bar{P}_i^{\text{true},24}} \tag{6-8}$$

$$\bar{P}_i^{\text{true},24} = \frac{1}{24} \sum_{i=1}^{24} P_i^{\text{true}}, \tag{6-9}$$

where MAPE_{day} is the daily error, P_i^{true} is the actual price for hour i, P_i^{est} is the predicted price for that hour, and $\bar{P}_i^{\text{true},24}$ is the average true price for the day.

The weekly error, $\text{MAPE}_{\text{week}}$ is computed as

$$\text{MAPE}_{\text{week}}(\%) = \frac{1}{168} \sum_{i=1}^{168} \frac{|P_i^{\text{true}} - P_i^{\text{est}}|}{\bar{P}_i^{\text{true},168}} \tag{6-10}$$

$$\bar{P}_i^{\text{true},168} = \frac{1}{168} \sum_{i=1}^{168} P_i^{\text{true}} \tag{6-11}$$

The reason for considering average true price in the denominator of (6-8) and (6-10) is that if the actual value is small, this will contribute a large term in MAPE even if the difference between the actual and forecasted values is small. In addition, if the forecasted value is small and the actual value is large, then absolute percentage error will be close to 100% [3]. Also, in order to avoid the adverse effect of prices close to zero [31], the new MAPE definitions as indicated in (6-8) and (6-10) have been adopted in this study.

To demonstrate the applicability of the proposed ANN model (Fig. 6-3), we performed simulations using two procedures: (i) forecasting the price curve based on averaging prices of similar days, that is, SD method, and (ii) forecasting the price curve based on averaging prices of SD plus ANN refinement.

Daily Price Forecasts   For the PJM electricity market, a single day has been selected for each month (January to May 2006) to forecast and validate the performance of the proposed model. Figures 6-6 to 6-10 show the daily price forecasts for three representative days (Friday, Saturday, and Sunday).

Figure 6-6 shows the day-ahead price forecasting results for January 20, 2006. The forecast results obtained from the proposed ANN are quite close to the actual LMP values. Using the MAPE definition as mentioned in (6-8), the prediction behavior of the proposed ANN technique for this day is very appropriate with a daily MAPE error of only 6.93%, which is much lower than that obtained using the SD approach (13.90%).

Similarly, day-ahead price forecasts for February 10 (Friday) and March 5 (Sunday), 2006 are shown in Figs. 6-7 and 6-8, respectively, where MAPE values obtained from the proposed ANN models are around 7%.

Figures 6-9 and 6-10 show the day-ahead price forecasts for the chosen days in the month of April and May, 2006. It can be seen in Fig. 6-9 that the prediction is particularly inaccurate for the morning and evening peaks of Friday (April 7). The MAPE value obtained from the proposed ANN is 9.02%, whereas, using SD approach, it is 14.14%. The MAPE value is found to be quite higher on this day. The PJM market experienced an average actual power demand of 32,486.11 MW in the month of April 2006. However, the power demand for this day only (April 7) was 34,589.84 MW. The increase in the MAPE error can possibly be because of the increase in power demand as price is driven by load. Also, price volatility increases with load. Moreover, the electricity price in the PJM electricity market is volatile, which can be observed from Fig. 6-11. Observe that the power demand shows a

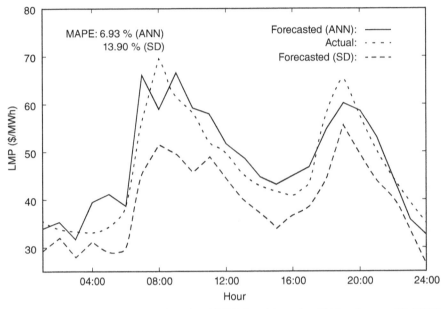

Figure 6-6   Actual and forecast day-ahead PJM electricity price (Friday, January 20, 2006).

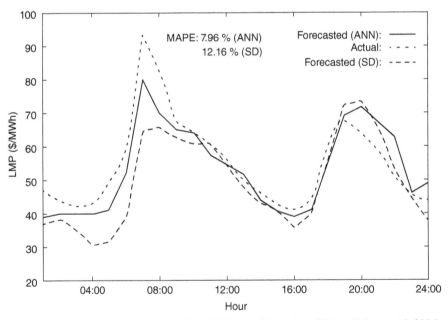

Figure 6-7   Actual and forecast day-ahead PJM electricity prices (Friday, February 10, 2006).

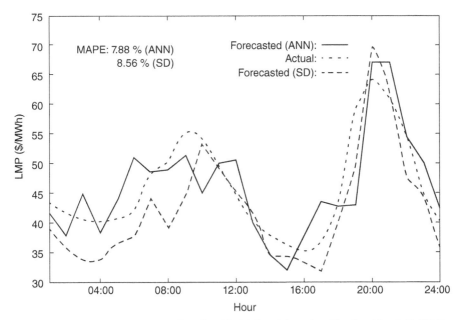

Figure 6-8   Actual and forecast day-ahead PJM electricity price (Sunday, March 5, 2006).

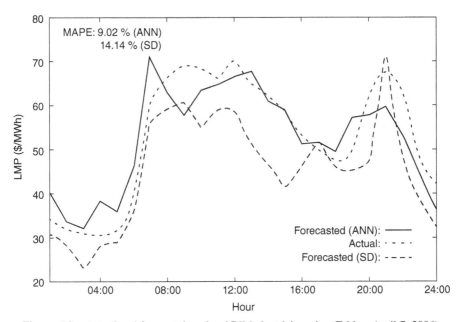

Figure 6-9   Actual and forecast day-ahead PJM electricity price (Friday, April 7, 2006).

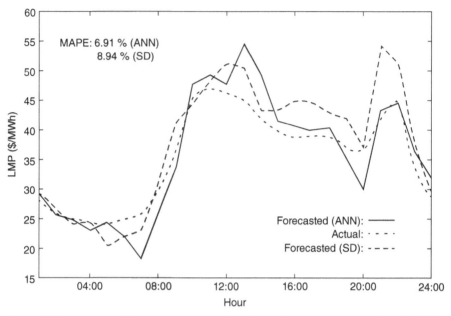

Figure 6-10   Actual and forecast day-ahead PJM electricity price (Saturday, May 13, 2006).

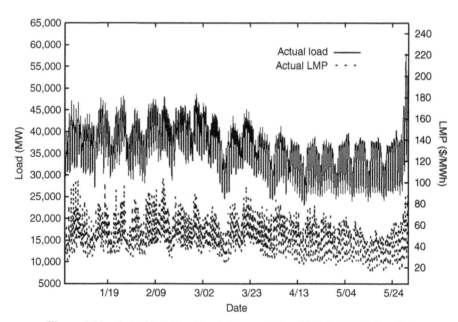

Figure 6-11   Actual L MP and load: January–May, 2006 in the PJM market.

similar pattern except at the end of May 2006. We can observe from Figs. 6-6 to 6-10 that the proposed ANN model can predict the peaks but not the really large ones. In all the above-mentioned cases, the ANN technique outperforms the SD method.

Weekly Price Forecasts   This study further deals with weekly price forecasting for the PJM day-ahead electricity market. Two weeks were selected to forecast and validate the performance of the proposed model. These forecasts are based on day-ahead and have been represented for a week. Figure 6-12 shows the weekly price forecasts during February 1–7, 2006, which is typically a low demand week (37,738.48 MW). It can be seen from Fig. 6-12 that the forecast results obtained from the proposed ANN are close to the actual LMP values. Also, it can be observed that when price spikes appear, our model does not forecast price jumps as in last three days of this week. Using the MAPE definition as mentioned in (6-10), weekly MAPE obtained from the proposed ANN approach is 7.66%, which is much lower than that obtained using the SD method (12.80%).

Similarly, the weekly price forecasts for the last week of February (February 22–28, 2006), which is a high demand week (40,345.28 MW), is shown in Fig. 6-13. In this case, weekly MAPE obtained from the proposed ANN is slightly higher (8.88%) than that obtained during the first week of February. It can be seen in Fig. 6-13 that the LMP forecasts in most days of this week, especially till 144 h, are close to the actual LMPs. The weekly MAPE obtained from the selected weeks show that the ANN technique outperforms the direct use of SD method.

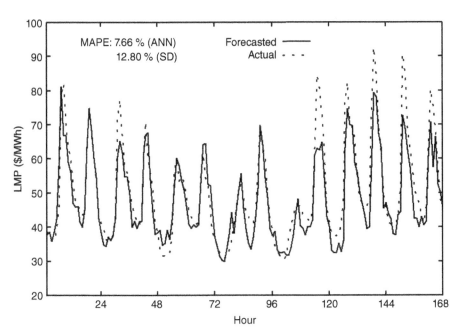

Figure 6-12   Weekly price forecasts during low demand week (Wednesday, February 1 to Tuesday, February 7, 2006).

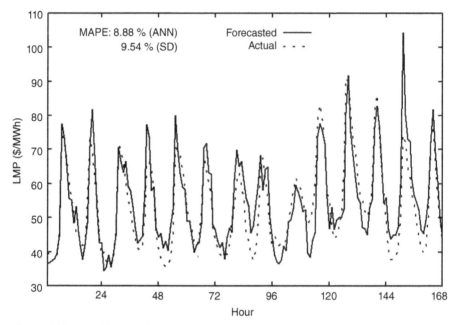

Figure 6-13 Weekly price forecasts during high demand week (Wednesday, February 22 to Tuesday, February 28, 2006).

Additionally, the forecast mean square error, FMSE, is computed [31]. This parameter is the square root of the average of either 24 (daily) or the 168 (weekly) square differences between the predicted prices and the actual ones, that is,

$$\text{FMSE}_{\text{day}} = \sqrt{\frac{1}{24} \sum_{i=1}^{24} (P_i^{\text{true}} - P_i^{\text{est}})^2} \qquad (6\text{-}12)$$

$$\text{FMSE}_{\text{week}} = \sqrt{\frac{1}{168} \sum_{i=1}^{168} (P_i^{\text{true}} - P_i^{\text{est}})^2} \qquad (6\text{-}13)$$

The forecasting error is the main concern for power engineers; a lower error indicates a better result. Table 6-1 provides a numerical overview of the prediction performance of the ANN and the SD method for the selected days where the values of average MAPE, MAE, and FMSE obtained from the ANN and SD approaches are compared. MAE provides an indication of error ranges. A comparison of the weekly forecasting performance of the proposed ANN method with SD method is presented in Table 6-2 where the values of maximum and minimum MAPE, average MAPE, FMSE, and MAE are compared. Table 6-3 presents the measure of uncertainty of the proposed model by means of statistical index, that is, variance [32]. The smaller

Table 6-1   Comparison of Daily Forecasting Performance of ANN with Similar Days Method

Year 2006	Avg. MAPE (%)		FMSE ($/MWh)		MAE ($/MWh)	
	ANN	SD	ANN	SD	ANN	SD
January 20	6.93	13.90	4.57	8.32	3.68	7.32
February 10	7.96	12.16	6.12	9.95	4.74	7.27
March 05	7.88	8.56	5.39	5.40	4.03	4.42
April 07	9.02	14.14	5.87	9.41	2.21	7.83
May 13	6.91	8.94	3.65	4.34	2.68	3.45

Table 6-2   Comparison of Weekly Forecasting Performance of ANN with Similar Days Method

	MAPE (%)						FMSE ($/MWh)		MAE ($/MWh)	
	Max. MAPE		Min. MAPE		Avg. MAPE					
Year 2006	ANN	SD	ANN	SD	ANN	SD	ANN	SD	ANN	SD
February 01–07	11.32	16.01	5.94	10.20	7.66	12.80	5.83	8.79	4.15	6.87
February 22–28	12.37	16.48	5.66	5.40	8.88	9.54	6.01	6.37	4.74	5.09

Table 6-3   Daily and Weekly Error Variances

	Error Variances	
Year 2006	ANN	SD
January 20	0.0034	0.0075
February 10	0.0050	0.0154
March 05	0.0061	0.0046
April 07	0.0038	0.0098
May 13	0.0049	0.0055
February 01–07	0.0066	0.0119
February 22–28	0.0047	0.0051

the variance, the more precise is the prediction of prices. Note that SD method is outperformed by the neural network technique in all the days and weeks under study.

Comparative Analysis   Reference [18] presented 1–6 h-ahead electricity price forecasting results based on ANN. If the method (Case 1) described in this work is

Table 6-4   Comparison of Case 1 Results with Other Method Based on the Recurrent Neural Network

	Reference [14]		Case 1	
	PJM Subarea 1	PJM Subarea 2	PJM Subarea	
	Weekday	Weekday	Weekday	Saturday
MAE ($/MWh)	2.98	3.75	2.21	2.68
Correlation value	0.92	0.96	0.92	0.95

applied to predicting 1–6 h-ahead prices, the error would be lower compared to day-ahead forecasting results obtained in this chapter. Because, as $\bar{P}_{t+1}$ increases (see Fig. 6-3), accumulated error that depends on the forecasting period (hourly, daily, weekly, monthly, etc.) will increase in case of predicting daily and weekly electricity prices.

Furthermore, in order to show the effectiveness of the proposed ANN method, the proposed price prediction method is compared with [14], where the authors have explored a technique of recurrent neural network-based LMP forecasting for the PJM deregulated market. In order to carry out the comparisons and assess the prediction capacity of the proposed ANN method, two statistical measures are used as shown in Table 6-4, that is, MAE and correlation values are calculated for the month of April and May 2006 based on Reference [14, equations (6) and (7)]. It can be seen from Table 6-4 that MAE ($/MWh) values obtained from the proposed ANN method are relatively lower than that of Reference [14], showing a better performance of our proposed ANN model. Moreover, correlation values compare favorably well with their values. The greater the correlation values, the more precise the prediction.

This case study contributes to solving an important problem of electricity price forecasting in which the authors have proposed a technique based on ANN model together with applying SD approach to forecasting day-ahead electricity price. Week-days and weekends are selected for daily and weekly forecasts and the test results obtained show that the proposed algorithm could provide a considerable improvement of forecasting accuracy for any day of the week very well. Case 1 provides reliable forecasts for day-ahead price forecasting as the results demonstrate the efficiency, accuracy, and high adequacy of the proposed ANN technique.

Price Forecasts Using Recursive Neural Network

In this study, Case 2 describes the price forecasting method using RNN, which is an extension work of Case 1, with better error analysis for a different data set of the PJM market to forecast hourly prices. Note that the SD technique is not the focus of this case, but the focus is on the integration of SD with the novel architecture of recursive neural network (RNN) to obtain better forecasting results compared to those obtained in Case 1. In addition to this, the proposed RNN, that is, a series of 24 cascaded MLPs, has been tested to examine the effect of seasonal variation on price curve forecasting and the test results obtained are compared with that of Case 1

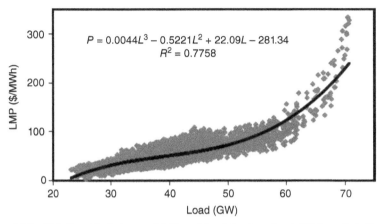

Figure 6-14   Relationship between LMP and load in day-ahead PJM market (January–December, 2006).

including other existing efficient techniques such as NN, ARIMA, transfer function [31], and neuro-fuzzy [33]. It is emphasized that the contributions of this section are the development of RNN and integrating SD with RNN to produce better results for forecasting the next 24-hour and 72-hour electricity prices.

For the selection of similar price days, we adopt the Euclidean norm equation as presented in (6-2). In order to minimize the number of input data, while maximizing the accuracy of the proposed algorithm, the authors did a comprehensive statistical analysis of historical data prior to the application of the Euclidean norm criterion. Figure 6-14 shows the correlation between demand and price for the period January–December 2006. The degree of correlated linearity can be tested by correlation coefficients. The correlation coefficient of determination (R^2) value is obtained as 0.7758, and the correlation coefficient (R) of 0.881 is reasonable to predict electricity price. The value of R^2 index obtained in Case 1 was relatively lower (0.6774) for the period January–May 2006.

In Case 2, the limit on the selection of SD differs from that of Case 1, that is, the time framework for the selection of similar price days corresponding to the forecast day is as shown in Fig. 6-15. The past 30 days from the day before a forecast day, and

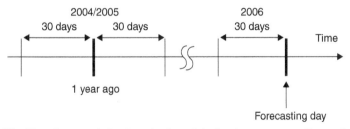

Figure 6-15   Time framework for the selection of similar days corresponding to forecast day (RNN approach).

the past 30 days before and after the forecast day in the previous year are considered for the selection of SD.

We select similar price days by calculating $\|D\|$ from (6-2). Price data of SD at time $t + 1$ are obtained from (6-2) and are assumed to be forecast similar price $\bar{P}_{t+1}$ at time $t + 1$. Then, it is assumed to be actual price and considered as one of the inputs to the proposed RNN model (Fig. 6-4).

Accuracy Measures   The proposed RNN model as shown in Fig. 6-4 was trained and tested using the hourly LMP and load data for the period January, 2004–December, 2006 [13]. The data set used in this case is different from that of Case 1. In this case, the forecasting technique has been tested for seasonal variation as all 12-months data set has been used for the year 2006.

Two forecasting procedures were analyzed: (i) forecasting the price curve based on averaging prices of similar days, that is, SD method, and (ii) forecasting the price curve based on averaging prices of SD plus RNN refinement. In Case 2, the forecasting year is 2006 and one day from each season of the year 2006 is chosen randomly to be studied for forecasting the next-24 h prices. Spring and summer are selected for predicting the next-72 h prices.

To evaluate the performance of the proposed RNN model, the MAPE as presented in (6-8) is adopted in general form as,

$$\text{MAPE } (\%) = \frac{1}{N} \sum_{i=1}^{N} \frac{\left| P_i^{\text{true}} - P_i^{\text{est}} \right|}{\bar{P}_i^{\text{true},N}} \tag{6-14}$$

$$\bar{P}_i^{\text{true},N} = \frac{1}{N} \sum_{i=1}^{N} P_i^{\text{true}} \tag{6-15}$$

where N is the number of data points, and $\bar{P}_i^{true,\,N}$ is the average true price for a day ($N = 24$) or three-day ($N = 72$). Since a set of nonstationary data is studied, a statistical analysis is more reasonable.

Additionally, FMSE, which is the square root of the average of either 24 (daily) or the 72 (three-days) square differences between the predicted prices and the actual ones, is computed and given as,

$$\text{FMSE} = \sqrt{\frac{1}{N} \sum_{i=1}^{N} (P_i^{\text{true}} - P_i^{\text{est}})^2} \tag{6-16}$$

In order to assess the prediction capacity of the proposed RNN method, MAE is calculated. Furthermore, the robustness of the proposed model can be measured by means of a statistical index, such as the variance of the error, which is calculated based on Reference [32, equations (6)–(11)].

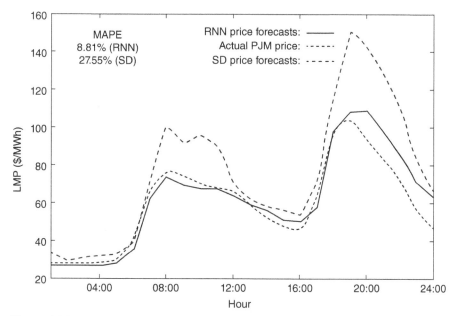

Figure 6-16   Actual and forecasted LMPs for PJM market in winter (Monday, December 4, 2006).

Case 2 uses the real market data to test the performance of the proposed RNN model. The results of price forecasting curve obtained from SD and RNN model are compared with the actual price curve shown in Figs. 6-16 to 6-19.

Daily Price Forecasts   Figure 6-16 shows the price forecasting result for a winter day (December 4, 2006). The daily price forecasts obtained from the proposed RNN are close to the actual price values, except at the evening peaks. The MAPE obtained from the proposed RNN approach is 8.81%, which is much lower than that obtained using the SD approach (27.55%). The average power demand for the month of December 2006 in the PJM market is 37,623.00 MW, whereas the PJM market experienced the power demand of 40,909.71 MW on December 4, which is higher than the average of this month. Moreover, at 7 p.m., the power demand and LMP values reach peaks of 49,324 MW and 105.42 $/MWh, respectively. The main factor driving the electricity price is the load. The MAPE value is slightly higher on this day, which can possibly be because of the increase in power demand. Price volatility also increases with load.

Figure 6-17 corresponds to a spring Saturday (May 13, 2006) with price forecasts computed using the SD method and the proposed RNN model. In this case, the prediction by the proposed RNN model and SD method is particularly inaccurate for the evening peaks, around 8 p.m. However, the prediction result obtained from RNN is very close to the true value during other hours. The MAPE value obtained

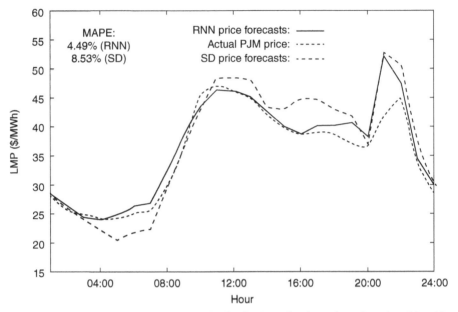

Figure 6-17 Actual and forecasted LMPs for PJM market in spring (Saturday, May 13, 2006).

from RNN in spring is more accurate and better (4.49%) than that obtained in winter. Using the SD approach, MAPE value is obtained as 8.53%.

Figure 6-18 shows the price forecasts obtained from the SD method and the proposed RNN model for a weekday in summer (July 10, 2006). It can be observed in Fig. 6-18 that the RNN projected curve is quite close to the true curve except during evening peaks. In the PJM market, the daily price curves are highly volatile. The volatility is higher in summer and autumn compared to other seasons. The price forecasts as shown in Fig. 6-18 means that our model is capable of forecasting the LMP values very accurately as the prediction behavior of the proposed RNN technique for this summer day is very appropriate with a daily MAPE error of only 5.11%.

The daily price forecasts for day-ahead PJM market during the autumn (November 2, 2006) is shown in Fig. 6-19 where it can be observed that the proposed RNN model predicted the values quite accurately other than during the morning and evening peaks. Using the MAPE definition as mentioned in (6-14), MAPE value obtained from the proposed RNN for this day is 7.72%, whereas using SD approach, it is 11.44%.

Note that the MAPE values obtained in all the daily forecasts as illustrated in Figs. 6-16 to 6-19 show that the RNN technique outperforms the direct use of SD approach. We can also observe from Figs. 6-16 to 6-19 that the proposed RNN model can predict the peaks but not the really large ones. It is also observed that forecasting error is relatively lower during the weekend than that during weekdays. The daily PJM price curves are highly volatile. As demonstrated by simulation results, despite

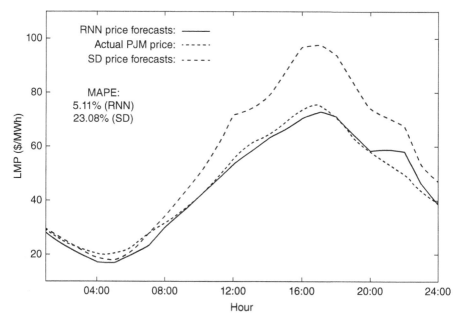

Figure 6-18 Actual and forecasted LMPs for PJM market in summer (Monday, July 10, 2006).

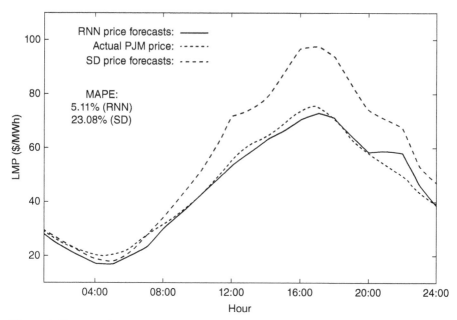

Figure 6-19 Actual and forecasted LMPs for PJM market in autumn (Thursday, November 2, 2006).

Table 6-5   Comparison of Daily Forecasting Performance of the proposed RNN with SD Method

	MAPE (%)		FMSE ($/MWh)		Error Variance		MAE ($/MWh)	
Year 2006	RNN	SD	RNN	SD	RNN	SD	RNN	SD
December 04	8.81	27.55	7.43	22.27	0.0077	0.0630	5.26	16.46
May 13	4.49	8.53	2.56	3.82	0.0032	0.0044	1.58	3.01
July 10	5.11	23.08	3.02	13.20	0.0017	0.0294	2.34	10.59
November 02	7.72	11.44	5.09	8.18	0.0047	0.0146	3.79	5.62

the volatility of the true price curves, the proposed RNN generated forecasts appeared to fit the curves relatively closer to the real values.

Table 6-5 summarizes the numerical results where the comparison of the daily forecasting performance of the proposed RNN model with SD method is presented. It can be observed from Table 6-5 that the daily MAPE, MAE, and FMSE of reasonably small values are obtained for the PJM data, which has coefficient of determination (R^2) of 0.7758 between load and electricity price. Table 6-5 also shows the robustness of the proposed model measured by the variance. The values of error variance demonstrate that the proposed forecasting algorithm functions reasonably well in predicting electricity price for any day of the week tested.

Three-Day Price Forecasts   Furthermore, to evaluate the accuracy of the proposed recursive neural network algorithm, we also performed a simulation to forecast the next-72 h prices in the PJM electricity market. Three-days in a row have been selected in spring and summer to forecast and validate the performance of the proposed RNN model. Figures 6-20 and 6-21 show the next-72 h price forecasts in spring (April 28–30) and summer (June 8–10), respectively. It is seen from both figures that when price spikes appear, our model does not forecast price jumps, which can be observed on the following days: April 29, June 8, and 9. The LMP forecasts during most hours of the selected days are close to the actual ones. The average MAPE for the next-72 h price forecasts obtained from the proposed RNN model is around 8% in both seasons. The MAPE value obtained implies that the proposed algorithm functions well even if any three days are taken in sequence arbitrarily. Table 6-6 presents the next-72 h price forecasting results obtained from the proposed RNN model and SD method. It is observed that MAE values of around 3 $/MWh only, and small values of FMSE and error variance are obtained for the PJM data. It is also seen in Table 6-6 that the maximum MAPEs for the selected days in spring and summer are around 10% only.

Comparative Analysis   This section presents a comprehensive comparison with existing methods reported in the literature. Forecasting results have been tested for seasonal variation and compared with the results obtained from the previous case

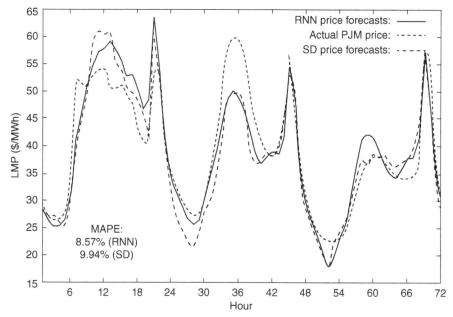

Figure 6-20 Actual and forecasted LMPs for PJM market in spring (Friday, April 28 to Sunday, April 30, 2006).

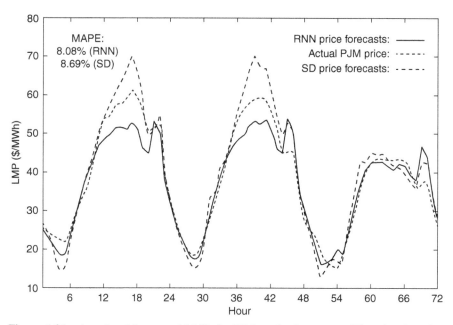

Figure 6-21 Actual and forecasted LMPs for PJM market in summer (Thursday, June 8 to Saturday, June 10, 2006).

Table 6-6 Comparison of Next-72 h Price Forecasting Performance of the Proposed RNN with SD Method

	MAPE (%)										
	Max. MAPE		Min. MAPE		Avg. MAPE		FMSE ($/MWh)		Error Variance		MAE ($/MWh)
Year 2006	RNN	SD	RNN	SD	RNN	SD	RNN	SD	RNN	SD	RNN	SD
April 28–30	10.20	12.95	6.64	6.53	8.57	9.94	4.59	5.01	0.0061	0.0062	3.38	3.92
June 8–10	9.72	10.40	5.84	7.46	8.08	8.69	3.98	4.21	0.0042	0.0045	3.10	3.33

(Case 1) and other efficient techniques. Better analysis has been done in this section for different data set of the PJM market to forecast hourly prices.

In order to show the effectiveness, and assess the prediction capacity of the proposed recursive NN model, Case 2 is compared with Reference [31], where the authors have proposed multi-layer perceptron with one hidden layer neural network along with other efficient methods, such as ARIMA and transfer function (TF) models, for forecasting electricity prices of the PJM market and same MAPE definition has been used. Table 6-7 presents the comparison of MAPE values obtained in this case (Case 2) with Reference [31] and Case 1. It can be observed from Table 6-7 that the MAPE values obtained for the weekday (winter) in Reference [31] range from 6.08 to 14.07% (NN), 5.06 to 10.76% (ARIMA), and 4.32 to 5.46% (TF), whereas in Case 2, the MAPE is around 8% for the weekday. The MAPE value obtained in this case, for the same day in spring (Saturday, May 13), is lower (4.49%) than that obtained in Case 1 (6.91%), which implies that the proposed algorithm using the recursive neural network is capable of improving the forecasting error effectively. In particular, the MAPEs during spring and summer are comparatively better than that of Reference [31]. The MAPE results show that the proposed recursive neural network model compares favorably with other methods during both weekend and weekdays.

Furthermore, the accuracy of the proposed method is verified by comparing the results obtained in spring with that of Case 1 as shown in Table 6-8 where weekday and Saturday resemble April 7 and May 13, 2006, respectively. Forecasting of the

Table 6-7 Comparison of Forecasting Performance of the Proposed RNN Model with Other Methods (MAPE comparison)

| | Reference [31] | | | Case 1 | Case 2 |
MAPE (%)	NN	ARIMA	TF	(ANN)	(RNN)
Winter (Weekday)	6.08–14.07	5.06–10.76	4.32–5.46	6.93–7.96	8.81
Spring (Weekday)	20.39–29.14	15.58–22.42	6.07–10.24	6.91	4.49
Summer (Weekday)	7.99–28.38	8.82–26.86	4.00–9.06	–	5.11
Autumn (Weekday)	6.45–16.81	5.77–16.50	4.18–6.70	–	7.72

Table 6-8 Comparison of Accuracy Measures in Spring

| | Case 1 | | Case 2 | |
| | PJM Subarea | | PJM Subarea | |
Spring	Weekday	Saturday	Weekday	Saturday
MAPE (%)	9.02	6.91	8.76	4.49
MAE ($/MWh)	2.21	2.68	4.62	1.58
FMSE ($/MWh)	5.87	3.65	5.32	2.56
Correlation value	0.92	0.95	0.96	0.96

next-24 h prices is performed on April 7 using the proposed RNN method and the MAPE value obtained is lower (8.76%) than that of Case 1 (9.02%). Correlation values are calculated based on Reference [14, equation (7)] and demonstrates the resemblance between the forecast curve and actual curve. FMSE and correlation values for Saturday are relatively better than that of Case 1. It is also seen that the MAE value for Saturday is obtained as 1.58 $/MWh only that shows a good performance of the developed recursive neural network algorithm. Moreover, FMSE and correlation values in weekday compare favorably with Case 1. The larger the correlation values, the more precise the prediction.

This section further compares its results obtained during summer with another literature [33], where the authors have proposed a neuro-fuzzy approach for price forecasting. The error standard deviation is calculated as [33]

$$\sigma = \sqrt{\frac{1}{N}\sum_{i=1}^{N}(e_i - \bar{e})^2}, \qquad (6\text{-}17)$$

where e_i is the difference between actual and forecasted price, and $\bar{e}$ is the mean of e_i, $i = 1, 2, ..., N$.

The values of error standard deviation are presented in Table 6-9. Note that the PJM price curves are highly volatile in summer compared to other seasons. It is seen from Table 6-9 that the values of the MAE and the error standard deviation

Table 6-9 Comparison of Accuracy Measures in Summer

| | Reference [33] | | | | Case 2 | |
| | PJM Subarea 1 | | PJM Subarea 2 | | PJM Subarea | |
Summer	Weekday	Saturday	Weekday	Saturday	Weekday	Saturday
MAE ($/MWh)	2.52	5.71	4.11	5.60	2.34	5.27
Error standard deviation ($/MWh)	5.41	7.37	8.95	7.46	2.91	2.67
Correlation value	0.9995	0.9879	0.9990	0.9867	0.9882	0.9851

obtained from the RNN model are comparatively lower than that of Reference [33]. These values show that the RNN-based prediction technique is efficient and provides a reliable forecast for short-term price forecasting.

This section analyzes an important problem of electricity price forecasting in which a technique based on recursive neural network model combined with SD method is applied to predict next-24 h and -72 h electricity prices of an electric energy market. The test results obtained for multiple seasons that include weekdays and weekends confirm that the prediction behavior of the developed RNN model is accurate.

CONCLUSIONS

This chapter presented a simple and easy-to-use technique for forecasting prices using neural networks based on the SD method in an energy market. Two major cases with different neural network architectures were discussed. These are Case 1 using integration of ANN and SD method, and Case 2 using integration of RNN and SD method. Appropriate examples based on data pertaining to the PJM electricity market were used to demonstrate the performance of the proposed neural networks. Detailed explanations of solution algorithm development including similar days and neural network approaches have been discussed. According to SD method, the price curves are forecasted by using the information of the days being similar to that of the forecast day. Two procedures were analyzed, for prediction based on averaging the prices of SD, and prediction based on averaging the prices of SD plus neural network refinement.

In Case 1, the proposed ANN model was applied to generate daily and weekly price forecasts. The factors affecting electricity price forecasting, including time factors, demand factors, and historical price factor are discussed. It is confirmed that demand is the most important variable influencing electricity price. For days and weeks studied, the daily and weekly MAPE values obtained in Case 1 indicate promising results in the PJM electricity market.

In Case 2, the developed recursive neural network model was used to generate the next-24 h and -72 h electricity prices for multiple seasons of the year. Different performance criteria, such as MAPE, MAE, FMSE, and error variance were used to characterize the prediction behavior of the proposed RNN model. In order to show the effectiveness of the proposed RNN model, a comparative analysis was carried out with respect to other methods reported in the literature, and the statistical measures, such as correlation value and error standard deviation were also calculated. The values obtained show that the proposed RNN model compares favorably with other studies.

In both cases, the MAPE, MAE, and FMSE values of reasonably small were obtained for the PJM data. MAPE values obtained from all the forecasts show that the proposed neural networks techniques outperform the direct use of SD method. The obtained smaller value of error variance implies the robustness of the developed price forecasting algorithm.

This chapter contributed to solving an important problem of short-term electricity price forecasting as the test results obtained through the simulation demonstrate that the developed algorithm is significantly accurate, efficient, highly adequate, and performs well in multiple seasons.

Future work would consider other influential aspects such as transmission congestion, outages, and spinning reserves, along with studying volatility, and risk in electricity price forecasting.

APPENDIX

Calculation of Weighting Factor

The weighted factor $\hat{w}_i (i = 1 - 4)$, is determined by least square method using multiple regression model:

$$P_{t+1} = w_0 + w_1 L_t + w_2 L_{t-1} + w_3 P_t + w_4 P_{t-1} \tag{6-18}$$

where, L_t is load at t; L_{t-1} is load at $t - 1$; P_t is price at t; and P_{t-1} is price at $t - 1$; and P_{t+1} is an estimated value of price using actual data of L_t, L_{t-1}, P_t, and P_{t-1}.

Equation (A6-18) is simplified using

$$y_\alpha = w_0 + w_1 x_{\alpha 1} + w_2 x_{\alpha 2} + w_3 x_{\alpha 3} + w_4 x_{\alpha 4} \tag{6-19}$$

We define that $\hat{y}_\alpha$ is the predicted value of y_α as

$$\hat{y}_\alpha = \hat{w}_0 + \hat{w}_1 x_{\alpha 1} + \hat{w}_2 x_{\alpha 2} + \hat{w}_3 x_{\alpha 3} + \hat{w}_4 x_{\alpha 4} \tag{6-20}$$

where $\hat{w}_0 - \hat{w}_4$ are the estimated values.

In least square regression analysis, the objective is to minimize the sum of the squared distance of the errors (e) where errors are defined as the distance between the observed value and the predicted one.

$$e_\alpha = \sum_{\alpha=1}^n (y_\alpha - \hat{y}_\alpha)^2 = \sum_{\alpha=1}^n (y_\alpha - \hat{w}_0 - \hat{w}_1 x_{\alpha 1} - \hat{w}_2 x_{\alpha 2} - \hat{w}_3 x_{\alpha 3} - \hat{w}_4 x_{\alpha 4})^2$$

$$\tag{6-21}$$

The objective here is to minimize e_α, where (A6-21) is partially differentiated with respect to $\hat{w}_0, \hat{w}_1, \hat{w}_2, \hat{w}_3$, and $\hat{w}_4$. Finally, we obtain the values of $\hat{w}_1 - \hat{w}_4$.

REFERENCES

1. S. Stoft, *Power System Economics: Designing Markets for Electricity*, IEEE Press, John Wiley & Sons, Inc., New York, 2002.
2. M. Shahidehpour, H. Yamin, and Z. Li, *Market Operations in Electric Power Systems: Forecasting, Scheduling, and Risk Assessment*, John Wiley & Sons, Inc., New York, 2002.

3. H.Y. Yamin, S.M. Shahidehpour, and Z. Li, "Adaptive short-term electricity price forecasting using artificial neural networks in the restructured power markets," *Electrical Power and Energy Systems*, vol. 26, pp. 571–581, 2004.

4. F. Gao, X. Guan, X.R. Cao, and A. Papalexopoulos, "Forecasting power market clearing price and quantity using a neural network," *IEEE PES Summer Meeting*, 2000, Seattle, WA.

5. R.C. Garcia, J. Contreras, M. Akkeren, and J.B.C. Garcia, "A GARCH forecasting model to predict day-ahead electricity prices," *IEEE Transactions on Power Systems*, vol. 20, no. 2, pp. 867–874, May 2005.

6. N. Amjady, "Day-Ahead price forecasting of electricity markets by a new fuzzy neural network," *IEEE Transactions on Power Systems*, vol. 21, no. 2, pp. 887–896, May 2006.

7. J.P.S. Catalao, S.J.P.S. Mariano, V.M.F. Mendes, and L.A.F.M. Ferreira, "Short-term electricity prices forecasting in a competitive market: a neural network approach," *IEEE Transactions on Power Systems*, vol. 77, no. 10, pp. 1297–1304, 2007.

8. N. Amjady and M. Hemmati, "Energy price forecasting-problem and proposals for such predictions," *Power and Energy Magazine, IEEE*, vol. 4, no. 2, pp. 20–29, March–April 2006.

9. B.R. Szkuta, L.A. Sanabria, and T.S. Dillon, "Electricity price short-term forecasting using ANN," *IEEE Transactions on Power Systems*, vol. 14, no. 3, pp. 851–857, August 1999.

10. F.J. Nogales, J. Contreras, A.J. Conejo, and R. Espinola, "Forecasting next-day electricity prices by Time Series Model," *IEEE Transactions on Power Systems*, vol. 17, no. 2, pp. 342–348, May 2002.

11. C.P. Rodriguez and G.J. Anders, "Energy price forecasting in the Ontario competitive power system market," *IEEE Transactions on Power Systems*, vol. 19, no. 3, pp. 366–374, Feb. 2004.

12. T. Mount, "Market power and price volatility in restructured markets for electricity," *Decision Support Systems*, vol. 30, no. 3, pp. 311–325, 2001.

13. PJM Web Site, *http://www.pjm.com*, Active March 2007.

14. Y.Y. Hong, C.Y. Hsiao, "Locational marginal price forecasting in deregulated electricity markets using artificial intelligence," *IEE Proceedings-Generation, Transmission, and Distribution*, vol. 149, no. 5, pp. 621–626, Sept. 2002.

15. J. Bastian, J. Zhu, V. Banunarayanan, and R. Mukerji, "Forecasting energy prices in a competitive market," *IEEE Computer Applications in Power*, vol. 12, no. 3, pp. 40–45, July 1999.

16. J. Contreras, R. Espinola, F.J. Nogales, and A.J. Conejo, "ARIMA models to predict next-day electricity prices," *IEEE Transactions on Power Systems*, vol. 18, no. 3, pp. 1014–1020, Aug. 2003.

17. A.M. Gonzalez, A.M.S. Roque, J. Gargia-Gonzalez, "Modeling and forecasting electricity prices with input/output hidden Markov models," *IEEE Transactions on Power Systems*, vol. 20, no. 1, pp. 13–24, 2005.

18. P. Mandal, T. Senjyu, and T. Funabashi, "Neural networks approach to forecast several hour ahead electricity prices and loads in deregulated market," *Energy Conversion and Management*, vol. 47, pp. 2128–2142, 2006.

19. H.S. Hippert, C.E. Pedreira, and R.C. Souza, "Neural networks for short-term load forecasting: A review and evaluation," *IEEE Transactions on Power Systems*, vol. 16, no. 1, pp. 44–55, 2001.

20. T. Senjyu, P. Mandal, K. Uezato, and T. Funabashi, "Next day load curve forecasting using hybrid correction method," *IEEE Transactions on Power Systems*, vol. 20, no. 1, pp. 102–109, 2005.

21. T. Senjyu, P. Mandal, K. Uezato, and T. Funabashi, "Next day load curve forecasting using recurrent neural network structure," *IEE Proceedings-Generation, Transmission, and Distribution*, vol. 151, no. 3, pp. 388–394, May 2004.

22. P. Mandal, T. Senjyu, N. Urasaki, and T. Funabashi, "A neural network based several-hour–ahead electric load forecasting using similar days approach," *International Journal of Electrical Power & Energy Systems*, vol. 28, no. 6, pp. 367–373, July 2006.

23. E. NI and P.B. Luh, "Forecasting power market clearing price and its discrete PDF using a Bayesian-based classification method," in *Proceedings of IEEE PES Engineering Society Winter Meeting*, Columbus, OH, 2001, vol. 28.

24. P. Doulai and W. Cahill, "Short-term price forecasting in electricity energy market," in *Proceedings of International Power Engineering Conference (IPEC2001)*, May 17–19, 2001, pp. 749–754.

25. L. Zhang, P.B. Luh, and K. Kasiviswanathan, "Energy clearing price prediction and confidence interval estimation with cascaded neural networks," *IEEE Transactions on Power Systems*, vol. 18, no. 1, pp. 99–105, 2003.

26. J. Guo and P.B. Luh, "Improving market clearing price prediction by using a committee machine of neural networks," *IEEE Transactions on Power Systems*, vol. 19, no. 4, pp. 1867–1876, 2004.

27. B. Ramsay and A.J. Wang, "A neural network based estimator for electricity spot-pricing with particular reference to weekend and public holidays," *Neurocomputing*, vol. 23, pp. 47–57, 1998.

28. P. Mandal, T. Senjyu, A. Yona, T. Funabashi, and A.K. Srivastava, "Price forecasting for day-ahead electricity market using recursive neural network," in *Proceedings of IEEE PES Engineering Society General Meeting*, Tampa, FL, June 24–28, 2007.

29. M. Negnevitsky, *Artificial Intelligence: A Guide to Intelligent Systems*, 2nd edition, Addison Wesley, Harlow, England, 2005.

30. S. Haykin, *Neural networks: A Comprehensive Foundation*. Macmillan College Publishing Company, Inc., 1994.

31. A.J. Conejo, J. Contreras, R. Espinola, and M.A. Plazas, "Forecasting electricity prices for a day-ahead pool-based electric energy market," *International Journal of Forecasting*, vol. 21, no. 3, pp. 435–462, 2005.

32. A.J. Conejo, M.A. Plazas, R. Espinola, and A.B. Molina, "Day-Ahead electricity price forecasting using the Wavelet Transform and ARIMA Models," *IEEE Transactions on Power Systems*, vol. 20, no. 12, pp. 1035–1042, May 2005.

33. Y.Y. Hong and C.F. Lee, "A neuro-fuzzy price forecasting approach in deregulated electricity markets," *Electrical Power Systems Research*, vol. 73, pp. 151–157, 2005.

Chapter 7

Estimation of Post-Storm Restoration Times for Electric Power Distribution Systems

Rachel A. Davidson, Haibin Liu, and Tatiyana V. Apanasovich

INTRODUCTION

Electric power systems throughout the Eastern United States have suffered repeated, often extensive damage from hurricanes and ice storms in recent years. In September 2003, for example, Hurricane Isabel caused 1.8 million (82%) of Dominion Virginia Power's 2.2 million customers to lose power for up to 2 weeks [1]. More than 10,700 power poles, 13,000 spans of wire, and 7900 transformers had to be replaced. In just 10 days, Dominion used a year's supply of poles, cross arms, and transformers, and about 4 years' worth of secondary wire and insulators. In December 2002, an ice storm caused 1.5 million (68%) of Duke Energy's 2.2 million customers to lose power for up to 9 days. That storm required 8500 field personnel and 4000 support personnel to restore power (personal communication with Duke Energy, 2003). Many similar examples exist. In fact, based on a survey of six utility companies' responses to 44 major storms from 1989 to 2003, it takes an average of 5.6 days to complete a major post-storm restoration [2].

As a result of storm-related damage to electric power systems, customers may experience business interruption and general inconvenience when they lose power. They may suffer irreversible loss of electronic data, food, or perishable goods. They may experience health impacts if, for example, power outages result in loss of heating during cold weather, as is often the case during ice storms. Even brief power interruptions may compromise security systems or financial transactions. A utility company's repair and recovery costs may be passed on to its customers as well.

Advances in Electric Power and Energy Systems: Load and Price Forecasting, First Edition.
Edited by Mohamed E. El-Hawary.
© 2017 by The Institute of Electrical and Electronics Engineers, Inc. Published 2017 by John Wiley & Sons, Inc.

Even those who are not power company customers may experience disruption due to power outages because of the indirect consequences they can have on business and government operations, water supply, traffic signaling, and other infrastructure systems. The reliance of water supply on electric power, for example, was clearly demonstrated when the August 2003 East Coast power failure led to the interruption of the water delivery system in Detroit, MI [3]. As society's dependence on and expectation of uninterrupted electric power increases, it becomes more and more critical to reduce these effects. Recognizing this growing need, regulators have increased their focus on power companies' natural disaster responses and post-disaster investigations by public utility commissions (e.g., [4]) have become increasingly common [5]. To the extent that a company knows ahead of time how long outages will last, it can better inform its customers, the public, and the state utility commission. Knowing the restoration time with some degree of certainty allows people to plan appropriately for the time without power. When outages occur, therefore, providing more, better, and earlier information can make customers more accepting of them [6].

Electric power companies typically use relatively simple deterministic models to estimate restoration times based on damage assessments reported from the field and a specified number of available crews. Because they depend on damage assessments, the models can be applied only after those assessments have been made. In this chapter, we present a new statistical approach to post-storm electric power restoration time estimation, and apply it for hurricanes and ice storms for three major electric power companies that together serve most of North Carolina, South Carolina, and Virginia—Dominion Virginia Power, Duke Energy, and Progress Energy Carolinas, which serve 2.2, 2.2, and 1.3 million customers, respectively (Fig. 7-1). Accelerated failure time (AFT) models, a type of survival analysis model, were fitted using an unusually large dataset that includes the companies' experiences in six hurricanes and eight ice storms (Tables 7-1 and 7-2). The models can be used to predict the duration of each probable outage in a storm, and by aggregating those estimated outage durations and accounting for variable outage start times, restoration times can be estimated for each county or other subregion of the service areas. The models can be used to better inform customers and the public of expected post-storm power restoration times.

This study focuses on hurricane wind (not storm surge) and ice storms because these hazards cause the most damage to power systems in the Eastern United States. It deals only with the distribution system because these storms primarily affect distribution. The same modeling approach could be applied to other infrastructure systems and hazards if data are available.

In Post-Storm Electric Power Outage Restoration Process, the real-life post-storm electric power outage restoration process is described. Previous work in restoration modeling for infrastructure systems in extreme events is summarized in Restoration Modeling Approaches. The data used in this analysis are described in Variable Definition and Data Sources, followed by a discussion of the AFT outage duration models in Accelerated Failure Time Outage Duration Models. The new method to estimate restoration times is presented in Restoration Time Estimation.

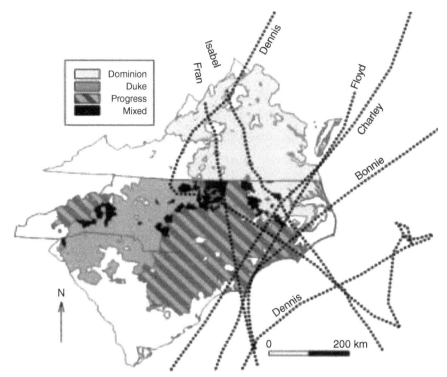

Figure 7-1   Dominion, Duke, and Progress service areas with recent hurricane tracks.

POST-STORM ELECTRIC POWER OUTAGE RESTORATION PROCESS

Strong winds and flooding are the main hazards that affect electric power distribution systems during hurricanes. In ice storms, ice accumulates on trees, lines, and electric

Table 7-1   Number of Outages by Company and Hurricane

Dominion		Duke		Progress	
Hurricane	Outages	Hurricane	Outages	Hurricane	Outages
Bonnie (8/98)	5079	Fran (8/96)	1818	Bonnie (8/98)	2746
Dennis (8/99)	2917	Floyd (9/99)	1061	Dennis (8/99)	1623
Floyd (9/99)	8186	Isabel (9/03)	2009	Floyd (9/99)	11,505
Isabel (9/03)	58,393			Isabel (9/03)	13,572
				Charley (8/04)	3470
Total	74,575	Total	4888	Total	32,916

Date is month/year of hurricane.

Table 7-2 Number of Outages by Company and Ice Storm

Dominion		Duke		Progress	
Ice Storm	Outages	Ice Storm	Outages	Ice Storm	Outages
12/23/98	10,125	12/24/02	13,370	1/25/04	7092
1/14/99	2597	2/27/03	3,277		
12/13/00	638				
12/4/02	1484				
12/11/02	1464				
12/24/02	468				
2/27/03	1095				
Total	17,871 Total		16,647	Total	7092

Date is first day of ice storm.

devices, often accompanied by strong winds. The most common associated damage modes in both cases are trees being uprooted and falling on lines, or tree branches breaking off and falling on lines. Other damage modes include poles falling over or breaking; cross-arms breaking; flying debris damaging lines, poles, or devices; cascading failure when failure of one span of line pulls down another; and in the case of ice storms, galloping-induced failure of lines due to the combination of wind and increased effective line surface area from accreted ice.

When a line breaks, a pole falls, or something else causes a permanent fault during a storm, the nearest protective device upstream is activated to isolate the damage (e.g., fuse melts or circuit breaker interrupts power flow). All customers on the isolated portion of the system lose power. In this chapter, such a scenario is considered a single *outage*, that is, an activation of a protective device caused by physical damage requiring repair by a crew. One outage, therefore, may be associated with a little or a lot of physical damage, and from one to many customers losing power.

Outages must be located and repaired to restore power following a storm. While each company conducts its post-storm restoration a little differently and each storm brings unanticipated conditions, the basic process of post-storm restoration is similar across companies and major events. Based primarily on interviews with emergency response personnel at Dominion Virginia, Duke, and Progress Carolinas, the process can be summarized in three main phases: (1) pre-storm preparations, (2) damage assessment, and (3) repair, tree clearing, and re-energizing.

Pre-storm Preparations

The company's internal meteorologist (or someone else if the company does not have one) continually monitors the development of any storms that may threaten the company's service area. When a storm develops that may have a major impact on the service area, the manager responsible for emergency response begins coordinating with the internal groups in charge of crew logistics, public affairs, resource procurement, and other relevant tasks. When the emergency response manager is reasonably sure that the company will be impacted and will need outside help for the restoration, that individual contacts other companies to request mutual aid in the form of patrollers (damage assessors), line crews, tree crews, and logistical support. This request may go through a professional organization like the Southeastern Electric Exchange. The number of crews initially requested typically depends on a simple deterministic calculation based on the hurricane category or predicted ice thickness and an estimated number of crews needed per impacted mile. The crews that are offered may be contractors or employees of other power companies. The amount of aid available depends on the extent to which other companies expect to be impacted and are willing to release their own employees. While a company's own employees tend to perform efficiently because they are familiar with the system, during major storm events, help from nonemployees may be required.

Companies may or may not deploy crews to the field and store materials at staging areas before a storm actually hits. Doing so requires some certainty about the severity and location of damage so that they can avoid moving crews into harm's way, ordering too many crews and materials, or positioning them in the wrong locations.

While the organization of each company differs, they all divide their service areas into smaller operating areas, each of which has an office that runs the local operations within that area. These local operations center typically coordinates a lot of the storm response, for example collecting damage reports from patrollers, dispatching line and tree crews, and reporting damage to central management.

Damage Assessment

After a storm actually impacts the company's service area, the company must locate outages that have occurred. Supervisory control and data acquisition (SCADA) systems locate outages in the transmission system, but they usually are not used in distribution. In distribution, outages are located primarily through a trouble call system. When a customer calls in to report a power outage, he or she provides information about the power outage, such as, whether it extends to the rest of the neighborhood, and whether there was any obvious cause that was observed by the customer (such as a fallen tree or loud noise from a fuse blowing). In normal times, 15% to 20% of customers without power call in to report it. During a storm, as outages persist, that percentage increases. The start time of the outage is recorded as the time (to the minute) that the first call associated with an outage is received. Based on the geographic distribution of incoming calls and the information obtained during the calls, the outage management system (OMS) uses an algorithm to locate the device (e.g., recloser, fuse) that is the suspected source of the outage. Crews are then dispatched to the site to perform restoration work. For major weather events, such as hurricanes and ice storms, patrollers (i.e., damage assessors) are sent out in advance of crews to review and document damages. They are sent to areas where the power is out, and they identify the protective device that has tripped off and document the extent of damages (e.g., number of broken poles, lines down). They record the materials and equipment needed for repair and send that information back to personnel at the local operations center, who then update the OMS (i.e., confirm or correct the suspected source of the outage). If the patrollers find an additional protective device activated downstream of the first, they notify the OMS and an additional outage is recorded. Customers affected by this newly discovered outage are removed from the first outage record and the starting time for the new outage record is usually listed as the time the new outage was entered, or possibly as the time at which the first, larger outage started.

Technical designers, engineers, and others serve as patrollers in a post-storm situation. Helicopters may be used to conduct high level damage assessment in particularly large events. Damage data are collected and used to determine the number of crews needed, and the materials and equipment needed. In general, this information is used to effectively direct the work of crews. Companies may collect damage data, but most record the impact only in terms of outages. While damage assessment and repair happen simultaneously throughout the storm, patrollers try to stay a day ahead of the line crews. Within 24 to 48 hours of the storm, based on damage assessments from the field, personnel at the local operations center estimates restoration times by county, feeder, or some similar unit, and upper management typically approves them. Those times are input into the voice response unit (VRU) so that when a customer calls in, he or she will hear an expected restoration time associated with his location.

Repairs, Tree Clearing, and Re-energizing

Crews are initially requested based on early damage assessments. Throughout the storm the managers repeatedly consider the amount of damage that has been reported, update the number of crews requested, estimate the restoration times given that damage and the number of crews available, reconsider the reported damage, and so on. It is an iterative process in which they alternate between the number of crews requested (which may or may not be the same as the number available) and restoration time estimates. Line and tree crews can vary from two to several people per crew.

Repair materials (e.g., cross arms, poles, transformers) are stored in staging areas that are located in shopping mall parking lots, fields, or any other open area the company can get permission to use. The number and locations of staging areas are a combination of pre-identified sites and *ad hoc* sites that vary from storm to storm depending on where they are needed. Materials are brought to the staging areas from the offices or warehouses where they are normally stored. Additional materials may be purchased and delivered by vendors, if necessary.

Repairs are scheduled as damage assessments come in from the field, OMS, and SCADA. Priority is often given to critical customers, such as police and fire stations, hospitals, schools, and water pumping stations, as indicated in the OMS. Beyond that, companies try to restore power to as many customers as quickly as possible, and thus they prioritize repair of outages based on how many customers each outage has affected and how easy it will be to restore.

Each crew is assigned one to several jobs at a time. It collects the necessary materials from one of the staging areas, goes out to locate and repair the damage as assigned. Once a repair is complete, the line crew will either call in a "switching order" to get the appropriate devices closed so that the circuit is re-energized, or close the devices itself in the field and call in to notify the operations center that they have. At that time, the line crew will also indicate whether or not the suspected outage location was correct, and if not, the record in the OMS database will be corrected. In some cases, a crew may wait until the end of the day to report the update to the operations center (although the outage end time should still be recorded as the time power was actually restored). Customers that previously called in to report an outage are then called back using an automated VRU to confirm that their power is actually back on. (Depending on the company, sometimes all customers affected by the outage are called). If a customer still does not have power, a new outage record is created with the time of the call-back used as a start time, and a new crew will be dispatched to repair it. Since mobile links to trucks often do not work during storms and many crews are not company employees, communication between the crews and operations center may be a challenge. During storms, tree crews will expedite the process by leaving trees they have cut off the lines at the site. Debris removal crews will come out later to remove them.

Possible Errors in Outage Start and End Times

Due to the way the outage information is collected, there is some potential for error in the recorded start and end times. First, if there are two locations of damage on the same circuit, the OMS algorithm may not identify the downstream location, which therefore, may not be identified until the patrollers or repair crews find it. For this reason, there is a possibility that the start time of the downstream outage may not be correct. Second, if a line crew does not report completion of a repair immediately, the outage end time may be inaccurate. Neither of these problems, however, is thought to cause large or widespread errors.

RESTORATION MODELING APPROACHES

Several methods have been used for modeling the restoration of infrastructure systems following extreme events (e.g., hurricanes, ice storms, earthquakes). They can be grouped into the following general approaches: (1) empirical curve fitting, (2) deterministic resource constraint modeling, (3) Markov modeling, (4) simulation, and (5) optimization. The output of these models is typically presented as either restoration curves (percentage of customers with power vs. time) or some version of the System Average Interruption Duration Index (SAIDI), a performance measure widely used in the electric power industry. Based on the potential uses of a restoration model and a detailed understanding of the real-life post-disaster restoration process, it is desirable for a restoration model to (based partly on [7]).

- Be usable in a predictive mode, before an infield damage assessment is complete.
- Include the utility company's decision variables explicitly, allowing exploration of their effects on the speed of the restoration. Possible decision

variables include number of response crews of different types, amount of repair materials of different types, and repair prioritization rules.

- Produce different restoration curves for each subregion within the service area rather than just one curve for the whole system.
- Represent the uncertainty in the restoration time estimates.
- Be based on and validated with real experiences and data.
- Limit the extent to which simplifying assumptions about the restoration process are required and ensure that any assumptions made are reasonable.

In Sections 3.1 to 3.6, the different approaches to the problem are briefly described and compared to each other and to the statistical approach proposed in this research, including a discussion of the extent to which each exhibits these desirable characteristics. This review draws on studies from the largely independent literatures on power distribution systems reliability and natural hazard lifeline engineering.

Empirical Curve Fitting

In the empirical curve fitting approach, data obtained from previous events and/or expert opinion is employed to fit restoration curves, and it is assumed that those curves represent future restorations. Either data are plotted and a restoration curve is fit to the data, or a functional form for the restoration curve is assumed and data are used to estimate the parameters for that function. This approach has been used in many previous studies, such as Reed [8], Reed et al. [9], ATC-25-1 [10], Shinozuka et al. [11], and Nojima et al. [12]. The restoration curves are derived directly from historical data, but the actual restoration process is not modeled explicitly. Usually, only one restoration curve is obtained for the entire lifeline system. Uncertainty and decision variables are not represented explicitly.

Deterministic Resource Constraint Modeling

In the deterministic resource constraint approach, the actual restoration process is modeled, but in a simplified way, typically using a set of simple equations and rules. Resource constraints are accounted for by specifying the number of repairs that can be made in any time period as a function of the number of repair personnel and/or materials available. This approach allows depiction of the progress of the restoration across time and space and enables investigation of the effect of some restoration time minimizing efforts, such as, prioritizing repairs or using mutual aid agreements [13]. This approach has been employed in Isumi and Shibuya [14], and HAZUS-water supply system section [15]. The Distribution Reliability Assessment Model uses this approach to estimate the Feeder Average Interruption Duration Index (FAIDI), but not for extreme events specifically [16–18]. Cooper et al. [19] and Cooper [20] use a deterministic resource constraint approach to estimate restoration times for each probable outage in a storm. Their method can be used in a predictive mode. The restoration time estimation models currently used by Dominion Virginia Power, Duke Energy, and Progress Energy Carolinas could be considered to use this approach as well. Billinton and Acharya [21] use a similar method that they call a reliability index segmenting approach. They first estimate the SAIFI and SAIDI using traditional reliability analysis based on component failure rates and repair times, and then disaggregate the indexes by weather state (normal, adverse, or extreme).

In these studies, the restoration process is modeled and resource constraints are considered, but the process is represented in a very simplified way that assumes, for example, that restoration involves only repair not damage assessment or other tasks. The models are deterministic and usually based on rough estimates of fixed repair rates. Estimates may be made for the entire service area or subregions within it.

Markov Modeling

Isoyama et al. [22] and Iwata [23] model individual lifeline's functional performance in the post-earthquake period using discrete-state, discrete-transition Markov processes. In later studies, such as Kozin and Zhou [24] and Zhang [25], a discrete-state, discrete-transition Markov process is employed to model the evolution of the restoration of various different lifelines together as a system. This modeling approach can produce spatially disaggregated restoration curves, include decision variables explicitly, and represent uncertainty in the restoration curves. One main disadvantage is that it requires data in the form of transition probabilities and state vectors, which can be difficult to obtain. The approach also requires many simplifying assumptions about the system and the restoration process. For example, the assumption in Zhang [25] and Kozin and Zhou [24] that lifelines compete with each other for available restoration resources is not typically true in reality.

Simulation

Brown et al. [26] and Balijepalli et al. [27] use Monte Carlo simulation to assess distribution system reliability in storms and lightning storms, respectively. They estimate failure rates and mean times to repair and switch, then simulate a simplified version of the storm restoration process. The results are presented as estimates of SAIDI and similar indices, for the whole system or by feeder. The method allows representation of uncertainty surrounding the SAIDI estimates, geographical disaggregation of restoration times, and use in a predictive mode. Broström [28] provides a comprehensive simulation-based approach to transmission system reliability in ice storms, including modeling winds and ice accretion, component vulnerability, and restoration times. The restoration time of each segment of the line is a function of the simulated times required to complete five tasks. Task times are assumed to be correlated across tasks and segments. Possible limitations are the relatively simplified representation of the restoration process and potential difficulty in estimating the required input data.

Newsom [29] presents early work on post-earthquake electric power restoration using discrete event simulation, but interestingly, no other studies could be found using that approach until almost 30 years later. Çağnan and Davidson [30] and Çağnan et al. [31] present a discrete event simulation model of the post-earthquake restoration process for the Los Angeles Department of Water and Power (LADWP) electric power system. Tabucchi and Davidson [32] do the same for the LADWP water supply system. These models explicitly represent the real-life restoration process including a high level of detail with few simplifications. The approach enables development of geographically disaggregated, quantitative restoration curves with uncertainty bounds, and explicit representation of the company's decision variables. Discrete event simulation models can be validated through comparison with restoration curves based on historical data or on statistical models like those presented in this research. Their key disadvantage is that it is very time consuming and data intensive to develop such models, and they are system-specific.

Optimization

While the first four categories of previous work focus on descriptively modeling the current restoration process, optimization aims to determine the "best" way to conduct a restoration process in terms of, for example, how to prioritize repairs, how many of each type of restoration crew to have, and how many of each type of repair material to have and where to store them. Xu et al. [33] summarize the literature in this area, including Wang et al. [34], Yao and Min [35], Wu et al. [36], Sato and Ichii [37], and Sugimoto and Tamura [38]. In general, restoration optimization models are presented in a relatively generic way, suggesting they can be applied equally well to different types of lifelines (e.g., electric power, roads) under different types of extreme events (e.g., earthquake, ice storm). They rely on simplified representations of the lifeline network(s), include applications only to relatively small hypothetical lifeline networks, and consider a single or small set of generally hypothetical initial damage conditions. They may or may not include uncertainty and geographically disaggregated results.

Statistical Regression

In this chapter, a statistical approach to modeling the restoration process is presented. Using the survival analysis AFT method and a large dataset of past hurricane- and ice-storm-related outages, models are developed that describe the relationship between the duration of each storm-related outage and a set of covariates that describe the outage, storm, and power system. The individual outage duration estimates resulting from those models are then aggregated, considering the variable outage start times, to develop estimated restoration times for each county or other subregion of the service areas. Sections 4 to 6 describe this new method.

Unlike empirical curve fitting, which is also based on historical data, this approach describes the relationship between outage durations and covariates, can be used in a predictive mode, provides geographically disaggregated restoration time estimates, and is based on rigorous statistical methods. A principal limitation of the statistical approach is that it does not explicitly represent the restoration process or power company decision variables, and thus cannot be used directly to examine the effects of changes in those variables. It could, however, be used to help calibrate a simulation approach that incorporates decision variables. The statistical approach also requires sufficient data from past events, and therefore, may not be possible for very rare events, like earthquakes.

VARIABLE DEFINITION AND DATA SOURCES

This section describes the covariates investigated for the new restoration model, why each was considered, and the data used to measure it. Table 7-3 summarizes the continuous variables considered in fitting the outage duration models. In addition, five categorical variables were considered. Hurricane (x_h), ice storm (x_{is}), and company affected (x_c), which indicate the particular storm and the company the outage was associated with, had six, eight, and three possible values, respectively, as listed in Tables 7-1 and 7-2. Type of device affected (x_{dev}) had four possible values (transmission side of substation circuit breaker; substation; protective device; service transformer/customer), and land cover type at outage location (x_{lc}) has six possible values: water; forest; grass; rock; wetlands; and residential/industrial/commercial/transportation.

The number of customers affected by the outage (x_{cus}), type of device affected by the outage (x_{dev}), population density of zip code outage (x_{pop}), and outage start time measured relative to the start time of the first outage in the storm (x_{start}) were considered because they all potentially affect the priority given to outages for repair. Companies typically prioritize restoration of outages that affect more customers or that affect a device that serves high priority customers, such as, hospitals. Outages that start earlier may last longer because there are many other outages early in the storm and restoration resources may not be fully mobilized yet. The total number of outages in the storm (x_{out}) was thought to be directly related to outage duration because storms that are larger overall are more likely to have inadequate resources and a more complex restoration process. Tables 7-1 and 7-2 list the 12 hurricane-company and 10 ice storm-company experiences, respectively, for which outage data were provided. A single outage may be caused by anything from a minor damage to a cross arm to several poles and spans of wire down. Unfortunately, the exact physical damage that causes each outage is not recorded. Maximum gust wind speed (x_w), duration of strong winds (d), rainfall (x_r), ice thickness (x_{it}), and land cover type (x_{lc}) are all potentially related to the amount of damage associated with an outage, and therefore, the time required to repair the damage and restore the outage.

The wind speed database contains the estimated maximum wind gust speed (in mph) at 10 m elevation at the centroid of each zip code during each hurricane. It also includes, for each zip code, the strong wind duration (in minutes), measured as the time during which the zip code experiences winds over 40 mph, including the passage of the hurricane's eye. For each hurricane of interest, a geographic information system (GIS)-based hurricane wind field simulation model [39] produced a preliminary wind

Table 7-3 Descriptive Statistics of Continuous Variables Considered in AFT Models

Variable	Hurricane					Ice storm				
	Minimum	Median	Mean	Maximum	Standard Deviation	Minimum	Median	Mean	Maximum	Standard Deviation
Outage duration, min. (T)	1	1294	2444	24,041	3058	1	1226	1934	13,507	2200
Maximum gust wind speed, mph (x_w)	5.72	41.77	41.43	81.16	12.60	NA	NA	NA	NA	NA
Duration of strong winds, min. (d)[a]	0	39	177.4	1746	239.58	NA	NA	NA	NA	NA
7-day rainfall, in. (x_r)	0.03	4.95	5.58	27.15	3.92	NA	NA	NA	NA	NA
Ice thickness, mm (x_{it})	NA	NA	NA	NA	NA	0	8.62	10.11	30.96	9.41
Total number of outages in storm (x_{out})	1061	58,393	34,787	58,393	25,165	468	10,125	9000	13,370	4308
Outage start time, min. (x_{start})	0	3886	5790	25,820	5403	0	3018	4118	15,733	3262
Number of customers affected by outage (x_{cus})	1	5	96.63	11,750	361.7	1	7	111.7	4815	337.7
Population density, people/sq. mi. (x_{pop})	2	163	519	5877	695.7	3	156	340	5877	467.7

[a] The covariate used in modeling is: $x_d = \ln(d)$.

field based on reconnaissance aircraft measurements, assumed surface-to-aircraft wind speed ratios, and an assumed decay of the storm as it moved inland. These preliminary model wind speed estimates were compared to wind speeds measured at various locations throughout the service areas, and the wind field was modified to ensure that the final wind speeds were consistent with the measured values. Using surface roughness information from Hagemann et al. [40], wind speeds were then adjusted based on land cover type, and the wind speed assigned to each 3 km × 3 km grid cell was the weighted average of adjusted wind speeds, where weights were the percentages of each land cover type appearing in the grid cell. The strong wind duration data were obtained from the wind model simulation.

For each hurricane, the rainfall database contains estimates of the total rainfall (in inches) experienced at the centroid of each zip code during the 7 days, preceding the time the hurricane reached its maximum intensity in the service areas. Rainfall may determine the extent to which the soil was saturated at the time of the strong winds, and therefore, the ease with which trees might be uprooted and fall on power lines. The appropriate 7-day rainfall totals were calculated for each rainfall station in the service areas using data from the National Climatic Data Center [41]. Seven-day rainfall totals at the centroids of each zip code were then estimated using Inverse Distance Weighted interpolation in GIS.

Ice accretion was estimated using a set of models executed in sequence [42]. The Weather Research and Forecasting (WRF) model [43, 44] was employed to produce 12-hour forecasts (initialized every 6 hours) of precipitation amount, wind speed, relative humidity, and temperature (wet-bulb and dry-bulb). A precipitation-type algorithm [45] was then executed using WRF output to determine where freezing rain (and potential ice accretion) has been forecasted. Finally, the simple ice accretion model [46] was applied to estimate ice accretion thickness at each model grid cell (about 12 km × 12 km).

National Land Cover Data (NLCD) was obtained from the Multi-Resolution Land Characteristics (MRLC) Consortium. The data, which are based on MRLC's Landsat 5 Thematic Mapper satellite data and ancillary sources, indicate the land cover class of each 30 m × 30 m grid cell in the coterminous United States. [47]. Of the 21 defined land cover classes, 15 appear in the three companies' service areas. Those 15 land cover types were recategorized into six main types as noted above, such that land cover types with similar outage patterns were grouped together.

For x_w, d, x_r, x_{it}, x_{lc}, and x_{pop}, the covariate value for each outage is the value for the grid cell that the outage is in. Characteristics of the soil and trees along power lines undoubtedly help determine how much physical damage is associated with each outage, and therefore, may be related to outage durations as well. However, since the limited data available to capture tree and soil characteristics were not found to be statistically significant in a related statistical outage count model [48], they were not considered. Similarly, power system age and condition were not included because of the lack of data. Information about available restoration resources (e.g., repair teams and materials) and the frequency and manner of tree trimming are important too. Further, including them would allow investigation of the effects of these variables— which unlike the others, are under the company's control—on restoration times.

Unfortunately, since the data required to include them are not available, they were not considered either.

ACCELERATED FAILURE TIME OUTAGE DURATION MODELS

AFT Model Formulation

AFT models were developed to estimate the duration of each probable storm-caused electric power outage. These outage durations were then used to estimate restoration times for each county. AFT is a type of survival analysis model, that is, statistical analysis specifically for time-to-event data, in this case, time until an outage is restored, or outage duration. The power outage restoration problem is a relatively simple application of survival analysis. The start and end times for every outage are known, so the data are not censored, and there is no possibility of zero or multiple events. The AFT model was chosen over the Cox proportional hazards (CPH) model because while the CPH model relates covariates to hazard rate (i.e., the conditional probability of an outage being repaired given that it has lasted until some specified time t), the AFT model describes a direct relationship between covariates and outage duration, the quantity of interest.

Let T_i be a random variable outage duration (survival time), x_i be a vector of covariates that describe the ith outage, and β be the regression parameter vector. The AFT model is [49]

$$\ln(T_i) = x_i^T \beta + \varepsilon_i \tag{7-1}$$

where the errors ε_i are typically assumed to be independent and identically distributed. The AFT model is similar to a linear regression model except that the ε_i do not necessarily follow a normal distribution. Estimation can be carried out by assuming a distribution for T_i. The distribution of T_i is often assumed to be exponential, Weibull, log-logistic or log-normal, corresponding, respectively, to extreme-value (one parameter), extreme-value (two parameters), logistic, and normal distributions of ε_i.

AFT Model Fitting

Using R software [50] and the maximum likelihood method, we fitted four AFT models, assuming exponential, Weibull, log-logistic, and lognormal distributions for outage duration T_i. For the hurricane models, all the covariates discussed in Variable Definition and Data Sources were considered except ice storm (x_{is}) and ice thickness (x_{it}); for the ice storm models, all except hurricane (x_h), maximum gust wind speed (x_w), duration of strong winds (x_d), and rainfall (x_r) were considered. Before the models were fitted, data from the most recent hurricane and ice storm (Hurricane Charley and the January 2004 ice storm) were withheld for model testing, as if the models were fitted before those events occurred, then applied when they did

(AFT Model Testing). Fitting the AFT models required: (1) defining the error term, (2) selecting the covariates and determining their functional form, and (3) identifying influential points and outliers. While the steps in fitting the models are intertwined and require iteration, in reality, they are presented sequentially here for ease of presentation.

Parametric Form   To determine the proper assumption for the errors, the AFT models were all fit using the training dataset and the fits were compared in a few ways. (The same statistically significant covariates were identified in all these initial AFT models.) First, for some parametric AFT models, a function of the survival function ($S(t) = P(T > t)$) can be found that is linear in $\ln(t)$ [51]. For instance, for the Weibull AFT model $\ln(-\ln S(t)) = \gamma \ln(\lambda) + \gamma \ln(t)$, where γ and λ are the shape and scale parameters, respectively. Therefore, using the Kaplan–Meier estimator of the survival function [49], under a certain AFT model, the plot of the estimate of the survival function against $\ln(t)$ should be a straight line if the model fits well. Figure 7-2 provides such graphs for the Weibull, log-logistic, and log-normal AFT hurricane outage duration models. The plots are similar for the ice storms models

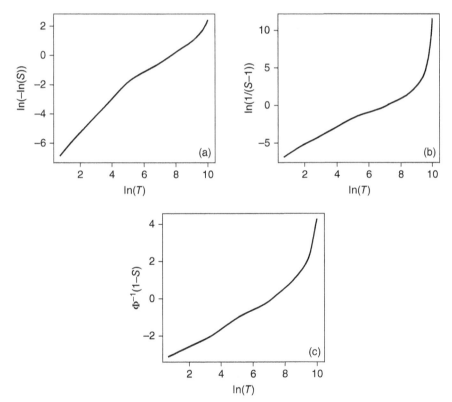

Figure 7-2   Function of survival function estimate versus $\ln(t)$ for (a) Weibull, (b) log-logistic, and (c) log-normal AFT models.

Table 7-4 Comparison of Parametric aft Outage Duration Models

	Hurricane		Ice Storm	
Model	Number of Parameters	AIC	Number of Parameters	AIC
Exponential	12	1,710,971	14	463,720
Weibull	13	1,710,188	15	463,647
Lognormal	13	1,735,359	15	480,899
Log-logistic	13	1,727,717	15	470,607

[48]. Since the plot for the exponential would be the same as that for Weibull but with a different y-intercept, it is not shown here. The graphs suggest that the Weibull distribution fits the data best of the three distributions for hurricane outage durations. The Weibull curve is reasonably linear, indicating that the Weibull distribution is an appropriate parametric model for these data.

Parametric AFT models can also be compared using the Akaike Information Criterion, defined as $AIC = -2logL + 2p$, where L is the likelihood and p is the number of parameters. For both hurricanes and ice storms, the Weibull distribution has the lowest AIC value indicating the best fit (Table 7-4). For nested AFT models like the exponential and Weibull, a log-likelihood ratio test can be used to compare model fits. The resulting p-value for the likelihood-ratio test was close to 0, confirming that the Weibull distribution is the better fit of the two.

Cox–Snell residuals [51], defined as $r_i^{CS} = -\ln \hat{S}(t_i|x_i)$, can also be used to assess goodness-of-fit for AFT models. If the distribution assumption is appropriate, plotting Cox–Snell residuals against a nonparametric estimate of $-\ln \hat{S}$, for example, the Kaplan–Meier estimator (i.e., the estimated cumulative hazard function $\hat{H}$) should produce a straight line through the origin with unit slope [49]. Plots for the Weibull regression models for hurricane and ice storm outage durations (Fig. 7-3) suggest that the model fits well, except at the upper end of the curve. At higher values, the Cox–Snell residuals are smaller than the Kaplan–Meier estimates, indicating that the model

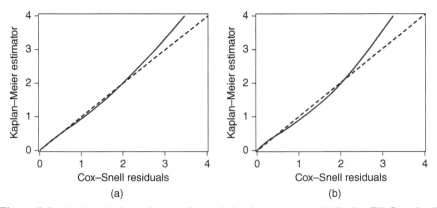

Figure 7-3 Kaplan–Meier estimate of cumulative hazard versus Weibull AFT Cox–Snell residuals for the (a) hurricane and (b) ice storm model.

overestimates the outage durations for the outages with the longest durations. Based on these analyses, the Weibull models were determined to provide the best fits, so they were chosen as the final models for both hurricane and ice storm outage durations.

Covariate Selection and Functional Form The best parametric forms for continuous covariates were investigated using generalized additive models (GAMs) [52]. In GAMs, the $\beta_j x_{ij}$ terms in the AFT models are replaced with smooth functions of the covariates, $f_j(x_j)$, in this case, splines. GAMs were fitted on $\ln(t)$ using the gam( ) procedure in R. We inspected the plots $f_j(x_j)$ versus $\ln(t)$ [48] and found the assumption of linearity to be acceptable for most of the continuous covariates. The number of customers (x_{cus}) and maximum gust wind speed (x_w) splines for the hurricane model exhibit some nonlinearity. However, this occurs primarily at the higher ends of the ranges where there are few observations, so the linear assumption still appears reasonable in general [48]. Since there are only 12 and 10 different values for the total number of outages in the storm (x_{out}) for the hurricane and ice storm models, respectively, and the relationship between (x_{out}) and $\ln(t)$ seems to be monotonic, we settled on the linear relationship for the sake of simplicity.

To determine which covariates to include in the model, several different models were compared using the AIC and log-likelihood values. Based on the results summarized in Table 7-5, Model 9 is the final recommended Weibull hurricane outage duration model. An "X" means that the corresponding variable is significant to the 0.0001 level, and an "OX" means that some of the land cover types are significant at the 0.0001 level and some are not. While Models 10, 11, and 12 are statistically better, practically they are not worth the additional complexity of adding one more covariate. Similarly, based on the results summarized in Table 7-6, Model 7 is the final recommended Weibull ice storm outage duration model.

Based on the initial results, for hurricane AFT models, outages associated with the transmission side of a substation circuit breaker were removed and the device indicator variable was recategorized into (1) substation and (2) protective device/service

Table 7-5 Weibull AFT Hurricane Outage Duration Model Comparison

Model	Intercept	x_{dev}	x_{cus}	x_{out}	x_c	x_w	x_d	x_h	x_{start}	x_{lc}	x_r	x_{pop}	Log-Likelihood	AIC
1	X												−888,505	1,777,013
2	X	X											−888,321	1,776,645
3	X	X	X										−887,923	1,775,851
4	X	X	X	X									−871,478	1,742,964
5	X	X	X	X	X								−870,443	1,740,899
6	X	X	X	X	X	X							−869,728	1,739,469
7	X	X	X	X	X	X	X						−869,482	1,738,981
8	X	X	X	X	X	X		X					−866,794	1,733,612
9[a]	X	X	X	X	X	X		X	X				−854,622	1,709,270
10	X	X	X	X	X	X		X	X	OX			−854,573	1,709,181
11	X	X	X	X	X	X		X	X		X		−854,586	1,709,200
12	X	X	X	X	X	X		X	X			X	−854,392	1,708,811

[a]Model 9 is the final recommended Weibull AFT hurricane outage duration mode.

Table 7-6   Weibull AFT Ice Storm Outage Duration Model Comparison

Model	Intercept	x_{dev}	x_{cus}	x_{out}	x_{ice}	X	x_{start}	x_{lc}	Log-Likelihood	AIC
1	X								−239,626	479,257
2	X	X							−239,494	478,999
3	X	X	X						−237,597	475,205
4	X	X	X	X					−242,474	484,962
5	X	X	X	X	X				−242,051	484,118
6	X	X	X	X		X			−238,337	476,702
7[a]	X	X	X	X		X	X		−231,808	463,647
8	X	X	X	X		X	X	OX	−231,801	463,641

[a]Model 7 is the final recommended Weibull AFT ice storm outage duration model.

transformer/customer. For the ice storm models, the original four device types were retained since they are all statistically significant in the final model.

Influential Points and Outliers   For each of the final models, scaled score residuals were used to identify influential observations (i.e., points whose removal from the dataset would cause a large change in the model fit), and deviance residuals were used to identify outliers (i.e., points that do not fit the current model). The scaled score residual for the kth covariate and ith observation approximates the change in the kth coefficient estimate if the ith observation was removed from the dataset and the model was re-estimated without that observation [49]. For example, Fig. 7-4a shows the plot of scaled score residuals versus the covariate *maximum gust wind speed* (x_w). There is one point on Fig. 7-4a with a scaled score residual value greater than 1.5, indicating that it has a great influence on the coefficient estimate for x_w. The observation is related to a substation device with relatively high maximum gust wind speed (63.6 mph) but short duration (<5 hours). Since there is no justification for removal of the data point, it was retained.

The deviance residual of observation i is defined as $r_i^D = \text{sgn}(t_i - \hat{t}_i)$ $(LL(t_i, \hat{t}_0; \hat{\lambda}) - LL(t_i, x_i\hat{\beta}; \hat{\lambda}))^{1/2}$, where $\hat{t}_0$ is the linear predictor for observation i in the model fitted with only this observation, and $LL(\ )$ are the log-likelihood functions of the respective models [51]. The values of $\hat{\beta}$ and $\hat{\lambda}$ were obtained from the overall fit of the corresponding AFT. For example, Fig. 7-4b shows the plot of deviance residuals versus the covariate *number of customers affected by outage* (x_{cus}). Among the observations with $|r_i^D|>4$, there are 16 observations that have an outage duration of less than 1 minute. They were considered to be transient faults and thus were removed from the training dataset. The removal resulted in less than 1% change in all coefficient estimates. As a result of the residual analyses, 23 observations were identified as probably being transient, not permanent, faults and thus were removed from the dataset before the final models were fitted.

Final AFT Models   The final outage duration models are given by model (1), where ε_i are two-parameter extreme value random variables, and parameter estimates

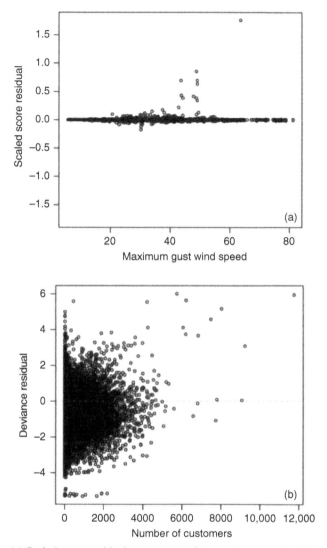

Figure 7-4 (a) Scaled score residuals versus covariate *maximum gust wind speed* (x_w) and (b) deviance residuals versus covariate *number of customers* (x_{cus}) for the final recommended AFT hurricane duration model.

are listed in Tables 7-7 and 7-8 for hurricanes and ice storms, respectively. Of the covariates related to prioritizing the restoration process, number of customers affected by the outage (x_{cus}), type of device affected in the outage (x_{dev}), and outage start time (x_{start}) were found to be important in both hurricane and ice storm models. Among the meteorological factors, only maximum gust wind speed (x_w) was selected. The total number of outages in the storm (x_{out}) was also important in determining the outage duration in both models.

Table 7-7 Final Weibull AFT Hurricane Outage Duration Model

Variable[a]	Mean	Standard Deviation	2.5%	97.5%
Intercept	5.3542	0.0172	5.3205	5.3879
Substation, $x_{dev} = 2$	−5.8315	0.0222	−5.8750	−5.7880
Protective device/transformer/customer ($x_{dev} = 3,4$)	0.0000	0.0000	0.0000	0.0000
Number of customers, x_{cus}	−0.0002	0.000008	−0.0002	−0.0002
Total number of outages in storm, x_{out}	0.00003	0.0000004	0.00002	0.00003
Maximum gust wind speed, x_w	0.0328	0.0004	0.0320	0.0336
Dominion, x_{c1}	0.7632	0.0115	0.7406	0.7857
Duke, x_{c2}	0.3066	0.0195	0.2683	0.3448
Progress, x_{c3}	0.0000	0.0000	0.0000	0.0000
Hurricane Floyd, x_{h1}	0.5170	0.0119	0.4936	0.5403
Hurricane Dennis, x_{h2}	−1.1467	0.0196	−1.1851	−1.1083
Hurricane Bonnie, x_{h3}	0.3704	0.0182	0.3347	0.4060
Hurricane Fran, x_{h4}	0.7452	0.0283	0.6897	0.8006
Hurricane Isabel, x_{h5}	0.0000	0.0000	0.0000	0.0000
Outage start time, x_{start}	−0.0001	0.000001	−0.0001	−0.0001
Log(scale)	−0.0735	0.0025	−0.0784	−0.0686

[a]All variables are significant at the 0.0001 level.

Table 7-8 Final Weibull AFT Ice Storm Outage Duration Model

Variable[a]	Mean	Standard Deviation	2.5%	97.5%
Intercept	7.3355	0.0238	7.2912	7.3798
Transmission, $x_{dev} = 1$	−1.3670	0.2800	−1.8805	−0.8535
Substation, $x_{dev} = 2$	−0.5568	0.0410	−0.6329	−0.4808
Protective device, $x_{dev} = 3$	−0.1579	0.0150	−0.1857	−0.1301
Transform./customer, $x_{dev} = 4$	0.0000	0.0000	0.0000	0.0000
Number of customers, x_{cus}	−0.00045	0.00002	−0.00049	0.00041
Total number of outages in storm, x_{out}	0.00018	0.00000	0.00017	0.00018
12/23/98 ice storm, x_{is1}	−0.4337	0.0305	−0.4905	−0.3768
1/14/99 ice storm, x_{is2}	−0.7330	0.0360	−0.7996	−0.6664
12/13/00 ice storm, x_{is3}	−1.7895	0.0585	−1.8989	−1.6801
12/4/02 ice storm, x_{is4}	−0.9542	0.0389	−1.0265	−0.8819
12/11/02 ice storm, x_{is5}	−1.0077	0.0400	−1.0817	−0.9336
12/24/02 ice storm, x_{is6}	−1.6820	0.0654	−1.8031	−1.5609
2/27/03 ice storm, x_{is7}	0.0000	0.00000	0.0000	0.0000
Outage start time, x_{start}	−0.00020	0.000002	−0.00020	−0.00019
Log(scale)	0.0427	0.0049	0.0330	0.0524

[a]All variables are significant at the 0.0001 level.

It was hoped that the storm indicator variables could be omitted because, when applying the models for an incoming storm, it is difficult to know what values to assign those indicators. Nevertheless, the storm indicator variable was included for both models as they are statistically significant and greatly improve the fit of the models. It is not clear which features of the individual storms are important in predicting outage durations and have not been captured by the other covariates. Possibilities include characteristics of the different land areas impacted by each storm (e.g., terrain, system maintenance, and tree trimming practices); age and size of the power system, and leaves on the trees at the time-each storm occurred (the storms span 8 years and a few months, see Tables 7-1 and 7-2); local wind bursts; storm surge in the hurricane model; or wind speed in the ice storm model. In an attempt to remove the hurricane indicator covariate, it was replaced with two others that are available in a predictive mode—central atmospheric pressure deficit at landfall in mb (x_{cp} = 1013–central pressure at landfall) and Saffir–Simpson category at landfall (x_{ss} = 0 to 5). The resulting restoration curve for the Hurricane Charley model test underestimated the restoration times by approximately the same percentage on average, (32%) as the model with the hurricane indicator covariate overestimated it (35%). The original model is recommended here to be consistent with the ice storm model and because x_{cp} and x_{ss} are only available once the hurricane actually makes landfall, but it would be reasonable to use the alternative model as well. See AFT Model Testing for the recommended way to apply the models in a predictive mode, given the inclusion of the storm indicator covariates.

AFT Model Testing

The final storm outage duration AFT models were applied to the model testing datasets (Hurricane Charley and the January 2004 ice storm). Without a better understanding of what the hurricane indicator covariate is capturing, it was assumed that Hurricane Charley was equally likely to be any hurricane in the dataset, the model was applied for each hurricane, and the resulting outage estimates were averaged. The same was done for the January 2004 ice storm.

Figure 7-5 shows a histogram of raw residuals (observed–predicted) for the Hurricane Charley test. The distribution has a median of −3.01 hours and a small positive skew. Plotting the raw residuals against number of customers affected by outage (x_{cus}) in Fig. 7-6 shows that prediction error decreases as more customers are involved and that the large prediction errors in Fig. 7-6 are associated with outages with very few customers. This may be because the relatively few outages that affect many customers are a high priority for repair, and thus, are the shorter duration outages, which are easier to estimate. The majority of outages affect no more than a few customers, and they are all of equally low priority, so it is difficult to estimate when each will be restored. The implication of Fig. 7-5 is that the models are more valuable than Fig. 7-4 suggests because, if all else is equal, power companies are more concerned with those outages that involve more customers than those that involve fewer. Figures 7-7 and 7-8 show similar results for the January 2004 ice storm. Tests

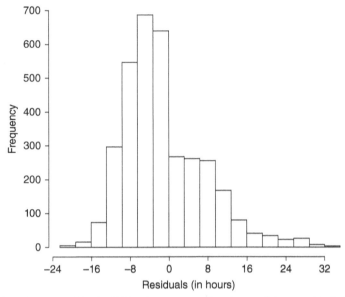

Figure 7-5   Histogram of final AFT hurricane outage duration model raw residuals (observed–predicted) in hours for the testing set (Hurricane Charley).

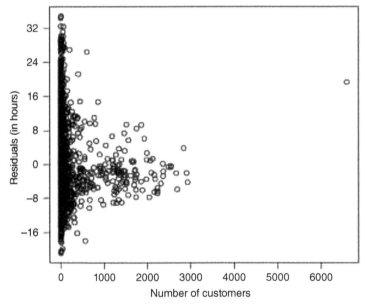

Figure 7-6   AFT hurricane outage duration model raw residuals (observed–predicted) in hours versus *number of customers affected by outage* (x_{cus}) for the testing set (Hurricane Charley).

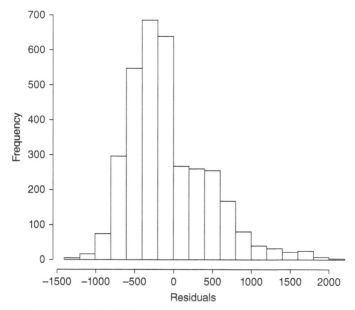

Figure 7-7   Histogram of final AFT ice storm outage duration model raw residuals (observed–predicted) in hours for the testing set (January 2004 ice storm).

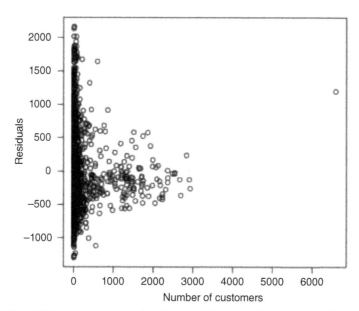

Figure 7-8   AFT ice storm outage duration model raw residuals (observed–predicted) in hours versus *number of customers affected by outage* (x_{cus}) for the testing set (January 2004 ice storm).

of the restoration time estimates—the quantities of real practical interest—for Hurricane Charley and the January 2004 ice storm are discussed in Restoration Time Estimation.

RESTORATION TIME ESTIMATION

Periodically during a storm, electric power companies tell their customers and the public when they expect the power to be restored. While each customer's power may be restored at a different time, companies typically make broad public announcements that provide a single approximate restoration time for each county or region. In the new method introduced in this study, these restoration estimates are quantified as the time by which a company-specified $Z\%$ of the customers in the county (say 90%) will have their power restored. In this research, restoration time estimates are developed by the county because companies want to be able to make estimates at that scale and because at that level of aggregation, errors for each individual outage cancel out providing reasonable accuracy (Results). The outage duration AFT models in Final AFT Models were used to estimate the duration of a particular outage based on characteristics of a storm and the outage itself. Those duration estimates were then combined with the varying start times and aggregated to counties.

Method

For simplicity, the restoration time estimation method is described for hurricanes and counties here, but the same method could be used for ice storms and other area units. The method is summarized in the flowchart shown in Fig. 7-9. First, using an outage count model like that in [53] or [54], estimate the number of outages in each county. (The negative binomial models from Liu [48] were used in this study.) Second, for each outage, the value of each covariate in the AFT outage duration model in model (1) must be estimated. The total number of outages in the storm (x_{out}) can be taken from the outage count model used in Step 1. The maximum gust wind speed at the outage location (x_w) can be estimated from the Huang et al. model [39] or other available weather forecasts. Based on the location of the outage, the company indicator covariate (x_c) is obtained. Since the incoming hurricane will not be in the historical data, it is difficult to know which value of the hurricane indicator covariate (x_h) to use. Without a better understanding of what that covariate captures, it can be assumed that the incoming storm is equally "like" every hurricane in the dataset, the model can be applied assuming each hurricane in turn, and the resulting outage duration estimates can be averaged.

For the remaining covariates—device affected (x_{dev}), number of customers affected by outage (x_{cus}), and outage start time (x_{start})—two approaches are possible. For each outage, one could use the actual x_{cus}, x_{dev}, and x_{start} values recorded in the OMS. Since new outages start throughout the storm, however, those values will not all be available until the storm is well underway and field damage reports are also becoming available, limiting the benefit of this method. Alternatively, the

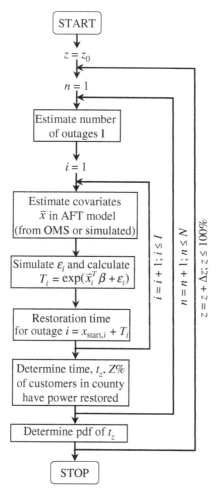

Figure 7-9 Flowchart of restoration time estimation method.

values may be sampled from empirical or fitted parametric distributions developed using historical hurricane data. Since x_{cus} and x_{dev} are correlated ($\rho = 0.31$), values of those two covariates were sampled together from the empirical joint distribution.

Since the range of outage start times varies considerably across hurricanes, with larger start times possible in hurricanes with more outages, in this analysis two separate distributions of outage start times were developed, using outages from hurricanes with $x_{out} \leq 4000$ and $x_{out} > 4000$, respectively. Liu [48] presents the cumulative distribution functions (CDFs) of outage start times. If a single distribution had been used, for a smaller hurricane, it would have been possible to sample an outage start time exceeding the hurricane duration, a situation that is impossible in reality. (For ice storms, three outage start time distributions were used, for $x_{out} \leq 4000$, $4000 < x_{out} \leq 8000$, and $x_{out} > 8000$.) An attempt was made to develop a different outage start time distribution for each county to capture geographic variability in x_{start}, but the results were not significantly better. In this study, both the *observed* (OMS) and *simulated* covariate approaches were applied. The former provides an indication of the best results that can be obtained with the current models, but the latter is how the method is more likely to be used in practice because the OMS will not provide data quickly enough to be useful for prediction.

The third step in the method is to simulate errors from the two-parameter extreme value distribution, then use the AFT model from Final AFT Models to calculate the outage duration. Fourth, calculate the restoration time for that outage (relative to the start time of the first outage) as the sum of the start time and duration. Fifth, repeat Steps 2 to 4 for every outage. (The wind speeds may or may not be resampled in each iteration. In this study, they were not.) For each county and the entire service area, using the observed/simulated outage restoration times and the number of customers associated with each outage, determine the time t_z at which $Z\%$ of the customers in the county (and service area) have their power restored. Sixth, repeat this process N times, where N is a large number, to obtain a probability density function (pdf) of the time at which $Z\%$ of the customers in the county will have their power restored. ($N = 20,000$ was used in this study since statistics stabilized beyond that). From that pdf, choose the value to be used in practice. One could use the median value or the upper limit of the $Y\%$ confidence interval around the time that $Z\%$ of customers in the county (and service area) will have their power restored. Repeating Steps 5 and 6 for different values of Z, one can develop a full restoration curve with uncertainty bounds.

Because this method estimates the duration of each individual outage, theoretically companies could use it to give an individualized restoration time estimate for each customer when he calls in. Due to the large variability in restoration times, however, restoration times cannot be predicted accurately enough for individual outages. When aggregated to a county, the estimates are more accurate.

Results

The restoration time estimation method was applied for Hurricane Charley and the January 2004 ice storm, both in the Progress Energy service area, using both the observed and the simulated covariate values. The resulting restoration model time estimates (using the median values from the 20,000 iterations of the simulation) were compared to the observed restoration times. Figure 7-10 presents the actual and estimated (based on both observed and simulated covariate values) restoration curves. The shape of the estimated restoration curves for Hurricane Charley matches the actual shape well, but the method overestimated the restoration times along the entire curve, especially when the simulated covariate values were used. When the observed covariate values were used, the restoration curve was shifted an average of 7.2 hours (35%) to the right. When the simulated covariate values were used, it was shifted an average of 19.0 hours (128%) to the right. The results for the January 2004 ice storm were much better. With the observed covariate values, the restoration curve was only an average of 2.8 hours (5%) different than the actual; and with

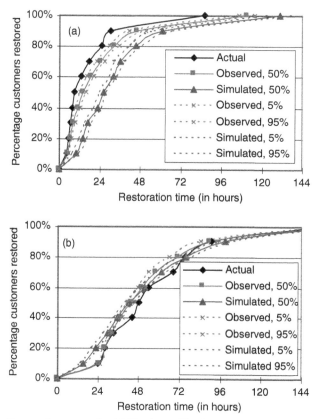

Figure 7-10   Restoration curves for (a) Hurricane Charley and (b) the January 2004 ice storm. Actual, predicted with observed covariates, and predicted with simulated covariates, median curves with 90% confidence intervals.

the simulated covariate values, it was only 0.1 hours (10%) different. Note that the confidence intervals appear small because they represent uncertainty due only to the error term in the AFT models (and for the simulate covariate version of results, due to uncertainty in the x_{dev}, x_{cus}, and x_{start} values).

The difference between the restoration curves with the observed values of x_{cus}, x_{dev}, and x_{start} and with the simulated values of those covariates is due only to uncertainty in those covariate values. Since the outage start time (x_{start}) is both an important covariate in the AFT outage duration models (Tables 7-5 and 7-6) and one of the two terms (with outage duration) that determines the restoration time, uncertainty in x_{start} is an important source of error in the final hurricane results. The results for Hurricane Charley may be worse than one would expect with other events because Hurricane Charley was somewhat unusual in having relatively short outage start times compared to other hurricanes with similar total numbers of outages. Therefore, the method estimated outage start times that were too long. The error that remains when observed values of x_{cus}, x_{dev}, and x_{start} are used is probably due to a combination of inclusion of the hurricane indicator covariate, which requires assuming that Hurricane Charley is an average of the historical hurricane experiences in the database, and errors in the models used to estimate x_w, x_{out}, and outage duration. An analysis of restoration time errors by the county showed no relationship between those errors and errors in outage counts, suggesting that the outage count model is not a key source of error. Comparing outage durations for Hurricane Charley with those for the other historical hurricanes suggest that it had relatively short outage durations, and therefore that the hurricane indicator assumption may be a main source of error. As more hurricane experiences are included in the database, the results should improve.

Nonetheless, the results should be useful in their current form. Imagine that before these two storms occurred, Progress Energy had planned to use the new method with the guideline to announce that power would be restored to most of the service area at the time the method predicted 90% of customers would have power restored (61.5 and 99.2 hours after the first outages began in Hurricane Charley and the January 2004 ice storm, respectively). In that case, 96% and 91% would have actually been restored by that time for the hurricane and ice storm, respectively.

The results can also be examined at the county level. Figures 7-11 and 7-12 are the scatterplots of the actual versus estimated times at which 90% of the customers restored for each county in the Progress Energy service area for Hurricane Charley and the January 2004 ice storm, using the observed and simulated covariate values, respectively. For results based on the observed covariate values (Fig. 7-11), the points lie close to the 45° line for both the hurricane and ice storm, although the model overestimates restoration times for most counties. This suggests that the shift in the restoration curves in Fig. 7-10 is due to errors not just in a few counties, but evenly distributed across counties. As expected, the results based on observed covariate values were better than those based on simulated covariate values (Fig. 7-12).

CONCLUSIONS AND FUTURE WORK

To the extent that a power company can estimate storm-caused power outage durations ahead of time, it can help its customers and the public plan for the outages and minimize the disruption they cause. Using a large dataset of historical outages from three major East Coast electric power companies, AFT models were developed to estimate hurricane- and ice-storm-related outage durations. A new method was also developed to use the outage duration estimates from the AFT models to estimate restoration times for each county in the company's service area. The method can be applied as a storm approaches, before field assessments of damage are available. When tested against observations in Hurricane Charley and the January 2004 ice storm, the restoration estimation method provided promising results, suggesting that it could offer a valuable tool to help power companies manage storms in the future.

There are several opportunities for future research, both in improving the method introduced herein and in building on the method and its application. A key limitation of the current method is that, because the necessary data could not be found, the outage duration models do not include potentially important tree-related or tree-trimming

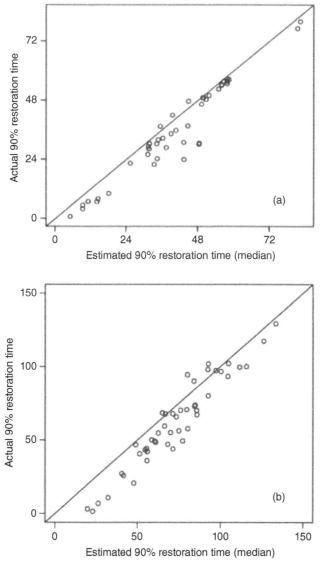

Figure 7-11   Actual time (hours) to restore 90% of customers versus median of time to restore 90% of customers from 20,000 simulated samples for (a) Hurricane Charley and (b) the January 2004 ice storm based on recommended AFT models using *observed* covariate values for x_{cus}, x_{dev}, and x_{start} (each point represents one county in the Progress Energy service area).

covariates or covariates describing restoration resources used (e.g., the number of line crews). Including such covariates would improve the models' predictive power and allow evaluation of the effect of changing tree-trimming practices or the amount of resources. Other avenues for possible refinement of the method are to try to

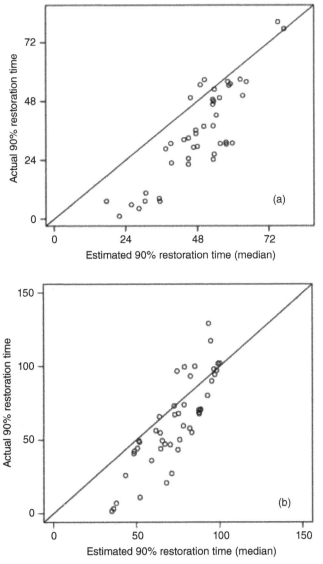

Figure 7-12   Actual time (hours) to restore 90% of customers versus median of time to restore 90% of customers from 20,000 simulated samples for (a) Hurricane Charley and (b) the January 2004 ice storm based on recommended AFT models using *simulated* covariate values for x_{cus}, x_{dev} and x_{start} (each point represents one county in the Progress Energy service area)

remove the storm indicator variables from the outage duration models (Final AFT Models) or further investigate the possibility of using GAMs to improve the fit of the outage duration models (Covariate Selection and Functional Form). In general, as more storms occur and additional data become available, the models can be refitted, improving the predictive power. Finally, since this new restoration time estimation approach depends on having a method to estimate the total number and geographic distribution of outages (e.g., [53]), any improvements in outage count models will improve the predictive power of this method as well.

In terms of extending this work (rather than refining the specific models), one could combine this method with long-term probabilistic hurricane and/or ice storm hazard models to assess the long-term restoration costs to the utilities, customers, and public. For hurricanes, for example, one could use a probabilistic hazard model to simulate the occurrence of many years of hurricanes, and for each event, use this approach to estimate the likely electric power restoration times. Probabilistically combining the results would provide, for example, the annual customer-hours without power due to hurricanes. That information might be useful in helping determine what long-term hurricane mitigation strategies would be cost-effective for a utility to undertake. As long as the required data were available, the approach could be applied to other regions, and possibly other hazards or lifelines.

ACKNOWLEDGMENTS

This material is based upon the work supported by the National Science Foundation under Grant No. CMS-0408525. The authors thank D. Rosowsky and K.H. Lee for the hurricane wind speed data; A. DeGaetano and B. Belcher for the ice storm data; and the power company personnel for data and consultation, especially L. Nunnery, D. Pickles, R. Meffert, M. Royster, E. Baine, C. Haller, and T. Kesler.

REFERENCES

1. Dominion Energy, "Hurricane Isabel: A Final Review," Dominion Energy Storm Center, Available at http://www.dom.com/news/isabel_review.jsp, November 2003.

2. B. Johnson, *Utility Storm Restoration Response*, Edison Electric Institute (EEI), Washington, DC, 2004.

3. Michigan Public Service Commission (MPSC), Report on August 14th Blackout, Available at www.michigan.gov/documents/mpsc_blackout_77423_7.pdf, 2003.

4. North Carolina Public Utility Commission (NCPUC) (2003), Response of Duke Power, A Division of Duke Energy Corporations to Data Request No. 1. Available at http://www.ncuc.commerce.state.nc.us/merger.pdf

5. E.O. Ferrell, "Dealing with Disasters: Utility Restoration of Service," Edison Electric Institute, Available at www.eei.org/meetings/nonav_2005-04-07-lm, July 2006.

6. J.D. Power and Associates, "Electric utility business customer satisfaction study," 2006. Available at http://businesscenter.jdpower.com/JDPAContent/CorpComm/News/content/Releases/pdf/2007043.pdf

7. Z. Çağnan, "Post-earthquake Restoration Modeling for Critical Lifeline Systems," *Ph.D. Dissertation, School of Civil and Environmental Engineering*, Cornell University, Ithaca, NY, 2005.

8. D. Reed, "Electric utility distribution analysis for extreme winds," *Journal of Wind Engineering and Industrial Aerodynamics*, vol. 96, no. 1, pp. 123–140, January 2008.

9. D. Reed, N. Nojima, and J. Park, "Performance Assessment of Lifelines," Proceedings of the 16th ASCE Engineering Mechanics Conference, Seattle, WA, 2003.

10. Applied Technology Council (ATC), "A Model Methodology for Assessment of Seismic Vulnerability and Impact Distribution of Water Supply Systems," Report No. ATC–25–1, Redwood City, CA, 1992.

11. M. Shinozuka, A. Rose, and R. Eguchi, Engineering and Socioeconomic Impacts of Earthquakes: An Analysis of Electricity Lifeline Disruptions in the New Madrid Area, Monograph No. 2, Multidisciplinary Center for Earthquake Engineering Research, Buffalo, NY, 1998.

12. N. Nojima, Y. Ishikawa, T. Okumura, and M. Sugito, "Empirical Estimation of Lifeline Outage Time in Seismic Disaster," Proceedings of US–Japan Joint Workshop and 3rd Grantee Meeting., US–Japan Cooperative Research on Urban Earthquake Disaster Mitigation, Seattle, WA, pp. 516–517, 2001.

13. S.E. Chang, M. Shinozuka, and W. Svekla, "Modeling Post-disaster Urban Lifeline Restoration," Proceedings of 5th US Conference on Lifeline Earthquake Engineering, Technical Council on Lifeline Earthquake Engineering, Monograph No. 16, ASCE, Seattle, WA, pp. 602–611, 1999.

14. M. Isumi, and T. Shibuya, "Simulation of post-earthquake restoration for lifeline systems," *International Journal of Mass Emerging Disasters*, vol. 3, no. 1, pp. 87–105, March 1985.

15. National Institute of Building Sciences (NIBS) Multi-hazard Loss Estimation Methodology, HAZUS®MH MR3, Technical Manual, FEMA Distribution Center, Jessup, MD, 1997. Available at https://www.nibs.org/?page=hazus

16. S. Goldberg, W. Horton, and V. Rose, "Analysis of feeder reliability using component failure rates," *IEEE Transactions on Power Delivery*, vol. 2, no. 4, pp. 1292–1296, 1987.

17. W.F. Horton, S. Goldberg, R.A. Hartwell, "A cost/benefit analysis in feeder reliability studies," *IEEE Transactions on Power Delivery*, vol. 4, no. 1, pp. 446–452, 1989.

18. C. Volkmann, S. Goldberg, and W. Horton, "A Probabilistic Approach to Distribution System Reliability Assessment," Proceedings of the 3rd International Conference on Probabilistic Methods Applied to Electric Power Systems, London, pp. 169–173, 1991.

19. L. Cooper, N. Schulz, and T. Nielsen, "A Computer Program for the Estimation of Restoration Times During Storms," Proceedings of 60th American Power Conference: Reliability and Economy—Technology Focus for Competition and Globalization, Chicago, IL, pt. 2, 2, pp. 777–780, 1998.

20. L Cooper, "A Computer Program for the Estimation of Restoration Times During Storms," M.S. Thesis, Department of Electrical Engineering, Michigan Technological University, Houghton, MI, 1998.

21. R. Billinton, and J. Acharya, "Weather-based distribution system reliability evaluation," *IEEE Proceedings on Generation, Transmission, and Distribution*, vol. 153, no. 5, pp. 499–506, September 2006.

22. R. Isoyama, T. Iwata, and T. Watanabe, "*Optimization of Post-earthquake Restoration of City Gas Systems*," Proceedings of Trilateral Seminar Workshop on Lifeline Earthquake Engineering, Taipei, Taiwan, pp. 43–57, 1985.

23. T. Iwata, "*Restoration Planning System for Earthquake-Damaged Gas Pipeline Network*," Proceedings of Trilateral Seminar Workshop on Lifeline Earthquake Engineering, Taipei, Taiwan, 1985.

24. F. Kozin, and H. Zhou, "System study of urban response and reconstruction due to earthquake," *Journal of Engineering Mechanics*, vol. 116, pp. 1959–1972, September 1991.

25. R. Zhang, "Lifeline Interaction and Post-earthquake Urban System Reconstruction," Proceedings of the 10th World Conference on Earthquake Engineering, Rotterdam, Netherlands, pp. 5475–5480, 1992.

26. R. Brown, S. Gupta, R. Christie, S. Venkata, and R. Fletcher, "Distribution system reliability assessment: Momentary interruptions and storms," *IEEE Transactions on Power Delivery*, vol. 12, no. 4, pp. 1569–1575, October 1997.

27. N. Balijepalli, S. Venkata, C. Richter, Jr., R. Christie, and V. Longo, "Distribution system reliability assessment due to lightning storms," *IEEE Transactions on Power Delivery*, vol. 20, no. 3, pp. 2153–2159, July 2005.

28. E. Broström, "Ice Storm Modelling in Transmission System Reliability Calculations," Licentiate Thesis, School of Electrical Engineering, Royal Institute of Technology, Stockholm, Sweden, 2007.

29. D. Newsom, *"Evaluating Lifeline Response to Earthquakes: A Simulation Methodology,"* Ph.D. Dissertation, Purdue University, West Lafayette, IN, 1977.

30. Z. Çağnan, and R. Davidson, "Discrete event simulation of the post-earthquake restoration process for electric power systems," *International Journal of Risk Assessment and Management*, vol. 7, no. 8, pp. 1138–1156, 2007.

31. Z. Çağnan, R. Davidson, and S. Guikema, "Post-earthquake restoration planning for Los Angeles electric power," *Earthquake Spectra*, vol. 22, no. 3, pp. 1–20, August 2006.

32. T. Tabucchi, and R. Davidson, *Post-earthquake Restoration of the Los Angeles Water Supply System, MCEER-08-0008*, Multidisciplinary Center for Earthquake Engineering Research, Buffalo, NY, 2008.

33. N. Xu, S. Guikema, R. Davidson, L. Nozick, Z. Çağnan, and K. Vaziri, "Optimizing scheduling of post-earthquake electric power restoration tasks," *Earthquake Engineering and Structural Dynamics*, vol. 36, no. 2, pp. 265–284, February 2007.

34. S. Wang, B. Sarker, L. Mann, Jr., and E. Triantaphyllou, "Resource planning and a depot location model for electric power restoration," *European Journal of Operational Research*, vol. 155, no. 1, pp. 22–43, 2004.

35. M. Yao, and K.J. Min, "Repair-unit location models for power failures," *IEEE Transactions on Engineering Management*, vol. 45, no. 1, pp. 57–65, 1998.

36. J. Wu, T. Lee, C. Tsai, T. Chang, and S. Tsai, "A Fuzzy Rule-Based System for Crew Management of Distribution Systems in Large-Scale Multiple Outages," Proceedings of the 2004 International Conference on Power System Technology—POWERCON, Singapore, pp. 1084–1089, 2004.

37. T. Sato, and K. Ichii, "Optimization of Post-earthquake Restoration of Lifeline Networks Using Genetic Algorithms," Proceedings of the 6th US–Japan Workshop on Earthquake Disaster Prevention for Lifeline Systems, Osaka, Japan, 1995.

38. H. Sugimoto, and T. Tamura, "Support System for Restoration Process of Disaster-Stricken Lifeline Networks by GA," Confronting Urban Earthquakes: Report of Fundamental Research on the Mitigation of Urban Disasters Caused by Near-Field Earthquakes, Kyoto University, Kyoto, Japan, pp. 610–613, 2000.

39. Z. Huang, D. Rosowsky, and P. Sparks, "Hurricane simulation techniques for the evaluation of wind speeds and expected insurance losses," *Journal of Wind Engineering and Industrial Aerodynamics*, vol. 89, no. 7–8, pp. 605–617, June 2001.

40. S. Hagemann, M. Botzet, L. Dümenil, and B. Machenhauer, *"Derivation of Global GCM Boundary Conditions from 1 km Land Use Satellite Data,"* Max-Planck-Institute for Meteorology, Hamburg, Germany, Report 289, 1999.

41. National Climatic Data Center (NCDC), Available at www.ncdc.noaa.gov, February 2006.

42. A. DeGaetano, B. Belcher, and P. Spier, "Short-term ice accretion forecasts for electric utilities using the Weather Research and Forecasting Model and modified precipitation type algorithm," *Weather and Forecasting*, vol. 23, pp. 835–853, 2008.

43. J. Michalakes, S. Chen, J. Dudhia, L. Hart, J. Klemp, J. Middlecoff, and W. Skamarock, "Development of a Next Generation Regional Weather Research and Forecast Model," Proceedings of Developments in Teracomputing, 9th ECMWF Workshop on the Use of High Performance Computing in Meteorology, Singapore, pp. 269–276, 2001.

44. W. Skamarock, J. Klemp, and J. Dudhia, "2001: Prototypes for the WRF (Weather Research and Forecast) Model," Proceedings of the 9th Conference on Mesoscale Processes, Fort Lauderdale, FL, pp. J11–J15, 2001.

45. J. Ramer, "An Empirical Technique for Diagnosing Precipitation Type from Model Output," Proceedings of the 5th International Conference on Aviation Weather Systems, Vienna, VA, pp. 227–230, 1993.

46. K. Jones, "A simple model for freezing rain ice loads," *Atmospheric Research*, vol. 46, no. 1–2, pp. 87–97, 1998.

47. Multi-Resolution Land Characteristics (MRLC) Consortium, National Land Cover Data, Available at http://www.epa.gov/mrlc/nlcd.html, July 2001.

48. H. Liu "Statistical Modeling of Electric Power Outage Counts and Restoration Times During Hurricanes and Ice Storms," *Ph.D. Dissertation, School of Civil and Environmental Engineering*, Cornell University, Ithaca, NY, 2006.

49. D. Hosmer, Jr., and S. Lemeshow, *Applied Survival Analysis: Regression Modeling of Time to Event Data*, John Wiley & Sons, New York, NY, 1999.

50. R Development Core Team, R: A Language and Environment for Statistical Computing, R Foundation for Statistical Computing, Vienna, Austria, Available at http://www.R-project.org, 2008.

51. J. Klein, and M. Moeschberger, *Survival Analysis: Techniques for Censored and Truncated Data*, Springer, New York, NY, 1997.

52. T. Hastie, and R. Tibshirani, *Generalized Additive Models*, Chapman and Hall, London, UK, 1990.

53. H. Liu, R. Davidson, D. Rosowsky, and J. Stedinger, "Negative binomial regression of electric power outages in hurricanes," *Journal of Infrastructure Systems*, vol. 11, no. 4, pp. 258–267, December 2005.

54. H. Liu, R. Davidson, and T. Apanasovich, "Spatial Generalized Linear Mixed Models of electric power outages due to hurricanes and ice storms," *Reliability Engineering. and System Safety*, vol. 93, no. 6, pp. 897–912, 2008.

Chapter 8

A Nonparametric Approach for River Flow Forecasting Based on Autonomous Neural Network Models

Vitor Hugo Ferreira and Alexandre P. Alves da Silva

INTRODUCTION

Artificial neural networks (ANNs) have been applied to many forecasting problems, motivating the organization of several competitions among research groups around the world. Besides the promising results obtained for large-scale forecasting problems involving lots of time-series, for example, short-term load forecasting [1], significant human intervention is still needed, mainly to define the input representation and complexity control of the models.

The ANN input representation and complexity control should not be treated separately, as it is a common practice when ANNs are applied to forecasting. The extent of nonlinearity required from an ANN is strongly dependent on the selected input variables. One of the advantages of neural network models is the universal approximation capability, that is, unlimited precision for continuous mapping. However, this theoretical advantage can backfire if data overfitting is not avoided. The main objective of model complexity control is to match data regularity with model structure, maximizing the generalization capacity.

A popular procedure for ANN complexity control is based on cross-validation with training early stopping, that is, the iterative updating of the connection weights until the error for the validation subset stops decreasing. This procedure is very heuristic,

Advances in Electric Power and Energy Systems: Load and Price Forecasting, First Edition.
Edited by Mohamed E. El-Hawary.

285

because it is not easy to detect the right iteration for interrupting the training process. Besides, although cross-validation has been successfully applied to neural classifiers design, serial dependence information can be lost when it is used in time-series forecasting. Shortcomings of cross-validation and early stopping are fully analyzed in Reference [2].

Input space representation is probably the most important subtask in forecasting. It has been shown that input variable selection based on linear auto- and cross-correlation analyses is not appropriate for nonlinear models such as ANNs. Feature extraction via multiresolution analysis, based on wavelets, has been proposed to overcome this problem [3]. However, a more ANN oriented input selection scheme is still needed to capture the important information about the linear and nonlinear interdependencies in the associated multivariate data. Reference [4] presents a feature extraction method based on curvilinear component analysis. As a filter method, results are not customized for the intended forecasting model.

This work presents a forecasting method based on appropriate techniques for controlling ANN complexity with simultaneous selection of explanatory input variables via a combination of filter and wrapper techniques. In order to automatically minimize the out-of-sample prediction error, the three levels of Bayesian inference applied to multilayered perceptrons (MLPs) have been exploited [5]. This training method includes complexity control terms in their objective function, which allow autonomous modeling and adaptation.

The methodology proposed in References [5] and [6] needs a specification of an initial set of possible inputs. To overcome this assumption, Takens' theorem [7] is applied in this work to select the lags to be used as the initial set of inputs, with the *time-delay* and *embedding dimension* estimated by methods such as minimal mutual information [8] and false neighbors [9], respectively. As usual, in nonlinear chaotic time-series analysis [10], local models are developed via the application of an automatic clustering algorithm based on the rival penalized expectation-maximization (RPEM) algorithm [11]. The technique presented in this chapter has been successfully applied to load forecasting [12]. In this chapter, the challenging problem of rainfall forecasting is employed for showing the robustness of the proposed technique in dealing with different time-series dynamics.

CHAOS INPUT SPACE RECONSTRUCTION

Chaos theory has been developed for the analysis of nonlinear dynamic systems. Given a state-space $\underline{X} \in \mathbb{R}^D$, a dynamic system $F(\underline{X}) : \mathbb{R}^D \to \mathbb{R}^D$ can be defined by:

$$\underline{X}(t + 1) = F[\underline{X}(t)] \tag{8-1}$$

Given the initial condition $\underline{X}(0)$, the state of the deterministic system $F(\underline{X})$ can be defined for any t. The set of initial conditions that asymptotically drives the system to a specific region of the state-space is named attraction base. This region is called attractor [10].

In Chaos theory, a time-series is considered as a compact representation of a dynamic system. Mathematically, a time-series $x(t) \in \mathbb{R}$, $t = 1, 2, \ldots, N$, is related to the corresponding dynamic system by the relation:

$$x(t) = s[\underline{X}(t)] + \eta(t) \tag{8-2}$$

where $s(\underline{X}) : \mathbb{R}^D \rightarrow \mathbb{R}$ is the measurement function and $\eta(t)$ the measurement noise.

Since $s(\underline{X})$ is unknown, the complete reconstruction of $\underline{X}(t)$ by means of $x(t)$ is impossible. To overcome this problem, Takens' theorem defines a new representation space $\underline{x}(t) \in \mathbb{R}^d$, in which the attractor is equivalent to the attractor of the original space $\underline{X}(t) \in \mathbb{R}^D$. This new space, named reconstructed space, can be obtained directly from the time-series $x(t)$ by the following equation [7]:

$$\underline{x}(t) = [\, x(t) \quad x(t - \tau) \quad \ldots \quad x(t - (d - 1)\tau)\,]^t \tag{8-3}$$

where d is the *embedding dimension* and τ is called the *time-delay*. Although the theorem implies that an arbitrary delay is sufficient to reconstruct the attractor, in practical problems with limited data, the specification of time-delay τ is critical. To unfold the attractor, the delay τ is usually estimated by means of the minimal mutual information criterion [8]. This criterion estimates τ based on the first minimum, along the τ axis, provided by the mutual information between $x(t)$ and $x(t - \tau)$. The minimal mutual information criterion aims to select near-independent input variables that allow phase-space reconstruction.

Since the rainfall databases considered in this work are related to seasonal time-series, the delay τ is also defined as half of the smallest seasonality period. The rationality behind this estimation is the following. As a strong series dependence is associated with a delay equal to the seasonality period, such an estimation for τ assures the synthesis of the seasonal pattern.

Another possibility for choosing the *time-delay* τ, according to Takens' theorem, is to make it equal to one as long as the embedding dimension d is set sufficiently large. The present work also investigates this possibility.

The reconstructed space might present special geometric aspects called strange attractors [7]. To develop a forecasting model, the time-series dynamics can be modeled by means of a global model or by a local one. The global model seeks the representation of the attractor as a whole, requiring a lot of data for this task. Local models attempt to represent specific aspects of the attractor, developing independent models for different regions of the reconstructed space. In practical applications with limited data, local models are recommended [7]. In the context of stochastic time-series, automatic clustering not only allows the application of local predictors, but also is helpful for identifying outliers. The reconstructed space is divided in this work via an automatic clustering algorithm.

AUTOMATIC CLUSTERING ALGORITHM

The automatic clustering algorithm adopted in this chapter is based on density mixture. Suppose that the N observations $X = (\underline{x}_1, \underline{x}_2, \ldots, \underline{x}_N)$ are independently and identically distributed from the following mixture model [11]:

$$p(\underline{x}, \underline{\Theta}^*) = \sum_{i=1}^{k^*} \sigma_i^* q(\underline{x}|\underline{\theta}_i^*) \qquad (8\text{-}4)$$

where k^*, $\underline{\sigma}^* = [\sigma_1^*, \ldots, \sigma_k^*]^t$, and $\underline{\theta}^* = \left[\underline{\theta}_1^{*t}, \ldots, \underline{\theta}_k^{*t}\right]^t$ are the model true parameters with

$$\sum_{i=1}^{k^*} \sigma_i^* = 1, \sigma_i^* > 0, i = 1, 2, \ldots, k^* \qquad (8\text{-}5)$$

Each mixture component in Eq. (8-4) is a probability density function (PDF) of $\underline{x}$ defined by a set of parameters $\underline{\theta}_i^*$ that defines $q(\underline{x}|\underline{\theta}_i^*)$. For Gaussian distributions, $\underline{\theta}_i^* = \{\underline{\mu}_i^*, \underline{\Sigma}_i^*\}$ with $\underline{\mu}_i^*$ and $\underline{\Sigma}_i^*$ representing the mean vector and covariance matrix, respectively.

The component $q(\underline{x}|\underline{\theta}_i^*)$ can be interpreted as the PDF of those points that form the corresponding cluster C_i, with the associated σ_i^* representing the proportion of points that lies in the C_i cluster. If the parameters $\underline{\theta}_i^*$ are known, then the posterior probability of $\underline{x}_j$ belonging to the C_r cluster can be calculated by:

$$h\left(C_r|\underline{x}_j, \underline{\theta}_r^*\right) = \frac{\sigma_r^* q\left(\underline{x}_j|\underline{\theta}_r^*\right)}{\sum_{i=1}^{k^*} \sigma_i^* q\left(\underline{x}|\underline{\theta}_i^*\right)} \qquad (8\text{-}6)$$

A test point $\underline{x}_j$ is assigned to the most probable cluster, that is, the cluster with largest $h(C_r|\underline{x}_j, \underline{\theta}_r^*)$.

In fact, the mixture parameters k^*, $\underline{\sigma}^*$, and $\underline{\theta}^*$ are unknown, and must be estimated from the data. In this chapter, the RPEM algorithm [11] is applied to estimate the mixture parameters, including an automatic model selection procedure for estimating the number of centers k^*. The RPEM algorithm is based on the expectation-maximization (EM) principle, making the centers compete along the iterative process. Therefore, the parameters of the winning center are updated and the ones from the rivals are penalized with a strength proportional to the corresponding posterior density probability. This rival penalization mechanism enables the RPEM to automatically select the number of densities k^*. More details about the RPEM algorithm can be found in Reference [11].

BAYESIAN INFERENCE APPLIED TO MLPs

Let $\underline{x} \in \mathbb{R}^n$ be the vector containing the input signals and $\underline{w} \in \mathbb{R}^M$ the vector including all the weights and biases of a single hidden layer, single output MLP, with m hidden neurons. Then, $M = mn + 2m + 1$, where n denotes the input space dimensionality and M is the number of connections in the MLP. Representing the biases of the hidden neurons' sigmoidal activation functions by b_k, with b representing the bias of the linear neuron of the output layer, w_k the weights connecting the hidden with the output layer, and w_{ik} the ones connecting the input with the hidden layer, the final output of the MLP is given by:

$$y = f(x, \underline{w}) = \sum_{k=1}^{m} \left[w_k \varphi \left(\sum_{i=1}^{n} (w_{ik} x_i) + b_k \right) \right] + b \qquad (8\text{-}7)$$

Given a dataset U with N input/output pairs, $U = \{X, D\}$ for $X = (\underline{x}_1, \underline{x}_2, \ldots, \underline{x}_N)$ and $D = (d_1, d_2, \ldots, d_N)$, where $d_j \in \mathbb{R}$ represents the desired outputs, the MLP training can be interpreted as the estimation of the conditional PDF of $\underline{w}$ given a dataset $p(\underline{w}|D, X)$ using Bayes' rule:

$$p(\underline{w}|D, X) = \frac{p(D|\underline{w}, X)p(\underline{w}|X)}{p(D|X)} \qquad (8\text{-}8)$$

Since X is conditioning all probabilities in Eq. (8-8), it will be omitted from this point on. Therefore, $p(D|\underline{w})$ is the likelihood of D given, $\underline{w}, p(\underline{w})$ is $\underline{w}$'s *a priori* PDF, and $p(D) = \int p(D|\underline{w}) \, p(\underline{w}) d\underline{w}$ is a normalizing factor enforcing $\int p(\underline{w}|D) \, d\underline{w} = 1$.

It is initially assumed that $\underline{w}$ presents a Gaussian distribution with zero mean and diagonal covariance matrix equal to $\alpha^{-1}\underline{\underline{I}}$, where $\underline{\underline{I}}$ is the $M \times M$ identity matrix, that is:

$$p(\underline{w}) = \frac{1}{Z_{\underline{w}}(\alpha)} e^{-\left(\frac{\alpha}{2}\|\underline{w}\|^2\right)}, \text{ where } Z_{\underline{w}}(\alpha) = \left(\frac{2\pi}{\alpha}\right)^{\frac{M}{2}} \qquad (8\text{-}9)$$

The desired outputs can be represented by $d_j = f(\underline{x}_j, \underline{w}) + \zeta_j$, where ζ is Gaussian white noise with zero mean and variance equal to β^{-1}. The regularization factors α and β (learning parameters, also called hyperparameters), contrary to the other regularization techniques, are estimated together with the parameters $\underline{w}$. Considering the previous hypotheses and assuming that the dataset patterns are independent,

$$p(D|\underline{w}) = \frac{e^{\left\{-\frac{\beta}{2}\sum_{j=1}^{N}[d_j - f(\underline{x}_j, \underline{w})]^2\right\}}}{Z_Y(\beta)}, \text{ where } Z_Y(\beta) = \left(\frac{2\pi}{\beta}\right)^{\frac{N}{2}} \qquad (8\text{-}10)$$

Consequently, based on Eq. (8-8),

$$p(\underline{w}|D) = \frac{e^{[-S(\underline{w})]}}{\int e^{-S(\underline{w})}d\underline{w}} \tag{8-11}$$

where

$$S(\underline{w}) = \frac{\beta}{2}\sum_{j=1}^{N}[d_j - f(\underline{x}_j, \underline{w})]^2 + \frac{\alpha}{2}\sum_{l=1}^{M}w_l^2 \tag{8-12}$$

Therefore, the maximization of the *a posteriori* distribution of $\underline{w}$, $p(\underline{w}|D)$, is equivalent to the minimization of $S(\underline{w})$. The regularization term in Eq. (8-12), known as *weight decay*, favors neural models with small magnitudes for the connection weights. Small values for the connection weights tend to propagate the input signals through the almost linear segment of the sigmoidal activation functions.

One of the advantages of the Bayesian training of an ANN is the embedded iterative mechanism for estimating the regularization parameters, that is, α and β, which avoids cross-validation. The advantage of avoiding cross-validation in forecasting problems has been extensively investigated in Reference [5]. For multivariate problems, the use of one single hyperparameter α for dealing with all connection weights is not recommended [5]. In this work, each group of connection weights directly related to an input variable receives a different α_i. The same idea is applied to the groups of weights associated with the biases (one α_i for the connections with hidden neurons and another for the output neuron connection). One last α_i is associated with all connection weights between the hidden and output layers. Therefore, for n dimensional input vectors $\underline{x}$, the total number of α_i's is $n + 3$. This procedure is known as automatic relevance determination (ARD), which has given rise to another functional $S(\underline{w})$:

$$S(\underline{w}) = \frac{\beta}{2}\sum_{k=1}^{N}[d_k - f(\underline{x}_k, \underline{w})]^2 + \frac{1}{2}\sum_{i=1}^{n+3}\left(\alpha_i \sum_{j=1}^{M_i}w_{ij}^2\right) \tag{8-13}$$

Automatic Input Selection

For a given model structure, the magnitudes of the α_i's can be compared to determine the relevance of the corresponding input variables (taken from the pre-selected set). As $p(_i\underline{w})$ is supposed to be normally distributed with zero mean and $\alpha_i^{-1}\underline{I}$ covariance, the largest α_i's lead to the smallest $\underline{w}_i$'s. For estimating the *a posteriori* PDF of $\underline{w}$, Bayesian training combines the *a priori* PDF with the information provided by the training set (Eq. 8.8). If an α_i is large, the prior information about $\underline{w}_i$ is almost certain, and the effect of the training data on the estimation of $\underline{w}_i$ is negligible. Another way to see the influence of α_i on $\underline{w}_i$ is through Eq. (8-13).

The impact on the output caused by input variables with very small $\underline{w}_i$'s, that is, very large α_i's, is not significant. However, a reference level for defining a very large α_i has to be established. Probe (i.e., uniformly distributed) random variables are applied to make the reference setting data driven. More details can be found in Reference [5]. After training the model with the pre-selected set of input variables, continuous and eventual dummy variables are separately ranked. For each rank, the variables with corresponding α_i's larger than α_{ref} (irrelevance level) are disregarded. After input selection, the ANN is retrained with the selected variables.

Model Structure Selection

Bayesian inference can also be employed to determine the best structure among a pre-defined set of possibilities, for example, $H = \{H_1, H_2, \ldots, H_K\}$, for which the corresponding inputs have been previously selected, that is,

$$P(H_h|D) = \frac{p(D|H_h)P(H_h)}{p(D)} \tag{8-14}$$

In Eq. (8-14), $P(H_h)$ represents the *a priori* probability of model H_h and $p(D|H_h)$ is given by:

$$p(D|H_h) = \int \int p(D|\alpha, \beta, H_h)p(\alpha, \beta|H_h) \, d\alpha \, d\beta \tag{8-15}$$

Using Gaussian approximation around the estimated hyperparameters (from training), analytic integration of Eq. (8-15) is possible, leading to:

$$\ln p(D|H_h) = -S(\underline{w}) - \frac{1}{2}\ln|\nabla\nabla S(\underline{w})| + \frac{1}{2}\sum_{i=1}^{n+3} M_i\alpha_i + \frac{N}{2}\ln\beta + \ln(m!) + 2\ln m$$

$$+ \frac{1}{2}\sum_{i=1}^{n+3}\ln\left(\frac{2}{\gamma_i}\right) + \frac{1}{2}\ln\left(\frac{2}{N-\gamma}\right) \tag{8-16}$$

where m denotes the number of hidden neurons in the ANN model H_h. Because all models, *a priori*, are assumed equally probable, H_h is selected by maximizing $P(D|H_h)$, which is equivalent to maximizing $\ln p(D|H_h)$. Consequently, Eq. (8-16) can be used for ranking and selecting among MLPs with different numbers of neurons in the hidden layer.

Extended Bayesian Training

The following steps describe the ANN structure and input selection via Bayesian inference.

Step 1. Set the minimum (N_{min}) and maximum (N_{max}) number of neurons in the hidden layer. In this work, $N_{min} = 1$ and $N_{max} = 10$.

Step 2. Make the number of neurons in the hidden layer $m = N_{min}$.

Step 3. Add the reference of irrelevance variables (probe(s)) to the pre-selected n-dimensional input vector, which can include dummy variables, as will be explained later in the paper. If dummy variables are used, the input set will contain $n = n + 2$ input variables. Otherwise, that is, if only continuous inputs are pre-selected, $n = n + 1$.

Step 4. Set $l = 0$ and initialize $\underline{w}(l) = [\underline{w}_1(l), \ldots, \underline{w}_{n+3}(l)]^t$, $\underline{\alpha}(l) = [\alpha_1(l), \ldots, \alpha_{n+3}(l)]^t$, and $\beta(l)$.

Step 5. Minimize $S(\underline{w})$ on $\underline{w}(l)$ to obtain $\underline{w}(l + 1)$.

Step 6. Calculate $\alpha_i(l + 1)$, $\beta(l + 1)$, and $\gamma_i(l + 1)$ using the following equations:

$$\nabla\nabla S(\underline{w})|_{w=w(l+1)} = \beta(l)\nabla\nabla E_s(\underline{w}, U)|_{w=w(l+1)} + \alpha(l)\underline{I} \qquad (8\text{-}17)$$

$$\underline{\underline{B}}_i(l + 1) = [\nabla\nabla S(\underline{w})|_{w=w(l+1)}]^{-1}\underline{\underline{I}}_i$$

$$\gamma_i(l + 1) = M_i - trace\{\underline{\underline{B}}_i(l + 1)\}$$

$$\alpha_i(l + 1) = \frac{\gamma_i(l + 1)}{\|\underline{w}_i(l + 1)\|^2}$$

$$\beta(l + 1) = \frac{N - \sum_{i=1}^{n+3} \gamma_i(l + 1)}{\sum_{j=1}^{N} [d_j - f(\underline{x}_j, \underline{w}(l + 1))]^2}$$

Step 7. Make $l = l + 1$ and return to Step 5 until convergence has been achieved. After convergence, go to the next step.

Step 8. Isolate in two lists the α_i's associated with the continuous input variables and the α_j's related to the dummy variables.

Step 9. For each list, select the inputs such that the corresponding $\alpha < \alpha_{ref}$, where α_{ref} stands for the hyperparameter associated with the added probe.

Step 10. Repeat Steps 4 to 7 using the inputs selected in Step 9, with n equal to the number of selected variables, to obtain the trained model H_m.

Step 11. Evaluate the log evidence of the hypothesis (ANN structure) H_m using Eq. (8-16).

Step 12. If $m = N_{max}$, then go to Step 13. Else, $m = m + 1$ and return to Step 3.

Step 13. Select the H_k with the largest log evidence.

In Eq. (8-14), $\underline{\underline{I}}_i$ is an $M \times M$ diagonal matrix with ones at the positions corresponding to the ith group of weights and with zeros otherwise. M_i is the number of connection weights in each group. Details on how to calculate the Hessian $\nabla\nabla E_s(\underline{w}, U)$ can be found in Reference [12].

FULLY AUTOMATIC BAYESIAN NEURAL FORECASTER

The fully automatic Bayesian neural forecaster (FABNF) can be summarized as follows [13]:

Step 1. Process the time-series to treat missing values and outliers via Gaussian filter. The parameters of the filter are estimated considering the seasonality involved.

Step 2. Specify the *time-delay* τ using the minimum mutual information criterion, or spectral analysis for determining $\tau/2$, or make τ equal to one.

Step 3. Estimate the *embedding dimension d* via false neighbors algorithm [9].

Step 4. Obtain the reconstructed space using Takens' theorem (Eq. 8-3).

Step 5. Add dummy variables (one of "n" representation) to the reconstructed space to represent seasonalities, if any.

Step 6. Apply the RPEM algorithm to cluster the data.

Step 7. Classify the forecasting pattern using the mixture density clustering model.

Step 8. Model the dynamics of the designated cluster using extended Bayesian training and make predictions.

RESULTS

River flow, for energy planning studies in Brazil, is usually estimated by distributed rainfall-runoff models using the output from a rainfall prediction model. The ETA model has been applied by the National System Operator in Brazil to provide rainfall forecasting. The ETA model is a state-of-the-art atmospheric model. The code is available for downloading at www.emc.ncep.noaa.gov/mmb/wrkstn_eta/. The proposed autonomous neural network model has been used as a corrector to ETA's daily predictions for 10 days ahead, that is,

$$C_i(k + i) = P_{real}(k + i) - P_{ETA}(k + i), i = 1, 2, \dots, 10 \qquad (8\text{-}18)$$

The grid points of reference in one of the six hydrographic basins are shown in Fig. 8-1. The grid points represent the resolution provided by the ETA model, and they are used to estimate the precipitations at the closest rainfall measurement stations. The idea of developing a corrector model based on ETA has proved to be a better solution than using the weather monitoring stations information, only.

The ANNs dedicated to correcting ETA's forecasting for each day ahead (10 different MISO models) use the following input variables:

- ETA's average precipitation forecasting for day $(k - 1)$;
- ETA's average precipitation forecasting for day k;

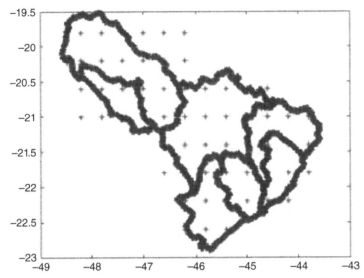

Figure 8-1 Grid coordinates of the Rio Grande basin, which has 122 rainfall monitoring stations.

- ETA's average precipitation forecasting for day $(k + 1)$;
- ETA's forecasts for 25 weather-related variables; and
- 12 binary variables for coding the month of the year.

The multi-input single-output (MISO) ANN model for one step ahead predictions has also been employed for covering the whole forecasting horizon. This is achieved via recursion, that is, feeding back the output as the new input. Figure 8-2 illustrates how the ANNs' training patterns are built.

The mean absolute error (MAE) has been applied to evaluate the forecasting models. The relative error (with sign) has also been employed to verify how biased the forecasts are. The MAE statistics has been calculated for the whole testing set, with more than 400 predictions for each point of interest. Results have indicated that when ETA forecasts a daily rainfall above 5 mm, the ANN model using recursion has been the dominant one for most of the basins. It has produced forecasting accuracy gains up to 22%. In fact, it has been observed that correction estimates by the ANNs are usually effective, unless precipitation falls below 2 mm per day. It has been interesting to verify that the stronger correlation among outputs assured by the ANN

ANN training			10 days ahead forecasting				
01/03/1996	...	12/06/2002	12/07/2002	12/08/2002	...	12/16/2002	
ANN retraining			10 days ahead forecasting				
01/03/1996	...	12/12/2002	12/13/2002	12/14/2002	...	12/22/2002	12/23/2002
ANN retraining			10 days ahead forecasting				
01/03/1996	...	12/18/2002	12/19/2002	12/20/2002	12/21/2002	...	12/29/2002	12/30/2002

Figure 8-2 ANNs' training adaptation and testing.

with recursion has paid-off. It has the additional advantage of reducing the training effort to one single model. Moreover, the rainfall correction estimates have also been useful to make the forecasts less biased.

CONCLUSION

This chapter presents an inductive learning procedure that combines several techniques to generate a fully data-driven forecasting model. Input selection is performed, without user intervention, by applying Chaos theory and Bayesian inference. Afterwards, neural network models are estimated, without cross-validation, relying on data partitioning and Bayesian regularization for complexity control. Automatic data clustering has been used for data partitioning. The proposed forecasting model has been successfully tested using rainfall data from six major hydrographic basins in Brazil.

REFERENCES

1. H.S. Hippert, R.C. Souza, and C.E. Pedreira, "Neural networks for load forecasting: A review and evaluation," *IEEE Transactions on Power Systems*, vol. 16, no. 1, pp. 44–55, February 2001.

2. S. Amari, N. Murata, K.R. Müller, M. Finke, and H. Yang, "Statistical theory of overtraining — Is cross-validation asymptotically effective?" In: *Advances in Neural Information Processing* Systems, vol. 8, pp. 176–182, MIT Press, 1996.

3. A.J.R. Reis and A.P. Alves da Silva, "Feature extraction via multi-resolution analysis for short-term load forecasting," *IEEE Transactions on Power Systems*, vol. 20, no. 1, pp. 189–198, February 2005.

4. A. Lendasse, J. Lee, V. Wertz, and M. Verleysen, "Forecasting electricity consumption using nonlinear projection and self-organizing maps," *Neurocomputing*, vol. 48, pp. 299–311, October 2002.

5. V.H. Ferreira and A.P. Alves da Silva, "Toward estimating autonomous neural network load forecasters," *IEEE Transactions on Power Systems*, vol. 22, no. 4, pp. 1554–1562, November 2007.

6. A.P. Alves da Silva, V.H. Ferreira, and R.M.G. Velásquez, "Input space to neural network based load forecasters," *International Journal of Forecasting*, vol. 24, no. 4, pp. 616–629, October/December 2008.

7. F. Takens, "Detecting Strange Attractors in Turbulence," In: D.A. Rand, L.-S. Young (eds.), *Dynamical Systems and Turbulence*, Lecture Notes in Mathematics, vol. 898, pp. 366–381, Springer-Verlag, 1981.

8. A.M. Fraser and H.L. Swinney, "Independent coordinates for strange attractors from mutual information," *Physical Review A*, vol. 33, no. 2, pp. 1134–1140, February 1986.

9. L. Cao, "Practical method for determining the minimum embedding dimension of a scalar time series," *Physica D*, vol. 110, no. 1–2, pp. 43–50, December 1997.

10. H. Kantz and T. Schreiber, *Nonlinear Time Series Analysis*, Cambridge Nonlinear Science Series, no. 7, Cambridge University Press, 1997.

11. Y-M. Cheung, "Maximum weighted likelihood via rival penalized EM for density mixture clustering with automatic model selection," *IEEE Transactions on Knowledge and Data Engineering*, vol. 17, no. 6, pp. 750–761, June 2005.

12. A.P. Alves da Silva and V.H. Ferreira, "Short-term load forecasting," In: *Electric Power Systems: Advanced Forecasting Techniques and Optimal Generation Scheduling*, pp. 301–322, CRC Press, 2012.

13. D.J.C. Mackay, Bayesian Methods for Adaptive Models, Ph.D. Dissertation, California Institute of Technology, Pasadena, CA, 1992.

Index

Advances in Electric Power and Energy Systems: Load and Price Forecasting, First Edition.
Edited by Mohamed E. El-Hawary.
© 2017 by The Institute of Electrical and Electronics Engineers, Inc. Published 2017 by John Wiley & Sons, Inc.

IEEE Press Series on Power Engineering

Series Editor: M. E. El-Hawary, Dalhousie University, Halifax, Nova Scotia, Canada

The mission of IEEE Press Series on Power Engineering is to publish leading-edge books that cover the broad spectrum of current and forward-looking technologies in this fast-moving area. The series attracts highly acclaimed authors from industry/academia to provide accessible coverage of current and emerging topics in power engineering and allied fields. Our target audience includes the power engineering professional who is interested in enhancing their knowledge and perspective in their areas of interest.

1. *Principles of Electric Machines with Power Electronic Applications, Second Edition*
M. E. El-Hawary

2. *Pulse Width Modulation for Power Converters: Principles and Practice*
D. Grahame Holmes and Thomas Lipo

3. *Analysis of Electric Machinery and Drive Systems, Second Edition*
Paul C. Krause, Oleg Wasynczuk, and Scott D. Sudhoff

4. *Risk Assessment for Power Systems: Models, Methods, and Applications*
Wenyuan Li

5. *Optimization Principles: Practical Applications to the Operations of Markets of the Electric Power Industry*
Narayan S. Rau

6. *Electric Economics: Regulation and Deregulation*
Geoffrey Rothwell and Tomas Gomez

7. *Electric Power Systems: Analysis and Control*
Fabio Saccomanno

8. *Electrical Insulation for Rotating Machines: Design, Evaluation, Aging, Testing, and Repair, Second Edition*
Greg Stone, Edward A. Boulter, Ian Culbert, and Hussein Dhirani

9. *Signal Processing of Power Quality Disturbances*
Math H. J. Bollen and Irene Y. H. Gu

10. *Instantaneous Power Theory and Applications to Power Conditioning*
Hirofumi Akagi, Edson H. Watanabe, and Mauricio Aredes